D0944890

Green Fluorescent Protein

METHODS IN MOLECULAR BIOLOGY™

John M. Walker, Series Editor

216. **PCR Detection of Microbial Pathogens:** *Methods and Protocols,* edited by *Konrad Sachse and Joachim Frey, 2003*
215. **Cytokines and Colony Stimulating Factors:** *Methods and Protocols,* edited by *Dieter Körholz and Wieland Kiess, 2003*
214. **Superantigen Protocols,** edited by *Teresa Krakauer, 2003*
213. **Capillary Electrophoresis of Carbohydrates,** edited by *Pierre Thibault and Susumu Honda, 2003*
212. **Single Nucleotide Polymorphisms:** *Methods and Protocols,* edited by *Piu-Yan Kwok, 2003*
211. **Protein Sequencing Protocols, 2nd ed.,** edited by *Bryan John Smith, 2003*
210. **MHC Protocols,** edited by *Stephen H. Powis and Robert W. Vaughan, 2003*
209. **Transgenic Mouse Methods and Protocols,** edited by *Marten Hofker and Jan van Deursen, 2002*
208. **Peptide Nucleic Acids:** *Methods and Protocols,* edited by *Peter E. Nielsen, 2002*
207. **Recombinant Antibodies for Cancer Therapy:** *Methods and Protocols.* edited by *Martin Welschof and Jürgen Krauss, 2002*
206. **Endothelin Protocols,** edited by *Janet J. Maguire and Anthony P. Davenport, 2002*
205. ***E. coli* Gene Expression Protocols,** edited by *Peter E. Vaillancourt, 2002*
204. **Molecular Cytogenetics:** *Protocols and Applications,* edited by *Yao-Shan Fan, 2002*
203. ***In Situ* Detection of DNA Damage:** *Methods and Protocols,* edited by *Vladimir V. Didenko, 2002*
202. **Thyroid Hormone Receptors:** *Methods and Protocols,* edited by *Aria Baniahmad, 2002*
201. **Combinatorial Library Methods and Protocols,** edited by *Lisa B. English, 2002*
200. **DNA Methylation Protocols,** edited by *Ken I. Mills and Bernie H, Ramsahoye, 2002*
199. **Liposome Methods and Protocols,** edited by *Subhash C. Basu and Manju Basu, 2002*
198. **Neural Stem Cells:** *Methods and Protocols,* edited by *Tanja Zigova, Juan R. Sanchez-Ramos, and Paul R. Sanberg, 2002*
197. **Mitochondrial DNA:** *Methods and Protocols,* edited by *William C. Copeland, 2002*
196. **Oxidants and Antioxidants:** *Ultrastructure and Molecular Biology Protocols,* edited by *Donald Armstrong, 2002*
195. **Quantitative Trait Loci:** *Methods and Protocols,* edited by *Nicola J. Camp and Angela Cox, 2002*
194. **Posttranslational Modifications of Proteins:** *Tools for Functional Proteomics,* edited by *Christoph Kannicht, 2002*
193. **RT-PCR Protocols,** edited by *Joseph O'Connell, 2002*
192. **PCR Cloning Protocols, 2nd ed.,** edited by *Bing-Yuan Chen and Harry W. Janes, 2002*
191. **Telomeres and Telomerase:** *Methods and Protocols,* edited by *John A. Double and Michael J. Thompson, 2002*
190. **High Throughput Screening:** *Methods and Protocols,* edited by *William P. Janzen, 2002*
189. **GTPase Protocols:** *The RAS Superfamily,* edited by *Edward J. Manser and Thomas Leung, 2002*
188. **Epithelial Cell Culture Protocols,** edited by *Clare Wise, 2002*
187. **PCR Mutation Detection Protocols,** edited by *Bimal D. M. Theophilus and Ralph Rapley, 2002*
186. **Oxidative Stress Biomarkers and Antioxidant Protocols,** edited by *Donald Armstrong, 2002*
185. **Embryonic Stem Cells:** *Methods and Protocols,* edited by *Kursad Turksen, 2002*
184. **Biostatistical Methods,** edited by *Stephen W. Looney, 2002*
183. **Green Fluorescent Protein:** *Applications and Protocols,* edited by *Barry W. Hicks, 2002*
182. **In Vitro Mutagenesis Protocols, 2nd ed.**, edited by *Jeff Braman, 2002*
181. **Genomic Imprinting:** *Methods and Protocols,* edited by *Andrew Ward, 2002*
180. **Transgenesis Techniques, 2nd ed.:** *Principles and Protocols,* edited by *Alan R. Clarke, 2002*
179. **Gene Probes:** *Principles and Protocols,* edited by *Marilena Aquino de Muro and Ralph Rapley, 2002*
178. **Antibody Phage Display:** *Methods and Protocols,* edited by *Philippa M. O'Brien and Robert Aitken, 2001*
177. **Two-Hybrid Systems:** *Methods and Protocols,* edited by *Paul N. MacDonald, 2001*
176. **Steroid Receptor Methods:** *Protocols and Assays,* edited by *Benjamin A. Lieberman, 2001*
175. **Genomics Protocols**, edited by *Michael P. Starkey and Ramnath Elaswarapu, 2001*
174. **Epstein-Barr Virus Protocols,** edited by *Joanna B. Wilson and Gerhard H. W. May, 2001*
173. **Calcium-Binding Protein Protocols, Volume 2:** *Methods and Techniques,* edited by *Hans J. Vogel, 2001*
172. **Calcium-Binding Protein Protocols, Volume 1:** *Reviews and Case Histories,* edited by *Hans J. Vogel, 2001*
171. **Proteoglycan Protocols,** edited by *Renato V. Iozzo, 2001*
170. **DNA Arrays:** *Methods and Protocols,* edited by *Jang B. Rampal, 2001*
169. **Neurotrophin Protocols,** edited by *Robert A. Rush, 2001*
168. **Protein Structure, Stability, and Folding,** edited by *Kenneth P. Murphy, 2001*
167. **DNA Sequencing Protocols,** *Second Edition,* edited by *Colin A. Graham and Alison J. M. Hill, 2001*
166. **Immunotoxin Methods and Protocols**, edited by *Walter A. Hall, 2001*
165. **SV40 Protocols**, edited by *Leda Raptis, 2001*
164. **Kinesin Protocols,** edited by *Isabelle Vernos, 2001*
163. **Capillary Electrophoresis of Nucleic Acids, Volume 2:** *Practical Applications of Capillary Electrophoresis,* edited by *Keith R. Mitchelson and Jing Cheng, 2001*
162. **Capillary Electrophoresis of Nucleic Acids, Volume 1:** *Introduction to the Capillary Electrophoresis of Nucleic Acids,* edited by *Keith R. Mitchelson and Jing Cheng, 2001*
161. **Cytoskeleton Methods and Protocols**, edited by *Ray H. Gavin, 2001*
160. **Nuclease Methods and Protocols,** edited by *Catherine H. Schein, 2001*
159. **Amino Acid Analysis Protocols,** edited by *Catherine Cooper, Nicole Packer, and Keith Williams, 2001*
158. **Gene Knockout Protocols,** edited by *Martin J. Tymms and Ismail Kola, 2001*
157. **Mycotoxin Protocols,** edited by *Mary W. Trucksess and Albert E. Pohland, 2001*
156. **Antigen Processing and Presentation Protocols,** edited by *Joyce C. Solheim, 2001*

METHODS IN MOLECULAR BIOLOGY™

Green Fluorescent Protein

Applications and Protocols

Edited by

Barry W. Hicks

Department of Chemistry,
United States Air Force Academy,
Colorado Springs, CO

Humana Press Totowa, New Jersey

999 Riverview Drive, Suite 208
Totowa, New Jersey 07512

www.humanpress.com

This publication is printed on acid-free paper. ∞
ANSI Z39.48-1984 (American Standards Institute)
Permanence of Paper for Printed Library Materials.

Cover illustration: The cover shows the crystal structure of the green fluorescent protein displayed using Rasmol software; the PDB ID number is 1EMG. The structure is shown in cartoon display with beta sheets in yellow, alpha helices in magenta, and turns in blue. The chromophore is colored green and depicted in space filling display.

Cover Design by Patricia F. Cleary.

Production Editor: Kim Hoather-Potter.

For additional copies, pricing for bulk purchases, and/or information about other Humana titles, contact Humana at the above address or at any of the following numbers: Tel.: 973-256-1699; Fax: 973-256-8341; E-mail: humana@humanapr.com or visit our Website: http://humanapress.com

Printed in the United States of America. 10 9 8 7 6 5 4 3 2 1

Library of Congress Cataloging in Publication Data

Green fluorescent protein : appplications and protocols / edited by Barry W. Hicks.
p. cm. -- (Methods in molecular biology ; v. 183)
Includes biolographical references and index.
ISBN 0-89603-905-6 (alk. paper)
1. Green fluorescent protein--Laboratory manuals. I. Hicks, Barry W. II. Series.

QP552.G73 G467 2002
572.8--dc21

2002068558

Preface

Could there be a better time to be a life scientist? In the past two decades, a host of new techniques have been added to the tool chests of biochemists and molecular biologists. A wonderful benefit of the basic scientific research that fueled the advances in these fields is the wide variety of direct applications in agriculture and medicine. Even with all of these advances, and with the accompanying explosion in computer and information technology, it is clear that the depth of our ignorance vastly exceeds the breadth of our knowledge about complex organisms at the molecular level. Any new techniques or materials that allow us to extend our research-based knowledge should be welcomed and utilized to their fullest potentials. With the cloning of the green fluorescent protein (GFP) from *Aequorea victoria* in 1992, another valuable tool was added to the arsenal. In *Green Fluorescent Proteins: Applications and Protocols* examples of how GFP can be utilized in a variety of fields are presented. Although the text has chapters that emphasize different areas of specialization, it is not meant to send molecular biologists to one section, botanists to another, and clinicians to still another. Perhaps the most valuable exchange for people in any discipline will come from seeing how others have been able to apply GFP in fields outside of their immediate areas of expertise.

GFP from *Aequorea victoria* is a fluorescent marker protein, and there are certainly other useful fluorophore markers. The wild-type GFP is not generally used by researchers today. In fact, the acronym GFP has become somewhat misleading because so many spectral variants are now available. All of the work described in this volume takes advantage of the mutant GFPs with altered spectral characteristics or with great cellular expression. It is also noteworthy that the first two chapters describe technique applied to other fluorescent markers: DsRed and other fluorescent proteins cloned from Anthozoans, and *cobA* and *CysG*, genes encoding for enzymes producing soluble red fluorescent markers. Although using the GFP marker to locate biomaterials remains the most often utilized application because of the advantages inherent in using GFP and the versatility offered by the many GFPs available, many more elegant methods have emerged, and several of these are demonstrated in this volume. Like all volumes in the Methods in Molecular Biology series, the text is designed to aid researchers who understand broad aspects of a topic to gain expertise in some narrow experimental portion of that topic. It might be most useful to postdoctoral researchers or graduate students who are actually performing the experimental work at the bench. In each chapter, methods with detail that go far beyond what is currently printed in most journals are provided and could aid in spreading GFP techniques to new laboratories.

Several groups and individuals deserve special attention for getting this text completed. Although the majority of the figures in the text are in black and white, I urge readers to take full advantage of the accompanying CD-ROM that was generously sponsored by Universal Imaging Corporation. The CD-ROM includes color figures and videos from over half of the chapters in this book. I would like to thank Dr. John Walker for allowing me the opportunity to edit this volume and further my own understanding of life science, which also allowed me to make research contacts with some fantastic people around the world utilizing autofluorescent proteins. Finally, I would like to thank my students at the US Air Force Academy for continuing to challenge me to stay abreast of the rapidly advancing discipline of biochemistry.

Barry W. Hicks

Contents

PART V. GREEN FLUORESCENT PROTEIN BIOSENSORS

PART VI. VIRAL APPLICATIONS OF GREEN FLUORESCENT PROTEIN

Contributors

PASCAL J. BAEHLER • *Institut de Biologie Cellulaire et de Morphologie, Lausanne, Switzerland*

JESSAMYN BAGLEY • *Transplantation Biology Research Center and Harvard Medical School, Boston, MA*

RICARDO M. BIONDI • *Division of Signal Transduction Therapy, University of Dundee, Dundee, UK*

DAVID L. BRAUTIGAN • *Center for Cell Signaling, University of Virginia Health Sciences Center, Charlottesville, VA*

ANTHONY W.S. CHAN • *Department of Human Genetics, School of Medicine, Yerkes Regional Primate Research Center, Emory University, Atlanta, GA*

KOWIT-YU CHONG • *Magee-Women's Research Institute, Pittsburgh Development Center (PDC), Department of Obstetrics, Gynecology and Reproductive Science, Pittsburgh, PA*

CHRISTOPHER COKER • *Department of Microbiology and Immunology, University of Maryland, Baltimore, MD*

NOBUHIDE DOI • *Department of Applied Chemistry, Faculty of Science and Technology, Keio University, Yokohama, Japan*

W. PAUL DUPREX • *School of Biology and Biochemistry, The Queen's University Belfast, Belfast, Northern Ireland*

MASILAMANI ELANGOVAN • *Department of Biology, W. M. Keck Center for Cellular Imaging, University of Virginia, Charlottesville, VA*

NICOLETTA ELIOPOULOS • *Lady Davis Institute for Medical Research, Division of Hematology-Oncology, Department of Experimental Medicine, McGill University, Montreal, Canada*

ELIZABETH ELLIOTT • *Center for Cell Signaling, University of Virginia Health Sciences Center, Charlottesville, VA*

JACQUES GALIPEAU • *Lady Davis Institute for Medical Research, Division of Hematology-Oncology, Department of Medicine, McGill University, Montreal, Quebec, Canada*

DONALD H. GILDEN • *Departments of Neurology and Microbiology, University of Colorado Health Sciences Center, Denver, CO*

RUDI GLOCKSHUBER • *Institut für Molekularbiologie und Biophysik, Eidgenössische Technische Hochschule Hönggerberg, Zürich, Switzerland*

MATTHEW D. HALFHILL • *Department of Crop Science, Williams Hall, North Carolina State University, Raleigh, NC; Department of Plant Sciences and Landscape Systems, Ellington Plant Sciences, The University of Tennessee, Knoxville, TN*

KLAAS J. HELLINGWERF • *Swammerdam Institute for Life Sciences, University of Amsterdam, Amsterdam, The Netherlands*

ROBERT M. HOFFMAN • *AntiCancer Inc., and Department of Surgery, University of California at San Diego, San Diego, CA*

HAIGEN HUANG • *Institute of Molecular Medicine and Genetics, Medical College of Georgia, Augusta, GA*

THOMAS J. HOPE • *Department of Microbiology and Immunology, University of Illinois at Chicago College of Medicine, Chicago, IL*

JOHN IACOMINI • *Transplantation Biology Research Center and Harvard Medical School, Boston, MA*

ALAN KIM JOHNSON • *Cardiovascular Center, Departments of Psychology and Pharmacology, University of Iowa, Iowa City, IA*

CARL H. JOHNSON • *Department of Biology, Vanderbilt University, Nashville, TN*

GEN KONDOH • *Department of Social and Environmental Medicine, Osaka University Graduate School of Medicine, Osaka, Japan*

SHUO LIN • *Institute of Molecular Medicine and Genetics, Medical College of Georgia, Augusta, Georgia*

JONATHAN E. LOEB • *Infectious Disease Laboratory, Salk Institute for Biological Studies, La Jolla, CA*

RAVI MAHALINGAM • *Department of Neurology, University of Colorado Health Sciences Center, Denver, CO*

TOMONAO MATSUSHITA • *Laboratory of Plant Physiology, Department of Botany, Kyoto University, Kyoto, Japan*

MIKHAIL V. MATZ • *Institute of Bioorganic Chemistry, Russian Academy of Sciences, Moscow, Russia*

REGINALD J. MILLWOOD • *Department of Plant Sciences and Landscape Systems, Ellington Plant Sciences, The University of Tennessee, Knoxville, TN*

HARRY L. T. MOBLEY • *Department of Microbiology and Immunology, University of Maryland, Baltimore, MD*

HAJIME MORI • *Department of Applied Biology, Kyoto Institute of Technology, Kyoto, Japan*

AKIRA NAGATANI • *Laboratory of Plant Physiology, Department of Botany, Kyoto University, Kyoto, Japan*

AMMASI PERIASAMY • *Department of Biology, W. M. Keck Center for Cellular Imaging, University of Virginia, Charlottesville, VA*

LASZLO PERLAKY • *Department of Pediatrics, Baylor College of Medicine, Houston, TX*

RHONDA PERRIMAN • *Center for Molecular Biology of RNA, Department of Biology, University of California at Santa Cruz, Santa Cruz, CA*

DAVID PISTON • *Department of Molecular Physiology and Biophysics, Vanderbilt University Medical School, Nashville, TN*

CHRISTOPHE D. REYMOND • *Institut de Biologie Cellulaire et de Morpologie, Lausanne, Switzerland*

BERT K. RIMA • *School of Biology and Biochemistry, The Queen's University Belfast, Belfast, Northern Ireland*

CHARLES A. ROESSNER • *Department of Chemistry, Texas A&M University, College Station, TX*

GUY A. RUTTER • *Department of Biochemistry, School of Medical Sciences, University of Bristol, Bristol, UK*

GERALD SCHATTEN • *Oregon Regional Primate Research Center, Oregon Health Sciences University, Beaverton, OR*

ARJEN SCHOTS • *Laboratory of Monoclonal Antibodies, Department of Plant Sciences, Wageningen University, Wageningen, The Netherlands*

EBRAHIM SHAFIZADEH • *Institute of Molecular Medicine and Genetics, Medical College of Georgia, Augusta, Georgia*

KEVAN SHOKAT • *Department of Chemistry, University of California, Berkeley, CA; Department of Cellular and Molecular Pharmacology, University of California, San Francisco, CA*

ROLAND H. STAUBER • *Institute for Medical and Clinical Virology, University of Erlangen-Nürenberg, Erlangen, Germany*

C. NEAL STEWART, JR. • *Department of Plant Sciences and Landscape Systems, Ellington Plant Sciences, The University of Tennessee, Knoxville, TN*

SIMON TOPELL • *Institut für Molekularbiologie und Biophysik, Eidgenössische Technische Hochschule Hönggerberg, Zürich, Switzerland*

SCOTT ULRICH • *Department of Cellular and Molecular Pharmacology, University of California, San Francisco, CA*

BENIGNO C. VALDEZ • *Department of Pharmacology, Baylor College of Medicine, Houston, TX*

MIGUEL VAN BEMMELEN • *Unité de Régulation Enzimatique des Activités Cellulaire, Institut Pasteur, Paris, France*

JAN M. VAN DER WOLF • *Plant Research International, Wageningen University and Research Centre, Wageningen, The Netherlands*

JASPER J. VAN THOR • *Laboratory of Molecular Biophysics, University of Oxford, Oxford, UK*
ANIKÓ VÁRADI • *Department of Biochemistry, School of Medical Sciences, University of Bristol, Bristol, UK*
ELISARDO CORRAL VASQUEZ • *Physiological Sciences Graduate Program, UFES & EMESCAM, Vitoria, Brazil*
MICHEL VÉRON • *Unité de Régulation Enzimatique des Activités Cellulaire, Institut Pasteur, Paris, France*
HENRY WEINER • *Biochemistry Department, Purdue University, West Lafayette, IN*
MATTHEW D. WEITZMAN • *Laboratory of Genetics, The Salk Institute for Biological Studies, La Jolla, CA*
YAO XU • *Department of Biology, Vanderbilt University, Nashville, TN*
HIROSHI YANAGAWA • *Department of Applied Chemistry, Faculty of Science and Technology, Keio University, Yokohama, Japan*
HUI ZHAO • *Department of Microbiology and Immunology, University of Maryland, Baltimore, MD*

I

Manipulation of Green Fluorescent Protein Structure at the Genetic Level

1

Amplification of Representative cDNA Samples from Microscopic Amounts of Invertebrate Tissue to Search for New Genes

Mikhail V. Matz

1. Introduction

Recently, we cloned six new green fluorescent protein (GFP)-like fluorescent proteins from five species of Antozoa *(1)*, including one red-emitting variant (DsRed), which is now commercially available. This project did not require expeditions and collection of animals on reefs: In all cases, the starting material was just several milligrams of tissue (e.g., a tentacle tip of a sea anemone), collected from a specimen in a private aquarium. This truly noninvasive kind of study was possible because the approach of total cDNA amplification, which is extensively applied to various tasks and biological models in our lab. This chapter outlines several year's of experience in this helpful technique.

The possibility of amplifying total cDNA obtained from small amounts of biological material is not yet routinely considered, despite the fact that obtaining amounts of material suitable for direct processing by standard methods is often time-consuming, expensive, and may be even impossible. Perhaps the most significant obstacle to the full appreciation of the technique is the widespread belief that polymerase chain reaction (PCR) amplification severely distorts the original cDNA profile, so that some cDNA species dramatically rise in abundance, but others diminish, and may even become completely lost. However, we found that there are just a few simple rules that should be followed to ensure that the amplified sample is minimally distorted and fully representative, i.e., contains all types of messages originally present in RNA, even the least abundant ones. This was demonstrated in experiments on differ-

From: *Methods in Molecular Biology, vol. 183: Green Fluorescent Protein: Applications and Protocols*
Edited by: B. W. Hicks © Humana Press Inc., Totowa, NJ

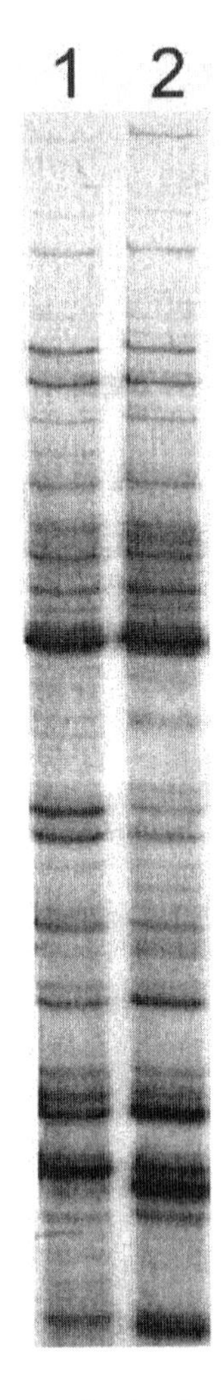

Fig. 1. Differential display patterns obtained according to **ref.** ***9*** for the same total RNA sample, either from the product of total cDNA amplification using a Klentaq/*Pfu* enzyme mixture (lane 1) or directly from nonamplified double-stranded cDNA (lane 2).

ential display (**Fig. 1**), and elsewhere in application of amplified cDNA as a probe for gene profiling by array technology *(2–5)*. According to our experience in gene hunting in various biological models, amplified cDNA can substitute for normal, nonamplified cDNA in virtually all tasks. Moreover, in PCR-based gene hunting techniques, such as rapid amplification of cDNA ends (RACE) *(6,7)*, subtraction *(8)*, or differential display *(9)*, the amplified cDNA usually outperforms the normal one, because all backgrounds are predictable, and can easily be kept under control.

1.1. Total RNA Isolation

We usually use the following procedure, rather than commercial kits, because this technique is suitable for virtually all animals. It is based on the well-known protocol of Chomczynski and Sacchi *(10)*, with one difference: All the procedures are performed at neutral pH, instead of acidic, as was originally suggested. Also, the step of RNA precipitation with lithium chloride (LiCl) is added, because it results in stable RNA preparations and considerably improves the consecutive procedures of cDNA synthesis. We have successfully applied the protocol to RNA isolation from representatives of 13 phyla of multicellular animals. As an alternative, a popular Trizol method (Gibco/Life Technologies) may be used in many cases, although it may not perform well on some nonstandard species, such as jellyfish. Kits for RNA isolation which utilize columns (such as Qiagen's RNeasy kit) are generally not recommended for nonstandard samples. The protocol is designed for rather large tissue samples (tissue vol 10–100 µL), which normally yield about 10–100 µg total RNA. The protocol for microscopic amounts of starting material (expected to yield ~1 µg RNA, or less) is the same but does not include second phenol–chloroform extraction (*see* **Subheading 3.1., step 4**) and LiCl precipitation (*see* **Subheading 3.1., step 6**). Additionally, the final pellet should be dissolved in 5 µL, instead of 40 µL of water and transferred directly to cDNA synthesis, omitting the agarose gel analysis..

1.2. cDNA Synthesis

Two alternatives are provided for preparing amplified total cDNA from the isolated RNA: method A and method B (**Fig. 2**). Both methods provide a possibility of amplifying a cDNA fraction corresponding to messenger (poly[A]$^+$) RNA, starting from total RNA. The fraction of ribosomal RNA in the amplified sample, as it was determined in an EST sequencing project based on amplified cDNA, is 15–20% represented mostly by small subunit RNA. This is the same figure that is normally obtained with standard methods of cDNA synthesis *(11)*.

Method A ("classical") is to synthesize a double-stranded cDNA by conventional means (employing DNA polymerase I/RNAse H/DNA ligase enzyme cocktail for second-strand synthesis), then ligate adaptors and amplify the sample, using adaptor-specific primers. The structure of the adaptors evokes a PCR-suppression effect *(12)*, and provides a method for selective amplification of only those cDNA molecules that contain both adaptor sequence and T-primer sequence, corresponding to the poly(A)$^+$ fraction of RNA. The principles behind this method are described *(13)*. The obvious advantage of this method is its high efficiency. A representative cDNA sample (with representation of 10^7 and higher) can be prepared from as little as 20–30 ng of total RNA. However, the method is rather laborious.

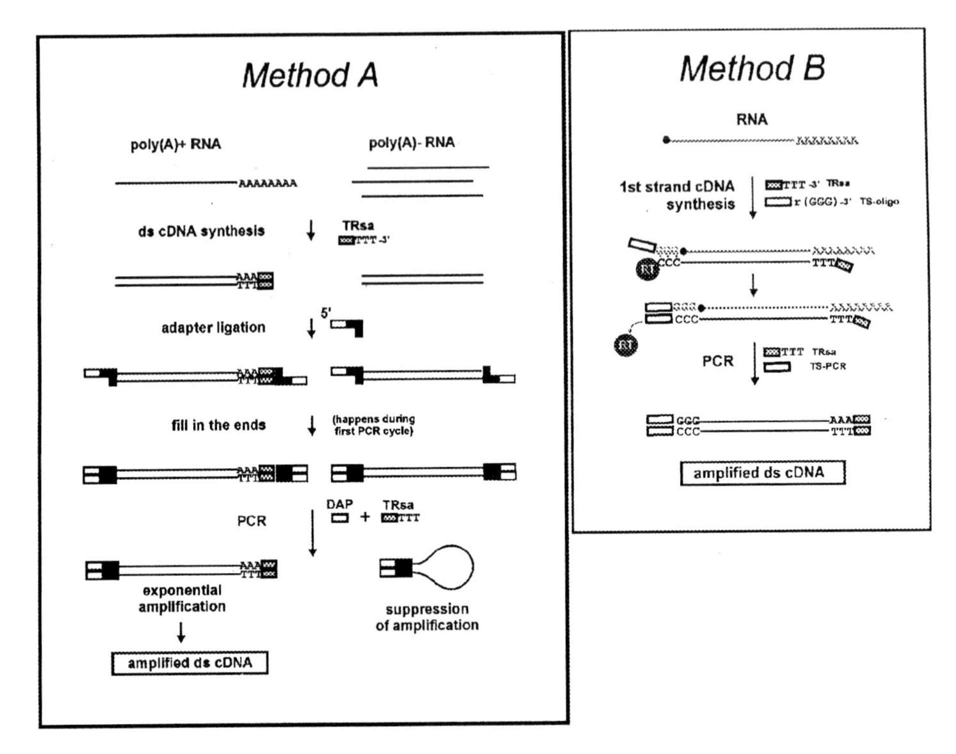

Fig. 2. Schematic outlines of cDNA amplification methods.

Method B is implemented in the SMART cDNA synthesis kit available from Clontech. It utilizes one surprising feature of Moloney murine leukemia virus reverse transcriptase (MMLV RT), its ability to add a few non-template deoxynucleotides (mostly C) to the 3' end of a newly synthesized cDNA strand, upon reaching the 5' end of the RNA template. Oligonucleotide containing oligo(rG) sequence on the 3' end, which is called "template-switch oligo" (TS-oligo), will base-pair with the deoxycytidine stretch produced by MMLV RT when added to the RT reaction. Reverse transcriptase then switches templates and continues replicating using the TS-oligo as a template. Thus, the sequence complementary to the TS-oligo can be attached to the 3' terminus of the first strand of cDNA synthesized, and may serve as a universal 5' terminal site for primer annealing during total cDNA amplification ***(14)***. Recently, an improvement to the original procedure was reported ***(15)***. Addition of $MnCl_2$ to the reaction mixture after first-strand synthesis, followed by a short incubation, increases the efficiency of nontemplate C addition to the cDNA, and thus results in higher overall yield following cDNA amplification.

Although method B is simpler and faster than method A, its reduced efficiency means that a cDNA sample of suitable representation (more than 10^6)

requires a minimum of 1 μg of total RNA. Both techniques (as they are described here) provide material not only for total cDNA amplification, but also for RACE, a procedure for obtaining unknown flanks of a fragment. This procedure is indispensable for cloning complete coding regions of proteins. Different RACE techniques are available for each of the methods of cDNA amplification described here (**refs. *6*** and ***7*** for methods A and B, respectively), both based on a PCR suppression effect ***(12)***.

2. Materials

2.1. Total RNA Isolation (see Note 1)

1. Dispersion buffer (buffer D): 4 *M* guanidine thiocyanate, 30 m*M* disodium citrate, 30 m*M* β-Mercaptoethanol, pH 7.0–7.5 (*see* **Note 2**).
2. Buffer-saturated phenol, pH 7.0–8.0 (Gibco-Life Technologies).
3. Chloroform–isoamyl alcohol mix (24:1).
4. 96% Ethanol.
5. 80% Ethanol.
6. 12 *M* LiCl.
7. Co-precipitant: SeeDNA reagent (Amersham) or glycogen.
8. Fresh MilliQ water.
9. Agarose gel (1%) containing ethidium bromide (EtBr).

2.2. cDNA Synthesis

2.2.1. Method A Using Conventional Second-Strand Synthesis (see **Note 3**)

1. SuperScript II reverse transcriptase, 200 U/μL (Life Technologies) or 20X PowerScript reverse transcriptase (Clontech) with provided buffer.
2. 0.1 *M* Dithiothreitol (DTT).
3. dNTP mix, 10 m*M* each.
4. 5X Second strand buffer: 500 m*M* KCl, 50 m*M* ammonium sulfate, 25 m*M* $MgCl_2$, 0.75 m*M* β-NAD, 100 m*M* Tris-HCl (pH 7.5), 0.25 mg/mL bovine serum albumin (BSA).
5. 20X Second-strand enzyme cocktail: 6 U/μL DNA polymerase I, 0.2 U/μL RNase H, 1.2 U/μL *Escherichia coli* DNA ligase.
6. T4 DNA polymerase (1–3 U/μL).
7. T4 DNA ligase 2–4 U/μL with provided buffer (New England Biolabs or equivalent).
8. T/M buffer: 10 m*M* Tris-HCl, pH 8.0, 1 m*M* $MgCl_2$.
9. Buffer-saturated phenol, pH 7.0–8.0 (Gibco-Life Technologies).
10. Chloroform–isoamyl alcohol mix (24:1).
11. Long-and-Accurate PCR enzyme mix (Advantage 2 polymerase mix by Clontech, LA-PCR by Takara, Expand *Taq* by Boehringer, or equivalent, *see* **Note 4**).
12. 10X PCR buffer: provided with the enzyme mix or, if Klentaq-based homemade mix is used: 300 m*M* tricine-KOH, pH 9.1, 160 m*M* ammonium sulfate, 30 m*M* $MgCl_2$, 0.2 mg/mL BSA.

Box 1

Invertebrate-optimized sets of oligos for total cDNA amplification and RACE

For Method A:

pseudo-double-stranded adaptor (two complementary oligonucleotides)

```
5'-CGA CGT GGA CTA TCC ATG AAC GCA ACT CTC CGA CCT CTC ACC GAG TAC G -3'
                                                    3'- TGG CTC ATG C -5'
```

DAP (Distal Adapter Primer)

```
5'-CGA CGT GGA CTA TCC ATG AAC GCA -3'
```

PAP (Proximal Adapter Primer)

```
5'- ACT CTC CGA CCT CTC ACC GA -3'
```

TRsa

$5'$-CGC AGT CGG TAC $(T)_{13}$ -3'

DAP-TRsa

$5'$-CGA CGT GGA CTA TCC ATG AAC GCA CGC AGT CGG TAC $(T)_{13}$ -3'

Oligos used in cDNA amlification:

TRsa - for 1st strand cDNA synthesis
adapter - to ligate to ds cDNA
DAP, TRsa - to amplify adapter-ligated cDNA

Oligos used in RACE from amplified cDNA (according to [6]):

DAP - universal primer for 5' RACE;
PAP - nested universal primer for 5' RACE
DAP-TRsa - universal primer for 3'-RACE
TRsa - nested universal primer for 3'-RACE

For Method B:

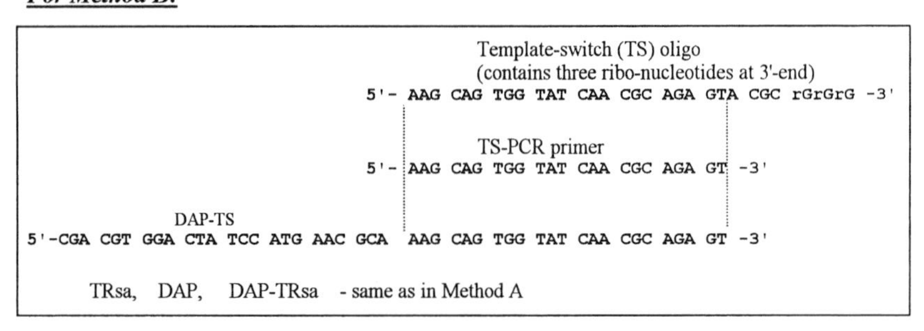

Oligos used in cDNA amlification:

TRsa, TS-oligo - for 1st strand cDNA synthesis
TRsa, TS-PCR - to amplify 1st strand cDNA

Oligos used in RACE from amplified cDNA (according to [7]):

DAP + DAP-TS ("step-out mixture") - universal for 5'-RACE
TS-PCR - nested for 5'-RACE
DAP-TRsa - universal primer for 3'-RACE
TRsa - nested universal primer for 3'-RACE

Box 1

13. Yeast tRNA, 10 μg/μL.
14. 3 *M* sodium acetate, pH 5.0.
15. Fresh MilliQ water.
16. Agarose gel (1%) containing EtBr.
17. Oligonucleotides: *see* **Box 1** and **Note 5**.

2.2.2. Method B Using the Template-Switching Effect

1. SuperScript II reverse transcriptase, 200 U/μL (Life Technologies) or 20X PowerScript reverse transcriptase (Clontech) with provided buffer.
2. 20 m*M* $MnCl_2$.
3. 0.1 *M* DTT.
4. dNTP mix, 10 m*M* each.
5. Long-and-Accurate PCR enzyme mix with buffer (*see* **Note 4**).
6. 10X PCR buffer: provided with the enzyme mix or, if Klentaq-based homemade mix is used: 300 m*M* tricine-KOH, pH 9.1, 160 m*M* ammonium sulfate, 30 m*M* $MgCl_2$, 0.2 mg/mL BSA.
7. Agarose gel (1%) containing EtBr.
8. Fresh MilliQ water.
9. Oligonucleotides: *see* **Box 1** and **Note 5**.

3. Methods

3.1. Total RNA Isolation

1. Dissolve the tissue sample in buffer D (*see* **Note 6**).
2. Spin the sample at maximum speed on table microcentrifuge for 5 min at room temperature, to remove debris. Transfer the supernatant to a new tube.
3. Put the tube on ice, add equal volume of buffer-saturated phenol, and mix. There will be no phase-separation at this moment. Add one-fifth vol chloroform–isoamyl alcohol (24:1), and vortex the sample. Two distinct phases will separate. Vortex 3–4× more with ~1-min intervals between steps. Incubate the tube on ice between steps. Spin at maximum speed on table microcentrifuge for 30 min at 4°C. Remove and save the upper, aqueous phase. Take care to avoid warming the tube with your fingers, or the interphase may become invisible.
4. Repeat **step 3**.
5. Add 1 μL co-precipitant, then add an equal vol of 96% ethanol and mix. Spin immediately at maximum speed on table microcentrifuge at room temperature for 10 min. The precipitate may not form a pellet, being instead spread over the back wall of the tube and thus being almost invisible even with co-precipitant added. Wash the pellet once with 0.5 mL 80% ethanol. Dry the pellet briefly, until no liquid is seen in the tube (do not over-dry).
6. Dissolve the pellet in 100 μL fresh MilliQ water. If the pellet cannot be dissolved completely, remove the debris by spinning the sample at maximum speed on table microcentrifuge for 3 min at room temperature. Transfer the supernatant to a new tube, then add equal volume of 12 *M* LiCl, and chill the solution at –20°C

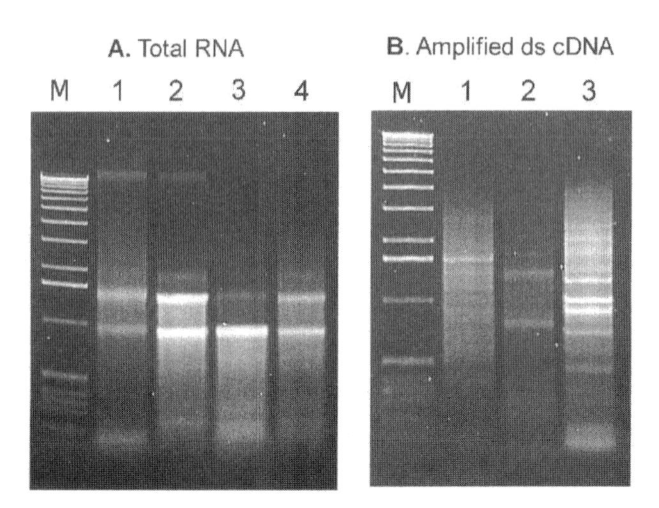

Fig. 3. **(A)** Nondenaturating agarose electrophoresis of total RNA from various invertebrate sources. Lane 1, unidentified sponge; lane 2, comb jelly *Bolinopsis infundibulum* (phylum Ctenophora); lane 3, planarian *Girardia tigrina* (phylum Platyhelminthes); lane 4, stony coral *Montastraea cavernosa* (phylum Cnidaria). M, 50 ng 1-kb DNA ladder (Gibco-Life Technologies). **(B)** Amplified total cDNA from various sources. Lane 1, comb jelly; lane 2, planarian; lane 3, mollusk *Tridacna sp.* M, 50 ng 1-kb DNA ladder (Gibco-Life Technologies). Product on lane 2 needs one more PCR cycle; product on lane 3 is already slightly overcycled (by 1–2 cycles), but is still well-suitable for further manipulations.

for 30 min. Spin at maximum speed on table microcentrifuge for 15 min at room temperature. Wash the pellet once with 0.5 mL 80% ethanol, and dry as previously done. The precipitated RNA is usually invisible, since co-precipitant does not precipitate in LiCl.

7. Dissolve the pellet in 40 µL fresh MilliQ water.
8. Load 2 µL solution onto a standard (nondenaturing) 1% agarose gel to check the amount and integrity of the RNA. Add EtBr to the gel to avoid the additional (potentially RNase-prone) step of gel staining. Load a known amount of some DNA on a neighboring lane to use as standard for determining the RNA concentration. Intact RNA should exhibit sharp band(s) of ribosomal RNA (*see* **Fig. 3A** and **Notes 7–10**).

3.2. cDNA Synthesis

3.2.1. Method A ("Classical")

3.2.1.1. First-Strand cDNA Synthesis

1. To 5 µL RNA solution in water (0.03–3 µg total RNA), add 1 µL of 10 µ*M* primer TRsa, and cover with mineral oil. Incubate at 65°C for 3 min, then put the tube on ice.

2. Add 2 μL 5X first-strand buffer (provided with reverse transcriptase), 1 μL 0.1 *M* DTT, 1 μL reverse transcriptase, 0.5 μL dNTP mix (10 m*M* each), and incubate at 42°C for 1 h, then put the tube on ice.

3.2.1.2. Second-Strand cDNA Synthesis

1. To the first-strand cDNA solution, add 49 μL of MilliQ water, 1.6 μL dNTP mix (10 m*M* each), 16 μL 5X second-strand reaction buffer, and 4 μL 20X second-strand enzyme cocktail (the total volume of the reaction mix is about 80 mL). Incubate at 16°C for 1.5 h, and then put the tube on ice.
2. Add 1 mL T4 DNA polymerase, incubate 0.5 h at 16°C to polish ends.
3. Stop the reaction by heating at 65°C for 5 min.
4. Take the reaction mix from under the oil, put in new tube, and add 0.5 vol phenol, then 0.5 vol chloroform–isoamyl alcohol (24:1). Vortex the solution and spin at maximum speed on table microcentrifuge for 10 min. Transfer the upper, aqueous phase into new tube.
5. Add carrier (SeeDNA, Amersham, or glycogen) and precipitate DNA by adding 0.1 vol (8 μL) 3 *M* sodium acetate (pH 5.0) and 2.5 vol (200 μL) 95% ethanol at room temperature. Spin immediately for 15 min at maximum speed on table microcentrifuge at room temperature.
6. Wash the pellet with 80% ethanol; air-dry the pellet for ~5 min at room temperature. Dissolve pellet in 6 μL water.

3.2.1.3. Adaptor Ligation

1. To the 6 μL double-stranded cDNA, add 2 μL of adaptor (10 μ*M*), 1 μL of 10X ligation buffer, 1 μL T4 DNA ligase, and incubate overnight at 16°C.
2. To the ligation mixture, add 90 μL MilliQ water and 10 μg yeast tRNA. Purify by QiaQuick PCR purification kit (Qiagen, follow manufacturer's instructions), elute with 40 μL T/M buffer. Alternatively, dilute the ligation mixture fivefold, by adding 40 μL MilliQ water to it (*see* **Note 11**).

3.2.1.4. cDNA Amplification

1. Prepare the PCR mixture (note that final concentration of primers is 0.1 μ*M*) as follows: Add 3 μL 10X PCR buffer, 1 μL dNTP mix (10 m*M* of each), 1.5 μL 2 m*M* TRsa primer, 1.5 μL 2 μ*M* DAP primer, 1 μL fivefold dilution of ligation mixture or 20 μL of QiaQuick purified sample of adapter-ligated cDNA, water to 30 μL, and Klentaq–Pfu homemade polymerase mixture, corresponding to 8 U Klentaq (*see* **Note 3**). When using commercial polymerase mixtures, follow manufacturer's recommendations.
2. Perform cycling: 94°C 30 s, 65°C 1 min, 72°C 2 min 30 s (block control), 95°C 10 s, 65°C 30 s, 72°C 2 min 30 s (tube control or simulated tube control). Check 2 μL of the product on a 1% agarose gel after 12 cycles, keeping the PCR tube at room temperature while the electrophoresis runs. If nothing is seen, put the tube back into thermal cycler and do five more cycles. If the product is barely visible,

do only three more cycles. It is very important to determine the minimal number of cycles required to amplify the product until it is readily detectable on an agarose gel with EtBr staining (*see* **Notes 12–15**).

3.2.2. Method B for cDNA Synthesis Using the Template-Switching Effect

3.2.2.1. First Strand cDNA Synthesis

1. To 4 µL RNA solution in water (1–3 µg total RNA), add 1 µL 10 m*M* primer TRsa, and cover with mineral oil. Incubate at 65°C for 3 min; put on ice.
2. Add 2 µL 5X first-strand buffer provided with reverse transcriptase, 1 µL 0.1 *M* DTT, 1 µL 5 µ*M* TS-oligo, 1 µL reverse transcriptase, 0.5 µL dNTP mix (10 m*M* each). Incubate at 42°C for 1 h, then add 1 µL 20 m*M* $MnCl_2$, and incubate for an additional 15 min at 42°C. Heat to 65°C, and incubate for 3 min to stop the reaction. The product can be stored at –20°C for several months.

3.2.2.2. cDNA Amplification

1. Prepare the PCR mixture (final concentration of primers is 0.1 m*M*) as follows: 3 mL of 10X PCR Buffer, 1 mL of dNTP mix (10 m*M* each), 1.5 mL of 2 m*M* TRsa primer, 1.5 mL of 2 m*M* TS-PCR primer, 1.5 mL of five-fold dilution of first-strand cDNA (from **step 2**), milliQ water to 30 mL, and KlenTaq/Pfu homemade polymerase mixture corresponding to 8 U of KlenTaq. When using commercial polymerase mixtures (*see* **Note 4**), follow manufacturer's recommendations.
2. Perform cycling: 94°C 30 s, 65°C 1 min, 72°C 2 min 30 s (block control); 95°C 10 s, 65°C 30 s, 72°C 2 min 30 s (tube control or simulated tube control). To determine the exact number of PCR cycles required to amplify cDNA, use the same strategy as described above for method A, but do 17 cycles before the first check on agarose gel. Typically, it takes ~17 cycles, if there was 1 µg total RNA at the start. If the number of cycles is 22–24, a 10^6-representation sample still can be accumulated by making amplification in 10-fold-larger volume (i.e., making it in 10 30-µL tubes instead of one), then pooling them. In this way, 10X more product of first-strand synthesis may be put into PCR, while avoiding the background problems caused by nonincorporated cDNA synthesis oligomers. However, this approach leads to slightly more distorted cDNA sample compared to direct less-than-20 amplification (*see* **Notes 12–15**).

4. Notes

1. There is widespread belief that RNA is very unstable and therefore all the reagents and materials for its handling should be specially treated to remove possible RNase activity. We have found that purified RNA is stable and, ironically, that too much anti-RNase treatment can become a source of problems. This especially applies to diethyl pyrocarbonate-treatment of aqueous solutions, which often leads to RNA preparations that are stable but completely unsuitable for cDNA

synthesis. Simple precautions, such as wearing gloves, avoiding speech over open tubes, using aerosol-barrier tips, and using fresh MilliQ water for all solutions, are sufficient to obtain stable RNA preparations. All organic liquids (phenol, chloroform, and ethanol) can be considered essentially RNase-free, by definition, as well as the dispersion buffer containing 4 *M* guanidine thiocyanate.

2. Normally, the dispersion buffer does not require titration. If pH comes out significantly lower than 7.0, try another batch of guanindine or disodium citrate. The buffer may be stored for years at 4°C in the dark.
3. For cDNA synthesis, we recommend the use of reagents (except oligonucleotides) provided in the Marathon kit (Clontech).
4. For LA-PCR enzyme mixtures, I strongly recommend enzymes based on Klentaq polymerase (Ab peptides) or its analogs (such as Advantaq polymerase, Clontech), instead of nontruncated *Taq* variants. In our experience, this enzyme produces the least distortion to the cDNA sample during amplification. The LA mixture can be prepared by adding 1 *Pfu* unit cloned *Pfu* polymerase (Stratagene) for every 30 U Klentaq polymerase. Calculate the required amount of mix, assuming that 25 U Klentaq are required for 100 µL PCR.
5. The set of oligonucleotides presented in Box 1 has been extensively tested on a number of various invertebrates and consistently produced good results. It is primarily designed for cDNA amplification and RACE, but can be also successfully applied to preparation of samples for suppression subtractive hybridization *(7)*, since the potentially interfering oligo-derived flanking sequences are removed by *Rsa*I digestion.
6. The volume of tissue should be not more than one-fifth vol buffer D. To avoid RNA degradation, tissue dispersion should be done as quickly and completely as possible, ensuring that cells do not die slowly on their own. To adequately disperse a piece of tissue usually takes 2–3 min of triturating, using a pipet, taking all or nearly all volume of buffer into the tip each time. The piece being dissolved must go up and down the tip, so it is sometimes helpful to cut the tip to increase the diameter of the opening for larger tissue pieces. Tissue dispersion can be done at room temperature. The dispersed samples can be stored at 4°C for several days (exceptions, such as *Balanoglossus* [acorn worm, phylum *Hemichordat*a] which contain high concentrations of highly reactive iodine in its tissues, are rare). The tissue dispersed in buffer D produces a highly viscous solution. The viscosity is usually caused by genomic DNA. This normally has no effect on the RNA isolation (except for dictating longer periods of spinning at the phenol–chloroform extraction steps), unless the amount of dissolved tissue was indeed too great. However, in some cases (e.g., freshwater planarians or mushroom anemones), mucus produced by the animal contributes to viscosity. This substance tends to co-purify with RNA, making it very difficult to collect the aqueous phase at the phenol–chloroform extraction step. It likewise lowers the efficiency of cDNA synthesis. The RNA sample contaminated with such mucus, although completely dissolved in water, does not enter agarose gel during electrophoresis. The EtBr-stained material stays in the well, probably because the mucus adsorbs RNA.

Including cysteine in buffer D can diminish the mucus problem. To buffer D add 0.1 vol solution containing 20% cysteine chloride and 50 m*M* tricine-KOH, pH 7.0 (requires a great deal of titration). The cysteine solution should be freshly prepared. After dissolving the tissue, incubate the sample for 2 h at 4°C, then proceed with the above protocol.

7. RNA degradation can be assessed using nondenaturing electrophoresis. The first sign of RNA degradation on the nondenaturing gel is a slight smear starting from the rRNA bands and extending to the area of shorter fragments, such as seen on **Fig. 3**, lanes 3 and 4. RNA showing this extent of degradation is still good for further procedures. However, if the downward smearing is so pronounced that the rRNA bands do not have a discernible lower edge, the RNA preparation should be discarded. The amount of RNA can be roughly estimated from the intensity of rRNA staining by EtBr in the gel, assuming that the dye incorporation efficiency is the same as for DNA (the rRNA may be considered a double-stranded molecule, because of its extensive secondary structure).
8. The rule for vertebrate rRNA, that in intact total RNA the upper (28s) rRNA band should be twice as intense as the lower (18s) band, does not apply to invertebrates. The overwhelming majority have 28s rRNA with a so-called "hidden break" ***(16)***. It is actually a true break in the middle of the 28s rRNA molecule, which is called "hidden," because under nondenaturing conditions the rRNA molecule is being held in one piece by the hydrogen bonding between its secondary structure elements. The two halves, should they separate, are each equivalent in electrophoretic mobility to 18s rRNA. In some organisms, the interaction between the halves is weak, so the total RNA preparation exhibits a single 18s-like rRNA band even on nondenaturing gel (**Fig. 3A**, lane 3). In others, the 28s rRNA is more robust, so it is still visible as a second band, but it rarely has twice the intensity of the lower one (**Fig. 3A**, lanes 1, 2, and 4).
9. Curiously, genomic DNA contamination is reproducible for a particular species, but varies between species. However, it never exceeds the amount seen at **Fig. 2A**, lanes 1 and 2, a weak band of high molecular weight. Such extent of contamination does not affect further procedures. In fact, the methods of cDNA amplification described here tolerate genomic DNA up to 50% of the total sample mass, without losing specificity or efficiency.
10. To store the isolated RNA, add 0.1 vol of 3 *M* sodium acetate and 2.5 vol of 96% ethanol to the RNA in water, and mix thoroughly. The sample may be stored for several years at –20°C.
11. Using the Qia-Quick purified ligation mixture removes excess nonincorporated adapter oligomers, and prevents them from interfering in the subsequent PCR. The step is necessary when the starting amount of RNA was lower than 0.3–0.5 µg, to make it possible to take all the generated cDNA into subsequent PCR. For higher initial amounts, the purification step may be replaced by fivefold dilution of the mixture, followed by PCR, starting with 1 µL of the dilution. Thus the nonincorporated oligos are simply diluted to a noninterfering concentration. In this case, only one-fiftieth part of the available adapter-ligated cDNA goes into

PCR, but, because of the excess of RNA at the start, this is usually enough to generate a representative cDNA sample. If you are not quite sure which variant to choose, start with dilution. If cDNA amplification requires too many cycles (more than 20, *see* below), purify the remaining ligation mixture by QiaQuick, and take it all into PCR. It is important to use T/M buffer, which contains 10 m*M* Tris-HCl and 1 m*M* $MgCl_2$, for elution. Elution with plain water leads to denaturation of DNA, caused by electrostatic repulsion of strands in low-salt conditions. This will decrease the specificity of amplification and promote background stemming from genomic DNA.

12. The number of PCR cycles required to amplify a visible amount of cDNA (i.e., ~5–10 ng/µL) is a key parameter to assess the representation of an amplified sample. There is a simple link between initial number of target DNA molecules and number of PCR cycles required to amplify the sample (**Box 2**), as it was empirically determined during the work on in vitro cloning ***(17,18)***. Using these guidelines, it can be calculated that a sample consisting of 10^6 molecules (a representation sufficient for most cDNA tasks) or more would require 20 or fewer PCR cycles to be amplified. In other words, if it took <20 PCR cycles to amplify the cDNA, this is a well-representative sample. In our practice, we prefer to achieve at least one-order-of-magnitude-higher representation (i.e., get robust cDNA product in 16–17 cycles) to ensure that we have even the most rare messages.
13. The amplified cDNA at agarose gel should look like a smear (which may contain some bands, corresponding to the most abundant cDNA species) with the average length ~1 kb (*see* **Fig. 3B**). If it comes out much less, this may be a sign of pronounced RNA degradation during cDNA synthesis (if the total RNA was confirmed to be intact), which is usually the result of poor quality of reverse transcriptase. Try another batch of it. Alternatively, something may be wrong with the PCR system. Probably the polymerase mixture is bad, but it is better to replace all the reaction components.
14. We recommend storing the product of amplification as a master sample. The unpurified PCR product produced by a Klentaq-based enzyme mixture can be stored at –20°C for several years. If a large amount of cDNA is required for further procedures (e.g., cloning), use aliquots of the master sample to amplify more material. Dilute the aliquot of the master sample 50-fold in deionized water, and add 1 µL dilution/each 20 µL of the PCR mixture, prepared as at **step 10**. Do exactly 10 PCR cycles, which will generate a product in a concentration equal to the master sample. Do not apply more cycles, attempting to generate more material, because overcycling produces the most pronounced distortions of cDNA profile. Instead, prepare a large volume of PCR mixture, distribute it into several tubes (30 µL/tube), and pool them after amplification is over. If pure DNA is required for further procedures, the amplified cDNA may be cleaned by QiaQuick PCR purification kit (Qiagen) according to the provided protocol, but using T/M buffer for elution, instead of the provided buffer.
15. If one intends to clone the product of cDNA amplification, it is necessary to perform a "chase" step after the product is amplified. The conditions for ampli-

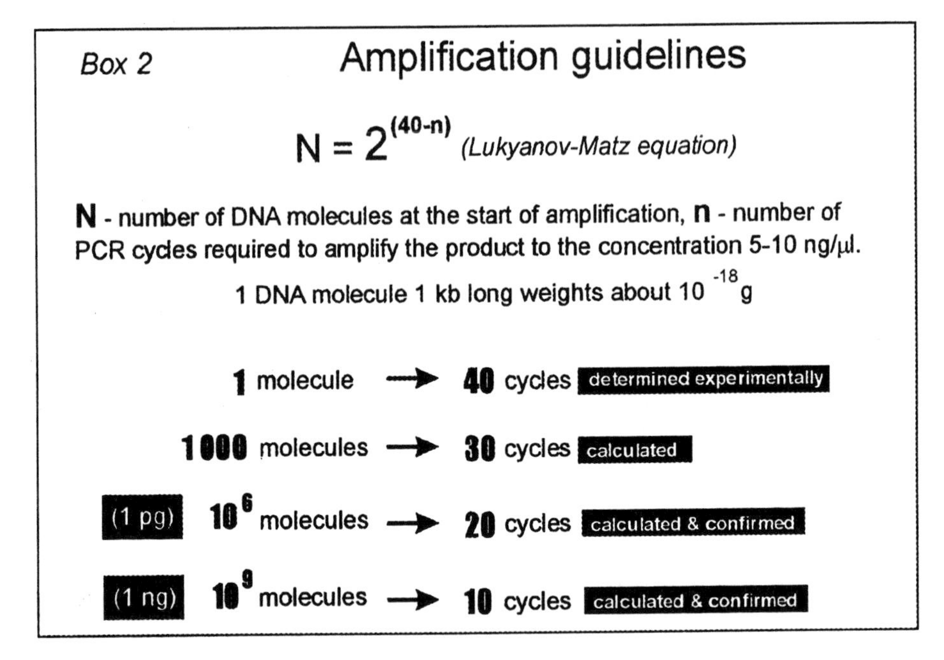

Box 2

fication recommended here include using low working concentration of primers (0.1 µ*M*), which greatly enhances the specificity of poly(A)$^+$-fraction amplification. However, there is a great chance that a substantial fraction of the sample will be denatured at the end of PCR, since there will already be no primers available to initiate the synthesis of complementary strand (especially, if slight overcycling occurred). Obviously, for cloning, it is highly desirable to have the entire PCR product double-stranded. To ensure this without sacrificing the specificity of amplification, do the following. Run the PCR with low primer concentration, as recommended, until the product is amplified; then, keeping the completed PCR reaction in the thermocycler at 72°C, inject an additional amount of primers there (up to 0.2 µ*M* more of each), and perform two nondenaturing chase cycles: 77°C 1 min, 65°C 1 min, 72°C 3 min. Purify the product by QiaQuick PCR purification kit (Qiagen) before cloning (use T/M buffer for elution).

References

1. Matz, M. V., Fradkov, A. F., Labas, Y. A., Savitsky, A. P., Zaraisky, A. G., Markelov, M. L., and Lukyanov, S. A. (1999) Fluorescent proteins from nonbioluminescent Anthozoa species. *Nat. Biotechnol.* **17,** 969–973.
2. Gonzalez, P., Zigler, J. S., Jr, Epstein, D. L., and Borras, T. (1999) Identification and isolation of differentially expressed genes from very small tissue samples. *Biotechniques* **26,** 884–886, 888–892.

3. Spirin, K. S., Ljubimov, A. V., Castellon, R., Wiedoeft, O., Marano, M., Sheppard, D., Kenney, M. C., and Brown, D. J. (1999) Analysis of gene expression in human bullous keratopathy corneas containing limiting amounts of RNA. *Invest. Opthalmol. Vis. Sci.* **40,** 3108–3115.
4. Wang, E., Miller L. D., Ohnmacht, G. A., Liu, E. T., and Marincola, M. (2000) High-fidelity mRNA amplification for gene profiling. *Nat. Biotechnol.* **18,** 457–459.
5. Livesey, F. J., Furukawa, T., Steffen, M. A. Church, G. M., and Cepko, C. L. (2000) Microarray analysis of the transcriptional network controlled by the photoreceptor homeobox gene Crx. *Curr. Biol.* **10,** 301–310.
6. Chenchik, A., Diachenko, L., Moqadam, F., Tarabykin, V., Lukyanov, S., and Siebert, P. D. (1996) Full-length cDNA cloning and determination of mRNA 5' and 3' ends by amplification of adaptor-ligated cDNA. *Biotechniques* **21,** 526–534.
7. Matz, M., Shagin, D., Bogdanova, E., Britanova, O., Lukyanov, S., Diatchenko, L., and Chenchik, A. (1999) Amplification of cDNA ends based on template-switching effect and step-out PCR. *Nucl. Acids Res.* **27,** 1558–1560.
8. Diatchenko, L., Lukyanov, S., Lau, Y. F., and Siebert, P. D (1999) Suppression subtractive hybridization: a versatile method for identifying differentially expressed genes. *Meth. Enzymol.* **303,** 349–380.
9. Matz, M., Usman, N., Shagin, D., Bogdanova, E., and Lukyanov, S. (1997) Ordered differential display: a simple method for systematic comparison of gene expression profiles. *Nucl. Acids Res.* **25,** 2541–2542.
10. Chomczynski, P. and Sacchi, N. (1987) Single-step method of RNA isolation by acid guanidinium thiocyanate-phenol-chloroform extraction. *Analyt. Biochem.* **162,** 156–159.
11. Lee, Y. H., Huang, G. M., Cameron, R. A., Graham, G., Davidson, E. H., Hood, L., and Britten, R. J. (1999) EST analysis of gene expression in early cleavage-stage sea urchin embryos. *Development* **126,** 3857–3867.
12. Siebert, P. D., Chenchik, A., Kellogg, D. E., Lukyanov, K. A., and Lukyanov, S. A. (1995) An improved PCR method for walking in uncloned genomic DNA. *Nucl. Acids Res.* **23,** 1087–1088.
13. Lukyanov, K. A., Diachenko, L., Chenchik, A., Nanisetti, A., Siebert, P. D., Usman, N. Y., Matz, M. V., and Lukyanov, S. A. (1997) Construction of cDNA libraries from small amounts.18 of total RNA using the suppression PCR effect. *Biophys. Biochem. Res. Comm.* **230,** 285–288.
14. Schmidt, W. M. and Mueller, M. W. (1999) CapSelect: a highly sensitive method for 5' CAP-dependent enrichment of full length cDNA in PCR-mediated analyses of mRNAs. *Nucleic Acid Res.* **27,** e31.
15. Chenchik, A., Zhu, Y. Y., Diatchenko, L., Li, R., Hill, J., and Siebert, P. D. (1998) *Gene Cloning and Analysis by RT-PC*R (Siebert, P. and Larrick, J., eds.) BioTechniques, Natick, MA, 305–319.
16. Ishikawa, H. (1977) Evolution of ribosomal RNA. *Comp. Biochem. Physiol. B* **58,** 1–7.

17. Lukyanov K. A., Matz M. V., Bogdanova E. A., Gurskaya N. G., and Lukyanov S. A. (1996) Molecule by molecule PCR amplification of complex DNA mixtures for direct sequencing: an approach to in vitro cloning. *Nucleic Acids Res.* **24,** 2194–2195.
18. Fradkov, A. F., Lukyanov, K. A., Matz, M. V., Diatchenko, L. B., Siebert, P. D., and Lukyanov, S. A. (1998) Sequence-independent method for in vitro generation of nested deletions for sequencing large DNA fragments. *Analyt. Biochem.* **258,** 138–141.

2

Use of *cobA* and *cysG*^{A} as Red Fluorescent Indicators

Charles A. Roessner

1. Introduction

This chapter is based on the observations *(1–3)* that *Escherichia coli* cells bearing the plasmid pISA417, for the overexpression of the *cobA* gene from the bacterium *Propionibacterium freudenreichii*, or the plasmid pEB1, for the overexpression of a truncated *cysG* (*cysGA*) gene of *E. coli*, exhibit bright red fluorescence (**Fig. 1**) when cultured on Luria-Bertani (LB) growth medium and illuminated with ultraviolet (UV) light. The genes both encode uroporphyrinogen III (urogen III) methyltransferases (referred herein to as CobA or CysG^A) which catalyze the methylation of urogen III, an intermediate in heme biosynthesis, using *S*-adenosyl-L-methionine as the methyl donor. Plasmid pISA417 was constructed by insertion of a DNA fragment bearing the complete *cobA* gene into pUC19 (**Fig. 2**) and was originally used for the characterization of urogen III methyltransferase *(1)*. During this study, it was noticed that *E. coli* colonies harboring pISA417 are brightly red fluorescent when illuminated with UV light. However, *E. coli* cells harboring pISA417 bearing a DNA insert that deletes or knocks out the expression of *cobA* are not fluorescent, thus providing the basis for its first use as a fluorescent indicator in selecting recombinant plasmids *(2)*.

The fluorescence is caused by the cytoplasmic accumulation of two polar fluorescent compounds derived by the methylation of urogen III at C-2, C-7, and C-12 (**Fig. 3**), to afford dihydrosirohydrochlorin (precorrin-2) and a fluorescent trimethylpyrrocorphin *(1,4)*. Precorrin-2 is oxidized to the fluorescent sirohydrochlorin (factor II) either by oxygen, or enzymatically, by CysG and NAD. In contrast to heme and siroheme, whose cellular concentrations are tightly regulated in *E. coli*, the fluorescent compounds are synthesized and

From: *Methods in Molecular Biology, vol. 183: Green Fluorescent Protein: Applications and Protocols*
Edited by: B. W. Hicks © Humana Press Inc., Totowa, NJ

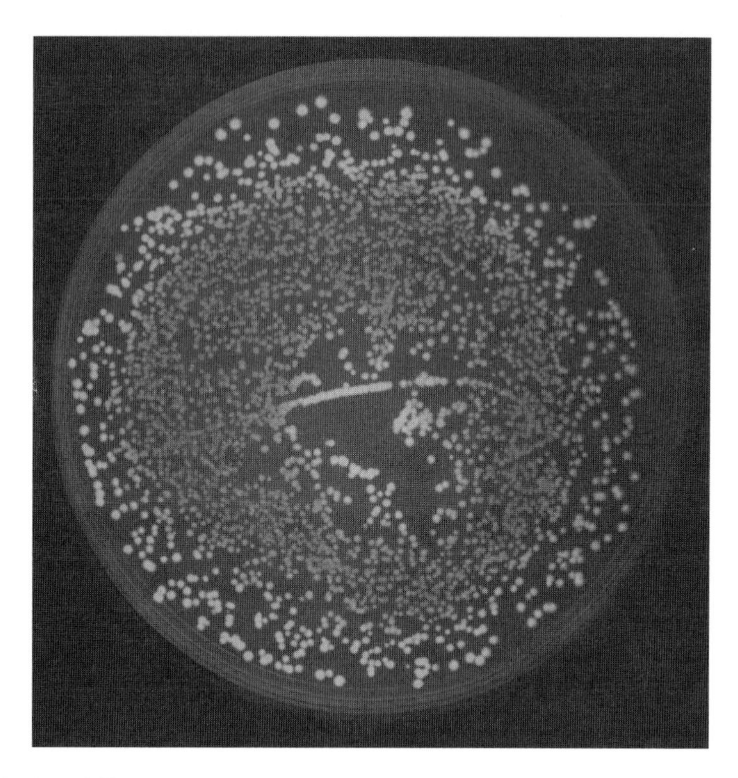

Fig. 1. A red fluorescent strain of *E. coli*. Strain CR417 (TB1 bearing pISA417) was grown on an LB-AMP plate and photographed with UV illumination. (For optimal, color representation please see accompanying CD-ROM.)

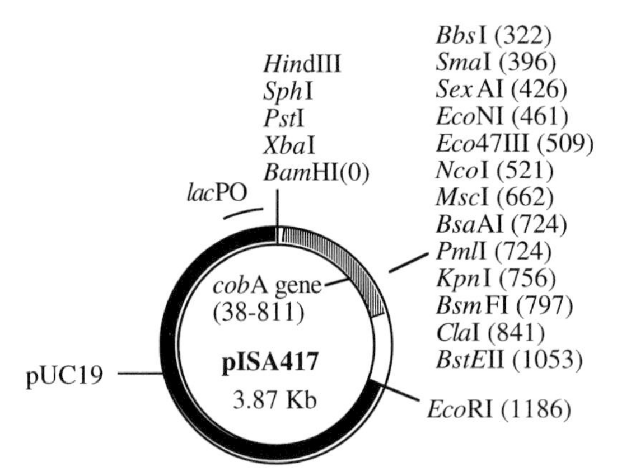

Fig. 2. The structure of pISA417, showing the location of the unique restriction sites derived from pUC19 and a 1.1-kb *Bam*HI-*Eco*RI insert bearing the *P. freudenreichii cobA* gene. The *Cla*I and *Bst*EII sites lie outside the *cobA* gene. The sites within *cobA* were predicted from the sequence (Genbank accession no. U13043). (Also on CD-ROM.)

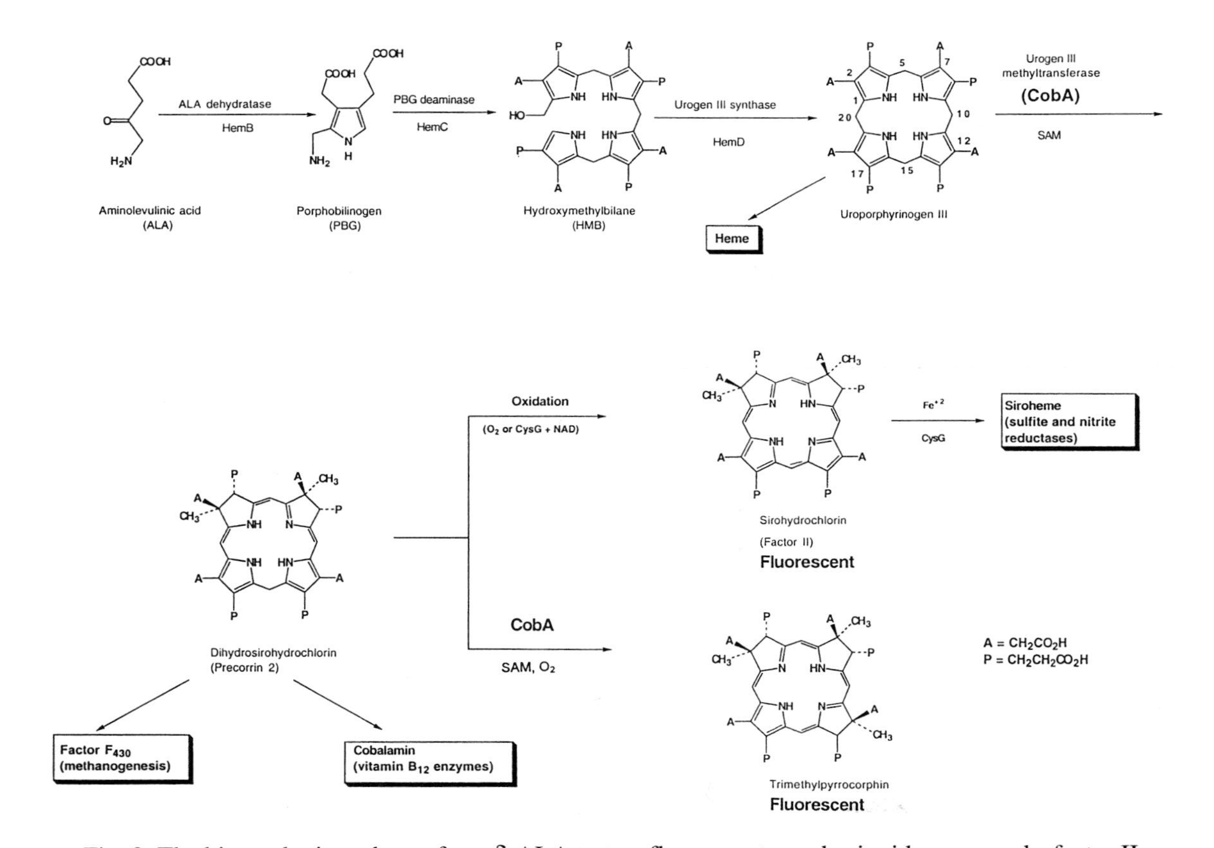

Fig. 3. The biosynthetic pathway from δ-ALA to two fluorescent porphyrinoid compounds, factor II and trimethylpyrrocorphin. (Also on CD-ROM.)

accumulate at relatively high levels, probably because of loss of feedback inhibition of aminolevulinic acid (ALA) synthesis in cells overexpressing urogen III methyltransferase ***(5)*** and stability of the products.

In *E. coli* and some other bacteria, such as *Salmonella typhimurium* and *Neisseria meningitidis*, urogen III methyltransferase is part of siroheme synthase (CysG), a multifunctional enzyme encoded by the *cysG* gene. CysG contains not only urogen III methyltransferase activity ($CysG^A$) in its C-terminal region, but also NAD-dependent oxidase and ferrochelatase activities ($CysG^B$) in its N-terminal region, which convert precorrin-2 to siroheme ***(3,6)***. Thus, overexpression of the complete *cysG* gene in *E. coli* leads to accumulation of siroheme, which is not fluorescent.

This chapter describes two methodologies: the use of pISA417, carrying the *cobA* gene as a red fluorescent indicator, for the selection of recombinant plasmids, and a protocol for the expression of the truncated *cysGA* gene. The procedure described for overexpression of *cysGA* in *E. coli* uses the polymerase chain reaction (PCR) and vector selection, to provide strong transcriptional and translational signals. Demonstrating its utility, this procedure has been adapted to construct plasmids for expressing *cobA* from *Pseudomonas denitrificans* and *UMP1* from *Arabidopsis thaliana* ***(5,7)*** to give red fluorescent *E. coli*. Since the *P. freudenreichii cobA* gene is derived from a high G-C, Gram-positive bacterium, it may not be suitable for expression in all organisms, and alternative sources of the gene may be desirable. However, similar technology has recently expanded the use of the *P. freudenreichii cobA* gene as a regulated red fluorescent reporter not only in bacteria but also in yeast (*Schizosaccharomyces pombe*) and cultured mammalian (Chinese hampster ovary) cells ***(8)***. In the latter case, *cobA* was expressed either by itself to provide red fluorescent cells, or in conjunction with the green fluorescent protein, to create cells that emitted both red and green fluorescence.

2. Materials

1. LB medium: 5 g/L yeast extract (Difco), 10 g/L tryptone (Difco), and 5 g/L NaCl.
2. LB agar: LB medium, add 15 g/L agar (Difco) before autoclaving. Add 50 μg/mL ampicillin (Sigma, sodium salt), after autoclaving. Add 10–20 μg/mL aminolevulinic acid (Sigma), after autoclaving, from a 10-mg/mL stock solution of ALA sterilized by filtration (ALA is destroyed by autoclaving).
3. *E. coli* K12 strain TB1 ***(9)*** is used throughout this work (*see* **Note 1**).
4. pISA417 is supplied (*see* **Note 2**) in strain CR417 (TB1 bearing pISA417).
5. pCR252 bearing the *E. coli cysG* gene ***(10)*** is isolated from TB1(pCR252).
6. pUC19 ***(11)*** is isolated from TB1(pUC19).
7. STET buffer: 8% sucrose, 50 m*M* Tris-HCl, pH 8.0, 50 m*M* EDTA, 0.5% Triton X-100; autoclave, and store at room temperature.

8. Egg white lysozyme (Sigma, 10 mg/mL in water).
9. Isopropanol.
10. 70% Ethanol, 100% ethanol.
11. TE buffer: 10 m*M* Tris-HCl, pH 8.0, 1.0 m*M* EDTA.
12. The insert DNA can be any DNA fragment of interest, whether a PCR product or a restriction fragment. In the example given here, a 0.6 kb blunt end PCR product is used.
13. Restriction enzymes and buffers: *Bam*HI, *Eco*RI, *Sma*I, and their 10X buffers (New England Biolabs).
14. T4 DNA ligase with 10X buffer (New England Biolabs).
15. *Taq* polymerase, 10X polymerase buffer (Mg free), and 25 m*M* $MgCl_2$ (Promega).
16. dNTPs for PCR (New England Biolabs). Dilute the four dNTPs (100 m*M*) to 10 m*M* with water, then a mixture is prepared by combining 50 µL of each dNTP with 200 µL water (1.25 m*M* final concentration). Store dNTP solutions at –20°C.
17. Phenol pH 8.0 (Ambion) is stored at –20°C. Prior to use, 8-hydroxyquinoline is added to 0.1%.
18. Autoclaved 7.5 *M* ammonium acetate.
19. Chloroform.
20. Sterile 10% glycerol.
21. Electroporation cuvets (1.0-mm gap) and an electroporator, e.g., the *E. coli* Pulser (Bio-Rad).
22. Recovery medium: 50 mL LB broth supplemented with 1.0 mL 20% glucose, 0.5 mL 1.0 *M* $MgSO_4$, and 0.05 mL 1.0 *M* $CaCl_2$.
23. For PCR, the following template and primers were used:
    a. Template DNA. pCR252 containing the complete *cysG* gene (***10***; *see* **Note 3**).
    b. PCR primers were synthesized on the 40-nmol scale. The *Bam*HI, RBS, and start codons are indicated:

    5' primer

    *Bam*HI RBS **Start** codons 211–220

    5'-CGCGCGGATCCAGGAAGGAATTTAAA**ATG***GAA*ACGACCGAAC AGTTAATCAACGAACCG-3'

    3' primer

    *Eco*RI **stop** anticodons 457–448

    5'-CGCCGGAATTC**TTA**ATGGTTGGAGAACCAGTTCAGTTTATCGCG-3'

    A 0.1-nmol/µL stock solution of the primers is prepared by dissolving 40 nmol of the primer in 400 µL of TE buffer, and stored at –20°C. Just prior to use, an aliquot of the stock is diluted to 0.01 nmol/µL with water.

3. Methods

3.1. Selection of Recombinant Plasmids, Using pISA417

The insertion of any DNA fragment into pISA417, as described here, using any of the unique sites shown in **Fig. 2** (*see* **Note 4**) will disrupt the *cobA* gene and result in nonfluorescent colonies.

3.1.1. Isolation of pISA417

1. Strain CR417 is usually received as filter disks that have been saturated with an overnight culture of the strain.
2. To recover the strain, place a filter disk on an LB-amp plate, streak for isolation, and incubate 16–20 h at 37°C.
3. Briefly illuminate the plate with a long-wavelength (302 nm) UV transilluminator (*see* **Note 5**), select a brightly fluorescent colony to inoculate into 50 mL LB-amp, and incubate overnight at 37°C, in a shaking water bath.
4. Fill a 1.5-mL microcentrifuge tube with the culture, pellet the cells in a microcentrifuge, and discard the supernatant. The cell pellet should be brightly fluorescent.
5. Resuspend the cells in 200 µL STET buffer (vortex vigorously), add 20 µL lysozyme solution, mix, and place the tube in a boiling water bath for 40 s.
6. Centrifuge at top speed in a microcentrifuge (≥10,000*g* for all microcentrifugations) for 15 min and remove the viscous pellet with a flat toothpick.
7. Add an equal volume of isopropanol (usually 150–200 µL), mix, and centrifuge for 10 min.
8. Remove the supernatant, add 0.5 mL 70% ethanol, vortex briefly, and centrifuge for 5 min.
9. Remove the supernatant, and dry the pellet under vacuum (Speed-Vac or lyophilizer).
10. Dissolve the pellet in 50 µL TE buffer (vortex vigorously), and store at –20°C. This procedure normally yields DNA concentrations of 100–200 ng/µL.

3.1.2. Restriction Enzyme Digestion

1. For this example, pISA417 is digested with *Sma*I in the following mixture, in a 0.5-mL microcentrifuge tube: 2 µL pISA417 (200 ng), 10 µL insert DNA (10–200 ng), 4 µL 10X *Sma*I buffer,1 µL *SmaI*I (10–20 U), sufficient water (23 µL) to make the total volume 40 µL.
2. Incubate the mixture 1 h at 25°C.
3. Extract the restriction digest with phenol to inactivate the enzymes. Add an equal volume of phenol to the digest and vortex for 1 min.
4. Centrifuge at top speed in a microcentrifuge and transfer the upper layer to a clean 0.5-mL tube. This layer should be clear but will sometimes appear milky, because of precipitation of phenol. The lower (phenol) layer will be yellow from the hydroxyquinoline.
5. To remove dissolved phenol from the DNA solution, add 40 µL chloroform, vortex briefly, centrifuge briefly, and remove the bottom (chloroform) layer with a micropipet. Perform the chloroform extraction a second time.
6. Precipitate the DNA by adding one-half vol 7.5 *M* ammonium acetate and 2 vol 100% ethanol. For example, if there is 30 µL DNA solution remaining after extraction with phenol and chloroform, add 15 µL ammonium acetate and 90 µL ethanol.

7. Mix and place the tube at –80°C for at least 30 min, centrifuge 10 min at top speed in a microcentrifuge, to pellet the DNA, and remove the supernatant.
8. To wash the pellet (usually not visible) add 200 µL 70% ethanol, vortex briefly, centrifuge 5 min, and completely remove the supernatant.
9. Dry the pellet for at least 1 h under vacuum to remove any remaining traces of the volatile ammonium acetate.

3.1.3. Ligation and Ethanol Precipitation of DNA Fragments

1. Dissolve the DNA pellet from **Subheading 3.1.2.** in 17 µL water, add 2 µL 10X ligation buffer, 1 µL T4 DNA ligase, and incubate 16–20 h at 16°C.
2. After ligation, ethanol precipitate the DNA as described in **Subheading 3.1.2.** and dissolve the pellet in 10 µL water.

3.1.4. Transformation of Electrocompetent TB1 Cells by Electroporation and Selection for Recombinant Plasmids

1. Produce electrocompetent cells by inoculating a colony of TB1 into 50 mL LB and incubate overnight at 37°C in a shaking water bath.
2. Inoculate two fresh 50-mL portions of LB with 0.5 mL of the overnight culture and grow the cells to an $A_{600} = 0.8$ at 37°C. Chill the cultures on ice and pellet the cells at 5000 rpm for 10 min in sterile 50-mL tubes in a Sorvall SS34 rotor or its equivalent. All centrifugations are done at 4°C.
3. Remove the medium and wash the cell pellets twice by gently resuspending them in 20 mL ice cold 10% glycerol and centrifugation as above. After the second wash, resuspend both pellets in a total of 1.0 mL 10% glycerol, and pellet the cells in a 1.5-mL microcentrifuge tube.
4. Resuspend the final pellet in 400 µL 10% glycerol, divide into 50-µL aliquots in microcentrifuge tubes on ice, and store at -80°C.
5. Thaw a tube of the electrocompetent TB1 cells on ice, and mix in 5 µL of the DNA solution.
6. Transfer the mixture to an ice-cold electroporation cuvet (1.0-mm gap), and incubate on ice for 5 min.
7. Thoroughly dry the outside walls of the cuvet, and electroshock the cells, using a setting of 1.8 kV on the electroporator.
8. Immediately add 1.0 mL recovery medium, and incubate the cells for 1.0 h at 37°C, to allow the cells to recover from the shock and allow expression of the ampicillin resistance gene.
9. Plate the cells by spreading on LB-amp plates (*see* **Note 6**) and incubate 16–20 h at 37°C. Several different amounts (1, 10, 100 µL) of cells should be plated, to ensure obtaining a plate that has isolated colonies. The smaller amounts should be added to 100 µL sterile water, before spreading.
10. Examine the plates with long-wavelength (302 nm) UV light (remove the Petri dish cover, and invert the plate over the light source), and select nonfluorescent colonies (**Fig. 4**) for further analysis.

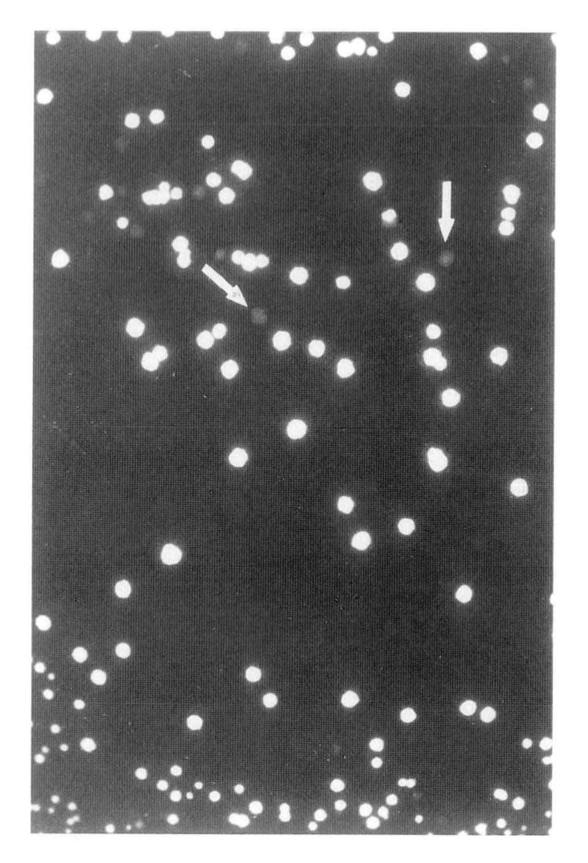

Fig. 4. *E. coli* TB1 that has been transformed with a ligation mixture prepared as described in **Subheading 3.1.** The photograph was taken with a Polaroid camera with an orange filter routinely used for photographing ethidium bromide-stained DNA gels. In black and white photographs, fluorescent colonies are bright white and nonfluorescent colonies are pale gray (*arrows*). (Also on CD-ROM.)

11. The presence of the insert is determined by preparing plasmid DNA from nonfluorescent cells, as described above, and analyzing for presence of the insert on a 1% agarose gel.

3.2. Using Genes Encoding Urogen III Methyltransferase as a Fluorescent Indicator: Overexpression of E. coli cysG^A Gene

This methodology is based on the use of PCR to amplify all or part of a gene, and, at the same time, provide optimal cloning, transcriptional, and/or translational signals for efficient expression of the gene, either through design of the PCR primers or selection of the vector into which the PCR product is inserted. In the example given, the portion of *cysG* encoding urogen III

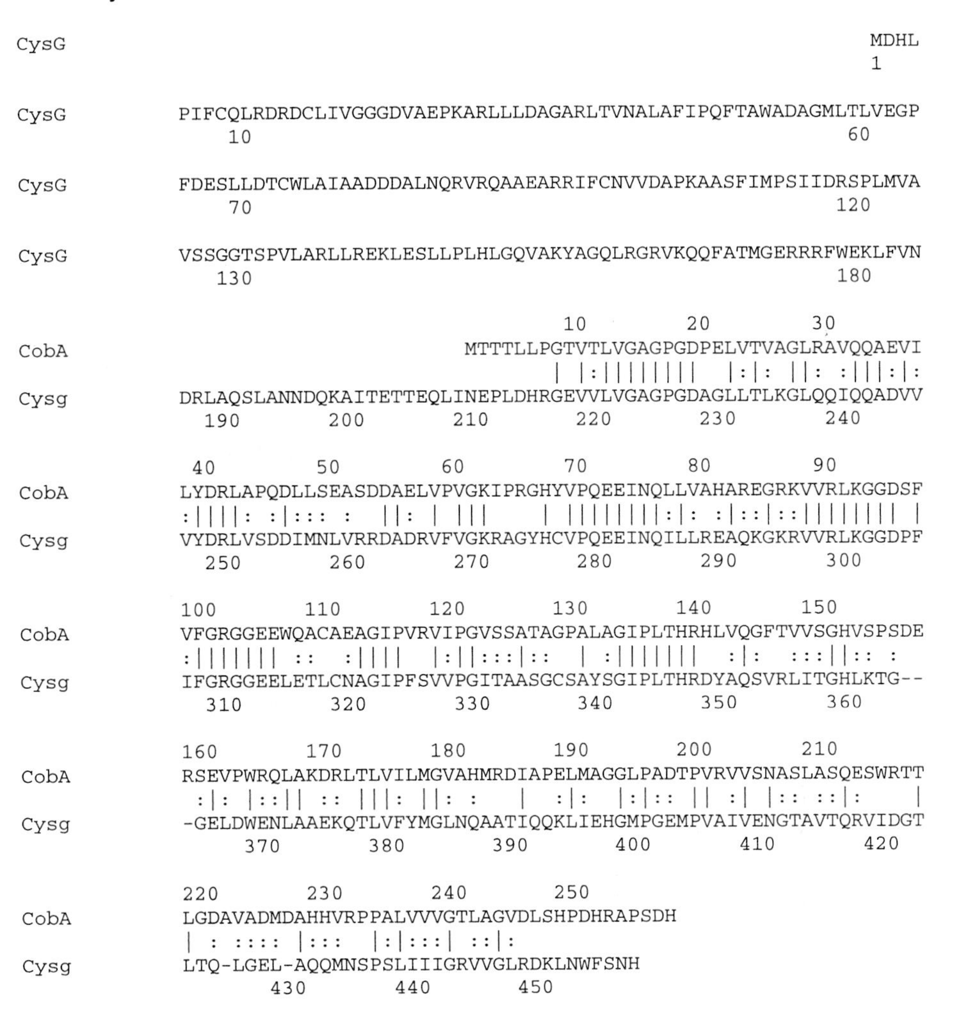

Fig. 5. Homology comparison between CobA and CysG, to determine where urogen III methyltransferase begins in CysG. There is 42.6% identity in a 237-amino acid overlap beginning with amino acid 211 (glutamate) of CysG. (Also on CD-ROM.)

methyltransferase was determined by a homology comparison of CysG with CobA from *P. freudenreichii* using the FASTA program (Genetics Computer Group, Madison, WI) with the result shown in **Fig. 5**. As can be seen, the region of overlap between CysG and CobA begins at Glu 211, and extends almost to the end, so the 5' PCR primer was designed to remove the first 210 codons of *cysG*. The vector chosen for expression of the truncated gene was pUC19, which provides the strong *lac* promoter, but no translational signals. Therefore, a ribosome-binding site, the ATG start codon, and codons for amino acids 211–220, in addition to a *Bam*HI restriction site (*see* **Note 7**), were incorporated into the 5' primer. The 3' primer was designed to provide the anticodons for the last 10 amino acids of CysG, a stop anticodon, and an *Eco*RI restriction site. Insertion of the PCR product (*cysGA*) into pUC19 results in a plasmid (pEB1), which, when transformed into TB1, affords red fluorescent cells indistinguishable from CR417 ***(3)***.

3.2.1. PCR of Methyltransferase Fragment of cysG^A

1. The following mixture is prepared in a 0.5-mL microcentrifuge tube for PCR amplification of the methyltransferase fragment of the *cysGA* gene: 16 µL water, 5 µL 10X buffer, 8 µL dNTP mix (1.25 m*M*), 5 µL 5' primer (0.01 nmol/µL), 5 µL 3' primer (0.01 nmol/µL), 5 µL $MgCl_2$ (25 mM, *see* **Note 8**), 1 µL pCR252 (100 ng/µL), 1 U *Taq* polymerase, for a total volume of 50 µL.
2. Overlay the mixture with 50 µL mineral oil, and perform 30 cycles of a sequence consisting of 94°C for 1 min, 55°C for 1 min, 72°C for 2 min.
3. At the end of the cycles, remove the mineral oil, and run 5 µL of the PCR mix on a 1% agarose gel.
4. If the product has been synthesized, extract the reaction mix with phenol and chloroform, and ethanol-precipitate the product as described in **Subheading 3.1.3.**
5. Dissolve the dried pellet in 50 µL TE buffer.

3.2.2. Restriction Enzyme Digestion, Ligation, Transformation, and Plating

These procedures are carried out by following all of the steps described above, except that the plasmid is pUC19, the insert is the PCR product, and two restriction enzymes, *Bam*HI and *Eco*RI (or others engineered into the insert by PCR), are used. After plating and an overnight incubation at 37°C, one should be able to observe fluorescent colonies that harbor the recombinant plasmid and express the *cysGA* gene.

4. Notes

1. In the examples given here, the host strain used is TB1 but any strain of *E. coli* that makes urogen III and *S*-adenosyl-L-methionine should work. If the strain overexpresses the *lac* repressor (*lacI^Q*), induction with isopropyl-β-D-thioglactoside may be required.
2. CR417 is available from the author (c-roessner@tamu.edu), and has also been submitted to the Belgian Coordinated Collections of Micro-organisms (http://www.belspo.be/bccm/lmbp.htm).
3. Sources of template DNA for amplification of urogen III methyltransferase genes from other organisms may include plasmids bearing the gene, prokaryotic genomic DNA, or genomic libraries, or cDNA libraries from eukaryotic organisms.
4. The *Bam*HI site, shown in **Fig. 2** apparently was lost during the construction of pISA417, therefore, pISA417 is not cut by *Bam*HI.

5. UV light causes thymidine dimer formation and can result in mutations and cell death. Therefore, exposure of the plates to UV light should be kept to a minimum at all times. Proper eye protection should be used to prevent UV damage to the retina.
6. Addition of ALA (10–20 μg/mL) to the medium may enhance the fluorescence of the colonies bearing nonrecombinant plasmids. However, it may also cause the colonies harboring recombinant plasmids to exhibit faint background fluorescence.
7. Care must be taken that the restriction sites chosen for cloning do not cut within the gene being inserted into the vector.
8. The most critical variable in PCR reactions is the magnesium ion concentration, which should be determined for each set of primers and template. Therefore, a series of concentrations (0.5, 1.0, 1.5, 2.0, 2.5, 3.0, 3.5, 4.0, 4.5, 5.0 m*M*) should be tested. Often, a difference of only 0.05 m*M* will have a drastic effect.

Acknowledgments

I would like to express deep gratitude to Professor Ian Scott for his continued support of a portion of the work described here through grants from the National Institutes of Health and the Robert Welch Foundation.

References

1. Sattler, I., Roessner, C. A., Stolowich, N. J., Hardin, S. H., Harris-Haller, L. W., Yokubaitis, N. T. (1995) Cloning, sequence and expression of the uroporphyrinogen III methyltransferase *cobA* gene of *Propionibacterium freudenreichii (shermanii). J. Bacteriol.* **177,** 1564–1569.
2. Roessner, C. A. and Scott, A. I. (1995) A fluorescence-based method for screening recombinant plasmids. *BioTechniques* **19,** 760–764.
3. Warren, M. J., Bolt, E. L., Roessner, C. A., Scott, A. I., Spencer, J. B., and Woodcock, S. (1994) Gene dissection demonstrates that the *Escherichia coli cysG* gene encodes a multifunctional protein. *Biochem. J.* **302,** 837–844.
4. Warren, M. J., Stolowich, N. J., Santander, P. J., Roessner, C. A., Sowa, B. A., and Scott, A. I. (1990) Enzymatic synthesis of dihydrosirohydrochlorin (precorrin-2) and of a novel pyrrocorphin by uroporphyrinogen III methylase. *FEBS Lett.* **261,** 76–80.
5. Roessner, C. A., Park, J.-H., and Scott, A. I. (1999) Genetic engineering of *E. coli* for the production of precorrin-3 in vivo and in vitro. *Bioorg. Med. Chem.* **7,** 2215–2219.
6. Spencer, J. B., Stolowich, N. J., Roessner, C. A., and Scott, A. I. (1993a) The *Escherichia coli cysG* gene encodes the multifunctional protein, siroheme synthase. *FEBS Lett.* **335,** 57–60.
7. Leustek, T., Smith, M., Murillo, M., Singh, D. P., Smith, A. G., Woodcock, S. C., Awan, S. J., and Warren, M. J. (1997) Siroheme biosynthesis in higher plants: Analysis of an S-adenosyl-L-methionine-dependent uroporpyrinogen III methyltransferase from *Arabidopsis thaliana. J. Biol. Chem.* **272,** 2744–2752.
8. Wildt, S. and Deuschle, U. (1999) cobA, a red fluorescent transcriptional reporter for *Escherichia coli*, yeast and mammalian cells. *Nat. Biotech.* **17,** 1175–1178.
9. Baldwin, T. O., Berends, T., Bunch, T. A., Holtzman, T. F., Rausch, S. K., Shamansky, L., and Treat, M. L. (1984) Cloning of the luciferase structural genes

from Vibrio harveyi and expression of bioluminescence in *Escherichia coli. Biochemistry* **23,** 3663–3667.

10. Warren, M. J., Roessner, C. A., Santander, P. J., and Scott, A. I. (1990) The *Escherichia coli cysG* gene encodes S-adenosylmethionine dependent uroporphyrinogen III methylase. *Biochem. J.* **265,** 725–729.
11. Yanisch-Perron, C., Vieira, J., and Messing, J. (1985) Improved cloning vectors and host strains: nucleotide sequence of M13mp18 and pUC19 vectors. *Gene* **33,** 103.

3

Circular Permutation of the Green Fluorescent Protein

Simon Topell and Rudi Glockshuber

1. Introduction

1.1. The Principle of Circular Permutation of Proteins

The green fluorescent protein (GFP) from the jellyfish *Aequorea victoria* has become one of the most important markers for studying gene expression and protein targeting in intact cells and organisms *(1–3)*. GFP represents the first genetically encoded reporter molecule that is detectable in the absence of an enzymatic substrate or cofactor, in a variety of cell types.

The vast majority of GFP variants generated in past years were the result of mutagenesis experiments on the natural, linear genetic sequence of GFP. Compared to these standard mutagenesis procedures, circular permutation is an alternative method of protein engineering that entails the manipulation of protein sequences to a much larger extent. In a thought experiment (**Fig. 1**), the concept can be considered as follows: A wild-type protein in which the termini are in close proximity is circularized by linking the N- and C-terminal ends directly, or via a short linker peptide. The circular protein is subsequently cleaved at another position in the sequence, generating new termini that are again in close proximity. In practice, this experiment is performed at the level of the genetic sequence encoding the corresponding polypeptide chain. This review summarizes the recently reported circular permutation experiments on GFP, and also gives a brief, general overview on the field of circular permutation of proteins.

Circular permutation has been extensively studied for its impact on protein structure, protein folding and conformational stability in the case of more than 20 different proteins. In general, two requirements must be fulfilled for a suc-

From: *Methods in Molecular Biology, vol. 183: Green Fluorescent Protein: Applications and Protocols*
Edited by: B. W. Hicks © Humana Press Inc., Totowa, NJ

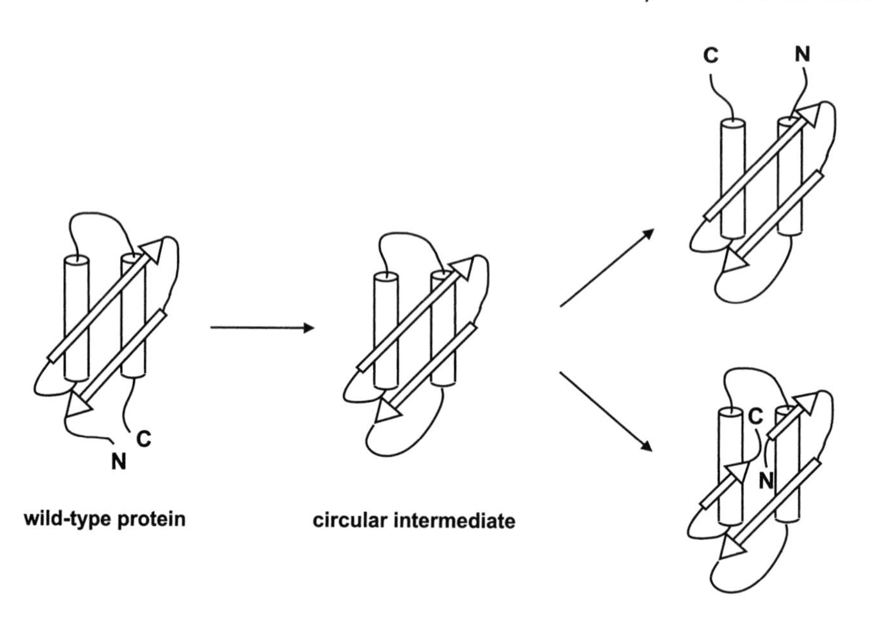

Fig. 1. The principle of circular permutation of a polypeptide chain. The N- and C-terminus of the wild-type protein are virtually connected, generating a circular protein intermediate. The protein backbone is then cleaved at a different position, yielding a circularly permuted variant of the protein with novel N- and C-termini.

cessful circular permutation experiment: The termini of the corresponding protein must be in close proximity, (a fact that is fulfilled by many proteins, including GFP *[4]*), and the introduction of the new termini may not affect folding of the protein to its functional tertiary structure. Folding and stability measurements have been reported for many engineered circularly permuted proteins. Those studies revealed that circular permutations are often tolerated by proteins without loss of functional tertiary structure, indicating that the position of the termini does not contain essential information for tertiary structure formation *(5,6)*. We will discuss several selected examples in the following.

In 1983, Goldenberg and Creighton performed the first circular permutation of a protein. They chemically cross-linked the natural termini of bovine pancreatic trypsin inhibitor (BPTI) with a water-soluble carbodiimide, thus creating a circular polypeptide chain. Subsequent incubation with trypsin yielded a linear protein with new termini in the trypsin-binding loop of BPTI between Lys15 and Ala16 of the wild-type protein. Permuted BPTI showed inhibitory activity against trypsin, and could be reconstituted in vitro after unfolding and reduction of the three disulfide bonds *(7)*.

The first circular permutation of a protein by genetic engineering was performed by Luger et al. ***(8)***, who created two circularly permuted variants of phosphoribosylanthranilate isomerase, with new termini located in surface-exposed loops. Both variants were similar to the wild-type in their three-dimensional structure and biologically active in vivo. The influence of circular permutations on the α-spectrin SH3 domain has been studied in great detail. All circular permutations of the protein with new termini in loop regions folded correctly, as probed by circular dichroism and nuclear magnetic resonance spectroscopy. However, their folding pathways significantly differed from that of the wild-type, which strongly supported the view that multiple folding pathways exist that lead to the thermodynamically most stable protein structure, i.e., the biologically active protein ***(9)***. Bacteriophage T4 lysozyme was among the first two-domain proteins to which circular permutation was applied ***(10)***. The two domains are connected by a long α-helix, with both termini in close vicinity to the interdomain surface on the side of the C-terminal domain.

Recent investigations on the role of interdomain interactions for folding and stability of the enzyme were performed by introducing new termini near the interdomain surface, generating a protein with a consecutive domain arrangement at the sequence level. Although this variant folded correctly, its stability was strongly reduced, compared to that of the wild-type. Another permutation experiment on a two-domain protein was reported by Wieligmann et al. ***(11)***, who introduced new termini into the linker sequence between the two domains of the homodimeric protein, βB2-crystallin. The resulting protein was the first example of a circularly permuted protein variant with altered quaternary structure. Probably because of improved interdomain interactions, the mode of domain interaction changed from intermolecular to intramolecular, so that the permuted protein became monomeric. Circular permutation was also reported for subunits of the hetero-oligomeric protein aspartate transcarbamoylase (ATCase), which is composed of six regulatory and six catalytic subunits. Circularly permuted variants of the catalytic subunit were able to assemble to stable, enzymatically active complexes with native regulatory chains ***(12)***.

In all these studies, the design of the permuted proteins was guided by the known three-dimensional structures of the proteins, in a way that new termini were exclusively introduced into loop regions between elements of regular secondary structure. To address the question of whether new termini in permuted proteins are also tolerated within regular secondary structures, a novel experiment was recently developed in which new termini were randomly introduced into the catalytic subunit of ATCase ***(13)*** and the disulfide oxidoreductase, DsbA ***(14)***. In both cases, random libraries of circularly permuted protein were generated with a method normally used for construction of cDNA libraries. Circular genes were constructed, randomly relinearized by limited digestion with DNase I, and cloned into expression plasmids. Appropriate screening systems were then developed that enabled identification of *Escherichia coli* cells that produced catalytically active permuted variants, and the active variants were isolated, sequenced, and investigated in detail.

These experiments showed an unexpected high tolerance for introduction of new termini into ATCase and especially into DsbA. Specifically, a large number of catalytically active circularly permuted variants with termini within regular secondary structures were found. In the case of DsbA, the active, permuted variants showed structural properties comparable to those of the wild-type. In certain regions of DsbA however, introduction of new termini always led to inactive variants. DsbA variants with termini in these “forbidden regions” showed altered overall folds and lacked catalytic activity. Analysis of active permuted DsbA variants also revealed that there was neither preference for new termini in solvent-accessible positions on the protein surface, nor for termini in segments with high mobility of the main chain (as judged from crystallographic *B*-factors).

Essentially the same observations were made, when Iwakura et al. ***(15)*** performed the first complete circular permutation study on a protein. They rationally designed every possible circularly permuted variant of dihydrofolate reductase (DHFR). Specifically, the regions in DHFR that could not be disrupted by introduction of new termini, without loss of folding competence, coincided with segments known to be involved in early folding events in DHFR.

1.2. Circular Permutations of GFP

From the systematic permutation studies on DsbA ***(14)*** and DHFR ***(15)***, clearly circular permutation is a sensitive method for identifying segments in a polypeptide chain that are important for folding and stability, and possibly represent essential folding nuclei. In this context, circular permutation of the 238-residue protein, GFP, appears to be an especially complex case because the folding pathway of GFP from the newly synthesized polypeptide chain to the native protein with intact fluorophore involves at least two defined intermediates and two posttranslational modifications. Thus, if the formation of any of these intermediates is affected by the introduction of new termini, the whole folding process could be influenced.

In the three-dimensional stucture of GFP, which consists of an 11-stranded β-barrel (**Fig. 2**), the *p*-hydroxybenzylideneimidazolidone chromophore is located within a helical segment in the center of the barrel, and is entirely shielded from the solvent. In vitro folding of unfolded GFP, isolated from bacterial inclusion bodies, and GFP expression studies in vivo suggest that newly synthesized GFP first forms an intermediate state, I1, which possibly already exhibits a native-like stucture. The main feature of I1 is that it brings

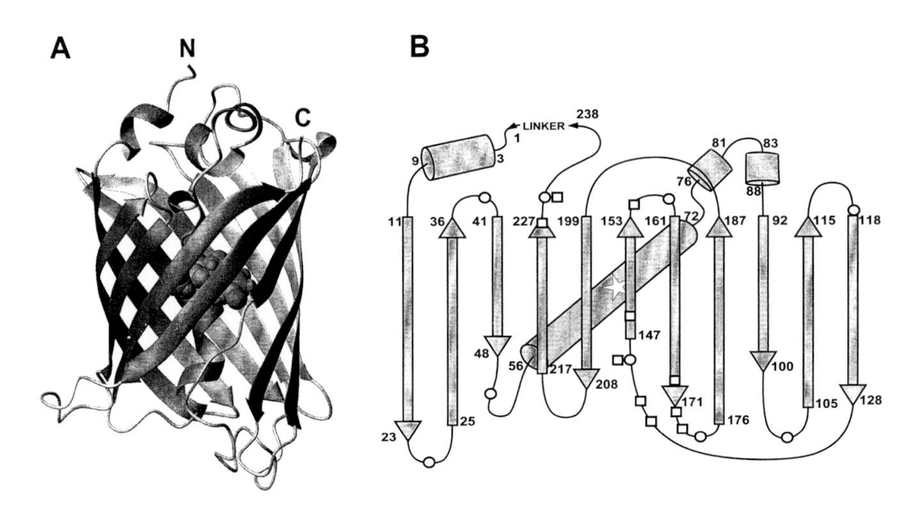

Fig. 2. **(A)** Ribbon diagram of the X-ray structure of wild type GFP from *A. victoria* ***(28)***. The figure was generated with the program MOLMOL ***(29)***. **(B)** Location of new termini in fluorescent, circularly permuted GFP variants. Termini of variants generated by random circular permutation are indicated by squares ***(17)***, and rationally designed permuted variants are indicated by circles ***(16)***. The linker had the sequence $(Gly)_2$-Thr-$(Gly)_2$-Ser, in the case of the random approach, and the sequence Gly-Ser-$(Gly)_2$-Thr-Gly, in the case of the rational approach. The first and last residues of regular secondary structures, as determined by Yang et al. ***(28)***, are indicated. The position of the chromophore in the central helical segment (56–72) is marked by a star.

the GFP segment around residues Ser65-Tyr66-Gly67 into a conformation that favors the spontaneous cyclization of the main chain between Ser65 and Gly67 with simultaneous release of a water molecule. This reaction, which does not occur spontaneously in unfolded GFP, yields the intermediate, I2. In the next folding step, I2 reacts with molecular oxygen, releasing a second water molecule, and generating the mature GFP chromophore. Whether another conformational rearrangement occurs in GFP after chromophore formation is unknown. Thus, the minimum scheme for the folding pathway of GFP in vivo can be described as shown in **Fig. 3** ***(15a)***.

Because none of the reaction steps in the folding pathway of GFP, i.e., the capability of the protein to form intermediates I1 and I2, may be affected in a circularly permuted GFP variant, without influencing the capability to form the chromophore, one might expect GFP to be more sensitive to circular permutation than other proteins. Recently, two independent investigations on circularly permuted variants of GFP have been reported which essentially support this view. Topell et al. ***(16)*** investigated the in vivo folding of 20 rationally designed, circularly permuted variants of the cycle 3 variant of GFP (also

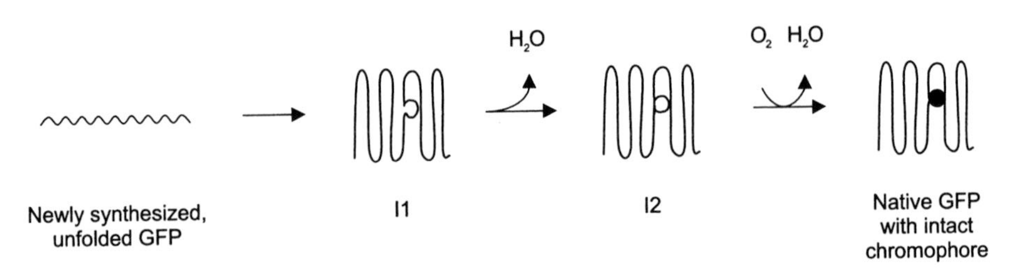

Fig. 3. The minimum steps necessary in the folding pathway, to produce active GFP.

termed "GFPuv") in *E. coli*. GFPuv differs from wild-type GFP by the replacements Phe99Ser, Met153Thr, and Val163Ala, and is one of the most frequently used GFPs, because it shows lower aggregation tendency than the wild-type. The 20 permuted variants were designed so that a hexapeptide linker connected the natural termini of GFPuv, and so that new termini were introduced into loop regions at the top and bottom of the 11-stranded β-barrel and within individual strands of the β-barrel scaffold (**Table 1**; **Fig. 2**). None of the variants with termini in strands were capable of folding into a native conformation and constitution with intact chromophore, indicating that the β-barrel is a highly cooperative folding unit in which each individual β-strand plays an important role for structural integrity of the barrel. A similar conclusion can be drawn from a random circular permutation experiment on the enhanced green fluorescent protein (EGFP) reported by Baird et al. ***(17)***, because new N-termini in β-strands could only be found at residue 148 at the beginning of the strand comprising residues 147–153, and at position 169 at the end of the strand, comprising residues 161–171 (**Fig. 2**).

GFP also shows a low tolerance toward introduction of new termini into loop regions. Rational introduction of 15 new termini into 12 different loops of GFPuv only yielded fluorescent variants in ~50% of the constructs, indicating that the loop regions of GFP also have an important role for folding and stability ***(16)***. All loops that were identified as being tolerant for the introduction of new termini by random circular permutation coincided with the tolerant loops identified by rational permutation of GFPuv ***(16,17)*** (**Fig. 2**).

Although GFPuv and EGFP exhibit similar fluorescence properties (emission maxima at 509 and 514 nm, respectively), their absorbance properties are significantly different. GFPuv has two absorbance maxima at 397 and 475 nm, but the first absorbance peak is almost completely lacking in EGFP (absorbance maximum, 488 nm). All circularly permuted GFPuv variants investigated so far showed spectroscopic characteristics that were surprisingly similar to those of the wild-type protein, with respect to absorbance and fluorescence maxima and fluorescence quantum yields (**Table 1**). The only exception is the permuted GFPuv variant, starting at Tyr145 and ending at Asn144. This variant showed an almost complete loss of the second absorbance peak at 475 nm, and a higher molar extinction coefficient at 397 nm (**Table 1**). The first and second absorbance maxima of GFP have been assigned to the protonated and deprotonated forms of the chromophore, respectively. Because both forms are not in equilibrium with the solvent, the ratio between the absorbance maxima reflects the apparent pK_a value of the chromophore in the environment of the folded GFP structure. The properties of the permuted variant, Tyr145-Asn144, can thus be interpreted in terms of an increase in the apparent pK_a of the chromophore due its altered environment in the permuted variant. Random circular permutation of EGFP yielded fluorescent variants, which, in contrast to EGFP wild-type, showed both absorbance maxima (**Table 1**). Only the permuted variants in which the peptide bond 144–145 was disrupted retained the single absorbance maximum at 488 nm. Thus, the circularly permuted variants of GFPuv and EGFP, starting at residue 145 and ending at residue 144, show exactly opposite absorbance properties, in that GFPuv 145–144 only exhibits the first and EGFP 145–144 only exhibits the second absorbance maximum ***(16–20)***.

Circularly permuted protein variants are thermodynamically less stable than the corresponding wild-type proteins ***(14,21,22)***. Analysis of the stability of five purified permuted GFPuv variants against thermal denaturation revealed that four of the variants essentially retained the extraordinary stability of wild-type GFPuv (T_m = 82°C), and that one of the variants (Thr50-Thr49) even showed an increased stability (**Table 1**; ***16***). Indeed, the permuted Thr50-Thr49 variant of GFPuv is the first example of a circularly permuted, monomeric protein that is more stable than the wild-type. Together with the almost unchanged spectroscopic properties of the permuted variants, these data indicate that interactions between the chromophore and side chains in the hydrophobic core of GFP strongly contribute to the overall stability of the protein. In accordance with this view is the observation that the only permuted GFPuv variant with significantly decreased thermal stability was the variant Tyr145-Asn144, in which, as discussed above, the local environment of the chromophore significantly differs from that in the wild-type.

What are possible applications of circularly permuted GFPs, in addition to answering general questions on protein folding, particularly GFP folding? Baird et al. ***(17)*** have used the fact that the permuted variant, Tyr145-Asn144, in which the new termini are in relatively close proximity to the chromophore, is capable of forming the chromophore to insert calmodulin and a zinc finger domain between residues 144 and 146 of EGFP and enhanced cyan, and yellow fluorescent proteins. Using these constructs, they were able to monitor

Table 1
Properties of Circulatory Permuted GFP Variants

N-terminal a.a.[a]	C-terminal a.a	Position of new termini in 3D structure	Absorbance/ Excitation maxima (nm)	Emission maximum (nm)	$A_{400}/A_{475\ (490)}$[d]	Quantum yield (%)	Apparent T_m[e] (°C)
		Rationally designed circularly permuted variants of GFPuv[b]					
		Fluorescent variants with intact three-dimensional structure					
GFPuv	w.t.	w.t.	397,475	506	2.16	79	81.9 ± 0.5
H25	H25-G24	loop β1-β2	397,475	506			
Y39	Y39-T38	loop β1-β3	397,475	506			
T50	T50-T49	loop β3-α2	397,475	506	2.04	76	83.3 ± 0.5
D103	D103-D102	loop β4-β5	397,475	506			
D117	D117-G116	loop β5-β6	397,475	506			
Y145	Y145-N144	loop β6-β7	397,475	506	14.9	47	68.3 ± 0.5
K158	K158-Q157	loop β7-β8	397,475	506	2.16	74	>75
G174	G174-D173	loop β8-β9	397,475	506	2.25	70	79.5 ± 1.0
I229	I229-G228	C-terminal tail	397,475	506	2.14	76	79.1 ± 0.5
		Non Flluorescent circularly permuted GFPuv variants					
G10	T9	loop α1-β1					
K45	L44	β3					
F64	T63	loop α2-α3					
Q69	V68	loop α2-α3					
D76	P75	loop α3-α4					
F83	D82	loop α3-α4					
T97	R96	β4					
K131	F130	loop β6-β7					
K166	F165	β8					
S205	Q204	β10					
E213	N212	loop β10-β11					

(Table 1 Continued)

Flluorescent circularly permuted variants of EGFP generated by random circular permutation[c]						
EGFP	w.t.	w.t.	488	508	0.15	60
E142M	N144LSE	loop β6–β7	404,494	514	1.09	
Y143N	N146LSE	loop β6–β7	404,494	512	0.31	
Y142I	N144LSE	loop β6–β7	404,494	512	1.16	
Y145I	N144	loop β6–β7	488	514	0.20	
Y145M	N144	loop β6–β7	487	512	0.23	
H148I	N149LSE	β7	488	510	0.08	
H148I	K162SE	β7, β8	398,490	512	0.35	
D155I	K156SE	loop β7–β8	400,496	514	1.47	
H169	N170LSE	β8	396,490	514	1.47	
H169I	N170LSE	β8	396,494	514	1.19	
E172M	I171DLSE	loop β8–β9	398,492	514	1.24	
D173I	D173LSE	loop β8–β9	398,494	514	1.38	
D173	E172SE	loop β8–β9	398,492	514	1.33	
A227	A227I	C-terminal tail	400,492	514	0.61	
I229	I229	C-terminal tail	396,492	514	1.21	

[a]Amino acid numbering according to ***(28)***, [b]From ***(16)***, [c]From ***(17)***, [d]A_{400}/A_{475} in the case of EGFP, [e]Determined at pH 7.4 ***(16)***.

metal binding-induced conformational changes in the inserted domains via a several-fold fluorescence increase of the chromophore in the GFP moiety ***(17)***. Insertions of proteins undergoing strong conformational changes upon ligand binding into GFP regions and tolerating introduction of new termini thus represent a promising alternative to measuring intracellular ligand-binding events by fluorescence resonance energy transfer, using fusions of blue fluorescent protein (BFP) and GFP to the termini of the corresponding protein ***(23,24)***.

2. Materials

1. The plasmid, pGFPuv-cyc, for circular permutation is based on GFPuv (Clontech), with a 6-amino-acid insert for the linker containing a *Kpn*I site, to span the distance between the N- and C-termini.
2. Expression vector: pRBI-PDI (accession code A22413) with *lac* promoter/operator ***(16)***.
3. Restriction enzymes and buffers: *Kpn*I.
4. Agarose gels, analytical and preparative.
5. T4 DNA ligase and 10X ligase buffer.
6. Polymerase chain reaction (PCR) primers for amplification of circularly permuted GFPuv (will vary for different permutations; those listed below are for the Y145-N144 permutation):
   a. Primer 1: 5'-CGACGCGAAT TCTAGATAAC GAGGGCAACA TATGTATAAC TCACACAATG TA-3'
   b. Primer 2: 5'-CGTGCGCCCG GGAGATCTTA GTTGTACTCG AGTTTGTG-3'
7. Lysis buffer for colony screening: 100 m*M* Tris-HCl, pH 7.8, 150 m*M* NaCl, 5 m*M* $MgCl_2$, 1.5% bovine serum albumin, 1 μg/mL pancreatic DNase I, 40 μg/mL lysozyme.
8. *E. coli* strains: XL1-Blue for amplification of plasmid DNA, JM 83 for expression of circularly permuted proteins.
9. Isopropyl-β-D-thiogalactoside (IPTG).
10. Qiagen plasmid mega kit (Qiagen, Basel, Switzerland) and Promega Wizard *Plus* SV Minipreps DNA purification system (Promega, Madison, WI) for plasmid purification.
11. Qiaex II and QiaQuick gel extraction kits (Qiagen) for isolation of DNA fragment from agarose gels.
12. Nitrocellulose filters (Millipore HATF, or similar).
13. Chloroform.
14. Western blot wash buffer: 10 m*M* Tris-HCl, pH 8.0, 150 m*M* NaCl.
15. Software for calculating protein extinction coefficients from given primary sequence is available from http://www.expasy.ch/tools/protparam.html (Swiss Institute of Bioinformatics).
16. 0.1 *M* NaOH, 0.2 *M* NaOH.
17. 5 m*M* sodium phosphate, pH 7.4.

3. Methods

3.1. Construction of a Circular GFP Gene

A length of six amino acids has turned out to be sufficient to span the distance between both termini of GFP *(16,17)* in circularly permuted GFP variants. **Figure 4** shows the region of the plasmid pGFPuv-cyc encoding the linker Gly-Ser-Gly-Gly-Thr-Gly, which was used for rational circular permutation of GFPuv by Topell et al. *(16)*.

1. The plasmid was constructed such that it can be used for generating a circular GFPuv gene with continuous reading frame, by cleavage with *Kpn*I and ligation, and as an expression plasmid of randomly circularly permuted GFP variants. The plasmid is freely available from the authors upon request.
2. Perform a plasmid preparation with a kit that yields at least 200 μg plasmid DNA, such as the Qiagen plasmid Maxi or Mega kit.
3. Digest an amount of pGFPuv-cyc or an analogous derivative with *Kpn*I in the appropriate buffer provided by the manufacturer, which will yield at least 10 μg of the 732-bp fragment (for pGFPuv-cyc, the amount is approx 60 μg).
4. Separate the digest on a 1% agarose gel.
5. Excise the 732-bp fragment from the agarose gel and extract it using, e.g., Qiagen Qiaex II kit.
6. Cyclize 10 μg of the gene fragment in a volume of 400 μL with 100 U T4 DNA ligase. Use 40 μL of 10X T4 DNA ligation buffer and dilute with the appropriate amount of water. Add the DNA ligase and the DNA. To minimize formation of oligomers, keep the concentration of the DNA below 2.5 μg/100 μL.
7. Run the ligase reaction for 18 h at 16°C (temperature cycle ligation may yield better results), (*see* **Note 1**).
8. Analyze the yield of the circular gene on a 1% agarose gel. The circular gene runs faster than the linear fragment (apparent length: ~400 bp). Dimers of the circular gene run slightly above the linear fragment (~800 bp). About 80% or more of the product should be the circular monomer.
9. Separate the products on a preparative 1% agarose gel.
10. Excise the band containing the circular monomer and extract it from the gel, using e.g., Qiagen Qiaex II kit.

3.2. Construction of Circularly Permuted GFPs by Rational Design

1. As outlined in **Fig. 5**, the circular GFP gene can be used as template to amplify any circularly permuted GFP gene by PCR. As an alternative to rational design, random permutations can be performed, but this procedure suffers from numerous drawbacks (*see* **Note 2**).
2. The corresponding PCR primers for rational permutation should be designed such that at least 12 bp at both 3' ends are exactly complementary to the template. We used the primers 1 and 2 and ran our PCR reactions under the following conditions: 95°C for 15 s, 45°C for 15 s, and 72°C for 70 s (repeat for 5 cycles); then 95°C for 15 s, 68°C for 15 s, and 72°C for 70 s (repeat for 25 cycles).

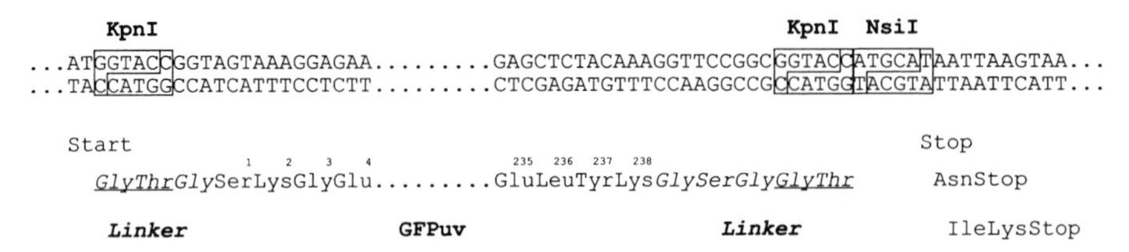

Fig. 4. Nucleotide sequence of a region of the plasmid pGFPuv-cyc. This plasmid has been used for the generation of the circular gene for GFPuv, i.e., the precursor of circularly permuted GFPuv genes *(16)*. The 732-bp fragment, containing the *gfpuv*-gene and the codons for the hexapeptide linker (Gly-Ser-Gly_2-Thr-Gly), is obtained by cleavage of pGFPuv-cyc with *Kpn*I. This fragment is then circularized with T4 DNA ligase, yielding a continuous GFPuv reading frame without start and stop codons (after cleavage with *Kpn*I and ligation, one of the two Gly-Thr codon couples [underlined] disappears). This fragment can then either be used as a template for PCR reactions, to generate rationally designed, circularly permuted GFPuv variants, or can be cleaved randomly with DNase I, yielding randomly circularly permuted GFP variants (*see* **Fig. 5**). Expression vectors for randomly circularly permuted GFP variants can also be derived from pGFPuv-cyc. Digestion with N*si*I, followed by digestion with *Kpn*I and subsequent removal of 3'-overhangs with T4 DNA polymerase, yields a vector fragment into which blunt-ended, circularly permuted GFPs can be cloned. The vector fragment ends with ATG, and starts with three consecutive stop codons, in all possible reading frames. Permuted GFP genes are cloned into an expression vector, and expressed under control of the *lac* promoter/operator sequence.

We recommend performing this two-step PCR reaction with ~5 cycles, at a temperature that allows primer annealing to the circular template, followed by ~20 cycles with a higher (more restrictive) annealing temperature.

3. Separate the amplified DNA fragment on a 1% agarose gel.
4. Isolate the PCR product from the gel.
5. Clone it into the expression vector (*see* **Note 3**). We digest the vector with *Xba*I and *Bam*HI, and the PCR products with *Xba*I and *Bgl*II. *Bam*HI and *Bgl*II give compatible cohesive ends that can be ligated with T4 DNA ligase.
6. Transform *E. coli* JM 83 cells by electroporation using, e.g., a Bio-Rad *E. coli* Pulser.

3.3. Colony Screening of E. coli for Expression of Fluorescent GFP Variants

In general, GFP variants may need substantially longer than the wild-type protein to fold and develop fluorescence in the cytoplasm of *E. coli*. According to our experience, growth of the bacteria on agar plates for 48 h at room temperature, and subsequent storage at 4°C, are conditions that allow detection of any GFP variant that is capable of forming the chromophore and which is not

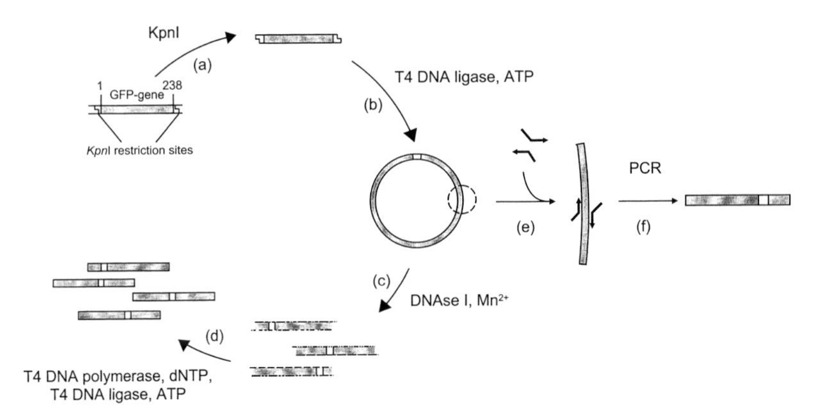

Fig. 5. Scheme for rational and random generation of circularly permuted variants of a protein. Both approaches utilize a circular gene that is constructed by cleavage at *Kpn*I restriction sites flanking the natural gene (a) and subsequent ligation with T4 DNA-ligase (b). For random circular permutation, the circular gene is partially digested with DNase I (c) and a library of relinearized genes is isolated. The library is treated with T4 DNA ligase and T4 DNA polymerase (d), yielding blunt-ended, repaired fragments that can be cloned into an appropriate expression plasmid. The circular gene can also be used as template for PCR amplification to generate rationally designed, circularly permuted variants (e,f).

degraded by *E. coli* proteases. In difficult cases, efficient development of the chromophore may take several days of incubation at 4°C. Agar plates should be screened on an UV transillluminator with a broad excitation spectrum, in the range of 360–520 nm. The time of exposure of the cells to UV light should be kept as short as possible (*see* **Note 4**).

3.4. Screening of E. coli for Production of Inactive GFPs

Some permuted GFP variants do not develop fluorescence during expression in *E. coli*, but adopt stable tertiary structures, because they are soluble and not proteolytically degraded *(16)*. To identify colonies producing nonfluorescent GFPs, a colony immunoblotting procedure can be applied according to the following protocol:

1. Place an autoclaved nitrocellulose filter (Millipore HATF, or similar) on top of an agar plate with single colonies and wait until the filter is completely soaked with medium. Mark the filter and the agar plate so that spots on the filter can later be assigned to the corresponding colonies.
2. Remove the filter, and place it, with the bacteria on top of the filter, onto a fresh agar plate containing IPTG (or a corresponding inducer), to start GFP expression. Incubate for 2–4 h at 42°C.

3. Remove the filter from the agar plate and place it onto a damp piece of paper in a covered glass container. Place an open vessel filled with chloroform beside the filters, and expose bacteria to the chloroform vapor for 15 min.
4. Transfer the filter into a container with lysis buffer, and shake for 12–16 h at room temperature.
5. Remove lysis buffer, wash filter twice with standard Western blot washing buffer (or comparable buffer), and wipe away all the remaining traces of lysed bacteria.
6. Treat the filter using conventional Western blot procedures, starting with the blocking step.
7. Detect GFP-expressing cells with polyclonal anti-GFP antibodies (*see* **Note 5**).

3.5. Determination of GFP Fluorescence Quantum Yields

Not all GFP variants generated by circular permutation will exhibit the same fluorescence characteristics as the wild-type. The following properties may be changed in GFP variants: the absorbance and emission maxima, the percentage of fully converted chromophore, and the fluorescence quantum yield. Once a new GFP variant has been isolated, the absorbance and emission wavelengths can easily be determined by fluorescence excitation and emission scans. The fluorescence quantum yield can be measured as outlined in the following subheading ***(16,26)***.

3.5.1. Determination of the Concentration of GFP Molecules with Intact Chromophore

This method only applies to circularly permuted GFP variants that contain a wild-type chromophore derived from the tripeptide sequence, S65-Y66-G67. The exact value for the extinction coefficient of the base-denatured chromophore is not available for all other chromophore sequences.

A certain fraction of a recombinant GFP variant may not show any fluorescence because of incomplete conversion of the chromophore. The exact concentration of GFP with intact chromophore can be determined by exploiting the absorbance characteristics of the base-denatured chromophore. The two typical GFP absorbance peaks in the range of 360–520 nm are converted into a single absorbance peak at 447 nm when the protein is denatured in 0.1 *M* NaOH. The extinction coefficient for this absorbance peak, $\varepsilon_{447nm} = 44{,}100\ M^{-1}/cm^{-1}$ ***(25)***, is then used to calculate the exact concentration of active GFP:

1. Adjust the protein concentration to ~10 µ*M* with 5 m*M* sodium phosphate, pH 7.4, by using its calculated extinction coefficient at 280 nm.
2. Dilute the protein solution 1:1 with 0.2 *M* NaOH, incubate for 3 min, and record an absorbance spectrum in the range of 350–500 nm. Measure the absorbance at 447 nm, correct for the volume increase, and calculate the concentration of active GFP, using $\varepsilon_{447nm} = 44{,}100\ M/cm$.

3. The molar extinction coefficient of the chromophore in native GFP variants at any wavelength, × nm, can be determined from the equation

$$\varepsilon_{X\text{nm}} = A^{\text{pH 7.4}}_{X\text{ nm}} \,/\, A^{0.1\,M\text{ NaOH}}_{447\text{ nm}} \cdot 44{,}100\ M/\text{cm}$$

3.5.2. Determination of Quantum Yields of GFP Variants

Variants of GFP may differ substantially in their fluorescence quantum yields. The quantum yield of a new variant can be determined by comparison with the fluorescence intensity and known quantum yield of the corresponding wild-type ***(26)***.

1. Prepare solutions of the variant and the wild-type, so that they exhibit the same absorbance at a wavelength where the absorbance spectra of both proteins overlap, and both proteins can be excited efficiently. The protein concentrations should be <1–2 μ*M* to avoid a possible dimerization of GFP, which might affect the fluorescence properties of the protein.
2. Excite both samples at the selected wavelength and record complete emission spectra.
3. Integrate the areas under the emission curves (AUC) for both scans.
4. The quantum yield (ϕ) of the variant is calculated by the equation

$$\phi_{\text{variant}} = (\text{AUC}_{\text{variant}}/\text{AUC}_{\text{wild-type}}) \cdot \phi_{\text{wild-type}}$$

5. The following quantum yields have been determined for frequently used GFPs:
   a. GFP wild-type: $\lambda_{ex} = 395$ nm; $\phi = 79\%$
   b. GFPuv: $\lambda_{ex} = 397$ nm; $\phi = 79\%$
   c. EGFP: $\lambda_{ex} = 488$ nm; $\phi = 60\%$
   d. EBFP: $\lambda_{ex} = 380$ nm; $\phi = 20\%$

4. Notes

1. Temperature cycle ligation has been shown to increase the efficiency of sticky- and blunt-end ligation reactions by a factor of 4–6 ***(27)***. A good protocol for ligation in a PCR thermocycler is given below:
   a. Incubate for 30 s at 10°C.
   b. Heat to 30°C with a heating rate of 0.2°C/s.
   c. Incubate for 30 s at 30°C.
   d. Cool to 10°C, with a cooling rate of 0.2°C/s.
   e. Repeat the previous steps 100×.

   This protocol is especially suited for ligation of complex libraries of randomly circularly permuted genes, via blunt-ended fragments into their expression vectors.
2. The circular template DNA can also be used for construction of randomly circularly permuted genes by limited digestion with DNase I in the presence of Mn^{2+} ions, and subsequent incubation with T4 DNA polymerase and T4 DNA ligase (**Fig. 5**). The corresponding protocol has been described in detail by Graf and Schachman ***(13)***. The yields of circularly permuted, relinearized genes after DNase I digestion, are generally extremely low, so that large quantities

(~10–50 μg) of the circular gene are required in order to end up with a reasonably complex library of permuted genes that can be cloned via blunt ends into an appropriate expression vector.

Another complication of the random circular permutation experiment is the fact that DNase I does not only create blunt ends, but also 5' and 3' protruding ends in the relinearized genes. Treatment with T4 DNA polymerase and deoxyribonucleoside triphosphate causes N/C-terminal elongations, in the case of 5' protruding ends, and N/C-terminal deletions in the case of 3' protruding ends. The deletions/elongations may be more than 100 bp long, but are generally much smaller (for a detailed discussion *see* **ref. *14***). Consequently, one generally has to expect not only permuted proteins with wild-type length, but also elongated variants and variants with terminal deletions. However, this is not necessarily a disadvantage of the random method, because one can address additional interesting questions, such as to what extent deletions are tolerated, and, in the case of elongations, whether the N- or C-terminal elongation is incorporated into the three-dimensional structure of the permuted protein (e.g., ***14***).

Finally, in generating circularly permuted variants of a protein with a random approach, there can be variations in the processing of the randomly cut fragment by T4 DNA polymerase and the subsequent blunt-end cloning of the fragment into the expression vector. Three possible reading frames result from the trimming of sticky ends, as well as two possible orientations of the fragment in the vector backbone, corresponding to a maximum theoretical yield of 17% of clones expressing active GFPs. Besides the already mentioned complications caused by elongated or shortened proteins, C-terminal extensions by 1–3 residues occur when DNase I cleavage of the circular gene creates overhangs of 1–2 bp (**Fig. 5**) ***(13,14,17)***.

3. The choice of the GFP expression system is worth some consideration. Since some variants may be very inefficient in forming the chromophore, a strong protein expression, possibly already in noninduced cells, is desirable. No reports have been made so far that GFP is in any form toxic to *E. coli* cells, so that a background expression in the absence of an inducer generally does not inhibit growth of *E. coli*. We have made good experiences with the high-copy plasmid, pRBI-PDI, containing the *lac* promoter/operator and lacking the *lac* repressor (*lac*I) gene ***(16)***.
4. For screening of bacterial colonies, use of an UV transilluminator or UV lamp is recommended (or possibly several UV lamps), to cover a broad excitation wavelength range. A range of 350–550 nm is ideal. If one only applies an excitation wavelength that is optimal for the wild-type before mutagenesis, one may fail to detect interesting new GFP variants.
5. When performing immunoblot procedures to detect circularly permuted variants of GFP, the use of monoclonal antibodies cannot be recommended. Since these antibodies only bind to one specific epitope that may be altered/disrupted in a circularly permuted variant, detection of this variant may fail with a monoclonal antibody. Polyclonal antibodies, on the other hand, recognize a whole range of

epitopes and it is highly unlikely that all of these epitopes disappear in one circularly permuted variant.

References

1. Pollok, B. A. and Heim, R. (1999) Using GFP in FRET-based applications. *Trends Cell Biol.* **9,** 57–60.
2. Goodwin, P.C. (1999) GFP biofluorescence: imaging gene expression and protein dynamics in living cells. Design considerations for a fluorescence imaging laboratory. *Meth. Cell Biol.* **58,** 343–367.
3. Ikawa, M., Yamada, S., Nakanishi, T., and Okabe, M. (1999) Green fluorescent protein (GFP) as a vital marker in mammals. *Curr. Top. Dev. Biol.* **44,** 1–20.
4. Thornton, J. M. and Sibanda, B. L. (1983) Amino and carboxy-terminal regions in globular proteins. *J. Mol. Biol.* **167,** 443–460.
5. Heinemann, U. and Hahn, M. (1995) Circular permutation of polypeptide chains: implications for protein folding and stability. *Prog. Biophys. Mol. Biol.* **64,** 121–143.
6. Rojas, A., Garcia-Vallve, S., Palau, J., and Romeu, A. (1999) Circular permutations in proteins. *Biologia* **54,** 255–277.
7. Goldenberg, D. P. and Creighton, T. E. (1983) Circular and circularly permuted forms of bovine pancreativ trypsin inhibitor. *J. Mol. Biol.* **165,** 407–413.
8. Luger, K., Hommel, U., Herold, M., Hofsteenge, J., and Kirschner, K. (1989) Correct folding of circularly permuted variants of a βα barrel enzyme in vivo. *Science* **243,** 206–210.
9. Viguera, A. R., Blanco, F. J., and Serrano, L. (1995) The order of secondary structure elements does not determine the structure of a protein but does affect its folding kinetics. *J. Mol. Biol.* **247,** 670–681.
10. Zhang, T., Bertelsen, E., Benvegnu, D., and Alber, T. (1993) Circular permutation of T4 lysozyme. *Biochemistry* **32,** 12,311–12,318.
11. Wieligmann, K., Norledge, B., Jaenicke, R., and Mayr, E.-M. (1998) Eye lens βB2–crystallin: circular permutation does not influence the oligomerization state but enhances the conformational stability. *J. Mol. Biol.* **280,** 721–729.
12. Yang, Y. R. and Schachman, H. K. (1993) Aspartate transcarbamoylase containing circularly permuted catalytic polypeptide chains. *Proc. Natl. Acad. Sci. USA* **90,** 11,980–11,984.
13. Graf, R. and Schachman, H. K. (1996) Random circular permutation of genes and expressed polypeptide chains: application of the method to the catalytic chains of aspartate transcarbamoylase. *Proc. Natl Acad. Sci. USA* **93,** 11,591–11,596.
14. Hennecke, J., Sebbel, P., and Glockshuber, R. (1999) Random circular permutation of DsbA reveals segments that are essential for protein folding and stability. *J. Mol. Biol.* **286,** 1197–1215.
15. Iwakura, M., Nakamura, T., Yamane, C.. and Maki, K. (2000) Systematic circular permutation of an entire protein reveals essential folding elements. *Nat. Struct. Biol.* **7,** 580–585.

15a. Cubitt, A. B., Heim, R., Adams, S. R., Boyd, A. E., Gross, L. A., and Tsein, R. Y. (1995) Understanding, improving and using green fluorescent proteins. *Trends Biochem. Sci.* **30,** 448–455.
16. Topell, S., Hennecke, J., and Glockshuber, R. (1999) Circularly permuted variants of the green fluorescent protein. *FEBS Lett.* **457,** 283–289.
17. Baird, G. S., Zacharias, D. A., and Tsien, R. Y. (1999) Circular permutations and receptor insertion within green fluorescent proteins. *Proc. Natl. Acad. Sci. USA* **96,** 11,241–11,246.
18. Chattoraj, M., King, B. A., Bublitz, G. U., and Boxer, S. G. (1996) Ultra-fast excited state dynamics in green fluorescent protein: multiple states and proton transfer. *Proc. Natl. Acad. Sci. USA* **93,** 8362–8367.
19. Cormack, B. P., Valdivia, R. H., and Falkow, S. (1996) FACS-optimized mutants of the green fluorescent protein (GFP). *Gene* **173,** 33–38.
20. Yang, T.-T., Cheng, L., and Kain, S. R. (1996) Optimized codon usage and chromophore mutations provide enhanced sensitivity with the green fluorescent protein. *Nucl. Acids Res.* **24,** 4592–4593.
21. Martinez, J. C., Viguera, A. R., Berisio, R., et al. (1999) Thermodynamic analysis of α-spectrin SH3 and two of its circular permutants with different loop lengths: discerning the reasons for rapid folfing in proteins. *Biochemistry* **38,** 549–559.
22. Llinas, M. and Marquese, S. (1998) Subdomain interactions as a determinant in the folding and stability of T4 lysozyme. *Protein Sci.* **7,** 96–104.
23. Heim, R. and Tsien, R. Y. (1996) Engineering green fluorescent protein for improved brightness, longer wavelengths and fluorescence resonance energy transfer. *Curr. Biol.* **6,** 178–182.
24. Miyawaki, A., Lopis, J., Heim, R., McCaffery, J. M., Adams, J. A., Ikura, M., and Tsien, R. Y. (1997) Fluorescent indicators for Ca^{2+} based on green fluorescent proteins and calmodulin. *Nature* **388,** 882–887.
25. Ward, W. W. (1981) Properties of the coelenterate green-fluorescent proteins, in *Bioluminescence and Chemiluminescence: Basic Chemistry and Analytical Applications.* (De Luca, M. and McElroy, D. W., eds.), Academic, New York, pp. 235–242.
26. Patterson, G. H., Knobel, S. M., Sharif, W. D., Kain, S. R. and Piston, D. W. (1997) Use of the green fluorescent protein and ist mutants in quantitative fluorescence microscopy. *Biophys. J.* **73,** 2782–2790.
27. Lund, A. H., Duch, M., and Pedersen, F. S. (1996) Increased cloning efficiency by temperature-cycle ligation. *Nucl. Acids Res.* **24,** 800–801.
28. Yang, F., Moss, L. G., and Phillips, G. N., Jr. (1996) The molecular structure of green fluorescent protein. *Nat. Biotech.* **14,** 1246–1251.
29. Koradi, R., Billeter, M., and Wüthrich, K. (1996) (MOLMOL: a program for display and analysis of macromolecular structures. *J. Mol. Graph.* **14,** 51–55.

4

Evolutionary Design of Generic Green Fluorescent Protein Biosensors

Nobuhide Doi and Hiroshi Yanagawa

1. Introduction

Protein-engineering techniques have been applied to the molecular design of protein-based biosensors that combine a molecular-recognition site with a signal-transduction function *(1)*. The optical signal-transduction mechanism of green fluorescent protein (GFP) is most attractive, because the fluorophore of GFP is intrinsic to the polypeptide chain, and thus easily applied to molecular imaging in living cells *(2)*. One of the useful methods for creating new molecular-recognition sites on GFPs is a combination of the insertional gene fusion technique with evolutionary biotechnology *(3)*.

Since our finding that a soluble domain accommodates insertions of large domain sequences with unexpectedly high frequency *(4)*, several fusion proteins have been produced by insertion of a globular domain into another domain *(5)*. One of the interesting features of such insertional fusion is that the function of one domain (e.g., fluorescence of GFP) is sensitively modulated by a conformational change of the insert domain, upon ligand binding. Thus, the insertional fusion of a binding domain and a reporter domain can be used to develop a new generation of molecular biosensors (**Fig. 1**). However, a desired binding protein does not always undergo dramatic structural change upon ligand binding, and thus a further procedure for improvement of sensor proteins is required.

Directed protein evolution, or screening of combinatorial protein libraries, has been used to design a number of proteins with novel or improved functions *(6–8)*. GFP is suitable for screening procedures, because the GFP signal can be easily detected by various methods. Directed evolution is, therefore, a power-

From: *Methods in Molecular Biology, vol. 183: Green Fluorescent Protein: Applications and Protocols*
Edited by: B. W. Hicks © Humana Press Inc., Totowa, NJ

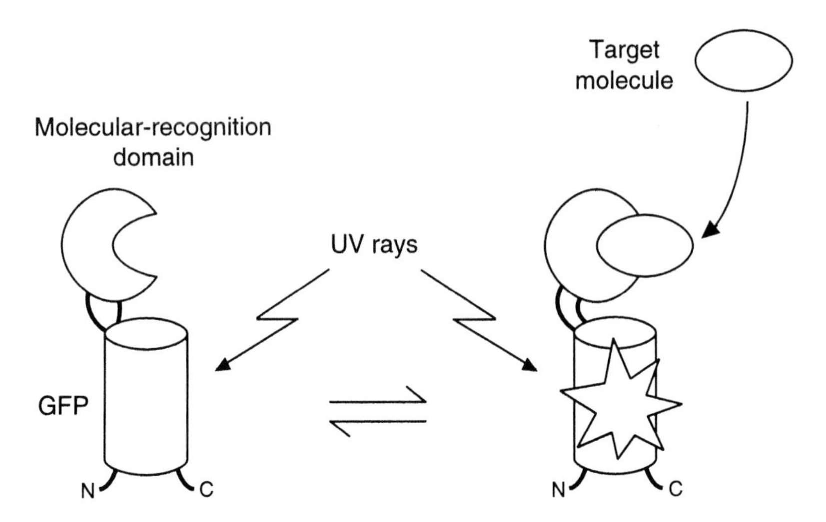

Fig. 1. Insertion of a binding domain as an internal fusion protein into a reporter domain (GFP) can be used to develop a new generation of molecular biosensors.

ful tool for improving generic GFP biosensors by selecting mutant proteins with greater intensity or with peak wavelength changes of fluorescence.

The scheme for creating desired molecular-recognition sites on GFPs comprises the following procedures. First, permissive sites for insertions are searched by linker-insertional mutagenesis of GFP. Next, a protein domain containing a desired molecular-binding site is inserted into a permissive site of GFP. Finally, if necessary, the insertional fusion protein is randomly mutated, and mutant proteins, which undergo changes in fluorescence upon binding of target molecules are selected from the random library.

This chapter describes a general technique for protein engineering and evolutionary design of GFP biosensors. Applications of individual GFP sensors to molecular imaging in living cells are described in the following chapters.

2. Materials

2.1. DNA Construction

1. Enzymes: *Ex Taq* DNA polymerase, restriction endonucleases, Ligation high (Toyobo).
2. *Escherichia coli* JM109 strain (e14$^-$(McrA$^-$), *recA1*, *endA1*, *gyrA96*, *thi-1*, *hsdR17*($r_K^- m_K^+$), *supE44*, *relA1*Δ(*lac-proAB*) [F' *traD36 proAB lacI*^{q} *Z*Δ*M15*]).
3. Luria-Bertoni (LB) plate: 10 g/L tryptone, 5 g/L yeast extract, 5 g/L NaCl, 15 g/L agar; sterilize by autoclaving ***(9)***.
4. Ampicillin stock solution: 100 mg/mL in sterile water. Store at –20°C for months ***(9)***.
5. Isopropyl thio-β-D-galactoside (IPTG): 20 mg/mL in sterile water.

2.2. Protein Overproduction and Purification

1. 2X YT medium: 16 g/L tryptone, 10 g/L yeast extract, 5 g/L NaCl; sterilize by autoclaving ***(9)***.
2. Kanamycin stock solution: 30 mg/mL in sterile water. Store at –20°C for months ***(9)***.
3. Ni-NTA agarose (Qiagen) or Talon metal affinity resin (Clontech).

2.3. Error-Prone Polymerase Chain Reaction for Random Mutagenesis

1. Enzymes: *Taq* DNA polymerase (Greiner).
2. 10X polymerase chain reaction (PCR) buffer: 0.67 *M* Tris-HCl, pH 8.8, 67 μ*M* EDTA, 0.166 *M* $(NH_4)_2SO_4$, 61 m*M* $MgCl_2$, 5 m*M* $MnCl_2$, 1.7 mg/mL BSA.
3. 10X Deoxyribonucleoside triphosphate (dNTP)s: 10 m*M* dGTP, 2 m*M* dATP, 10 m*M* dTTP, 10 m*M* dCTP.
4. 1 *M* 2-Mercaptoethanol (freshly prepared).
5. Dimethylsulfoxide.
6. Wizard PCR Preps (Promega) or QiaQuick PCR Purification Kit (Qiagen).

3. Methods

3.1. Linker Insertion Mutagenesis of GFP

To construct a GFP sensor, a binding protein for the target ligand is genetically inserted into GFP. The insertion site on GFP should be chosen according to the following considerations. First, suitable insertion sites must accept large domain insertions, without serious disturbance of the GFP structure and function. GFP consists of an 11-stranded β-barrel structure wrapped around a central helix, and solvent-exposed loops between β-strands seem to meet this criterion. Second, for sensitive signal-transduction from an inserted domain to the GFP fluorophore, favorable insertion sites may need to be three-dimensionally close to the fluorophore. So far, the Y145, Q157, E172, and L194 sites have been used as permissive sites for polypeptide insertions ***(3,10,11)***. For example, a linker-insertion mutant of GFP at the E172 site ***(3)*** is constructed as follows (*see* **Note 1**):

1. Prepare the N-terminal fragment (codons 1–172) of the GFP gene from pGFPuv ***(12)***, by PCR, with a sense primer containing an *Nhe*I site and a reverse primer containing *Hin*dIII-*Kpn*I linker sites. Similarly, prepare the C-terminal fragment (codons 173–238) with *Kpn*I-*Eco*RI linker sites at the 5' end and a *Sac*I site at the 3' end.
2. Digest the N-terminal fragment with *Nhe*I and *Kpn*I, and the C-terminal fragment with *Kpn*I and *Sac*I.
3. Ligate the two fragments simultaneously into the *Nhe*I-*Sac*I backbone vector pEOR ***(13)*** containing a *tac* promoter (*see* **Note 2**). Transform *E. coli* JM109 strain, plate transformants on an LB plate containing 100 μg/mL ampicillin and 40 μL 20 mg/mL IPTG, and incubate overnight at 37°C.

4. Pick up greenish colonies under an ultraviolet lamp, and test individual colonies for the presence of the desired insert by plasmid minipreps and *Kpn*I digestion.

The resulting plasmid contains the linker insertion corresponding to the *Hin*dIII-*Kpn*I-*Eco*RI sites between Glu172 and Asp173 of GFP. The multirestriction enzyme sites can be used for insertion of a desired domain.

3.2. Insertion of a Desired Binding Domain into GFP

A gene of a desired binding domain (designated as protein X) is inserted in-frame into a target site of the GFP gene.

1. Prepare a DNA fragment encoding protein X, minus the stop codon, by digestion of a plasmid or a PCR fragment with appropriate restriction enzymes (*see* **Note 3**).
2. Insert the DNA fragment into the vector with an appropriate insertion site on GFP (constructed as described in **Subheading 3.1.**), ligate the DNA, and transform *E. coli* JM109 strain. Plate transformants on LB medium containing 10 μg/mL kanamycin (assuming the plasmid carries the kanamycin resistance gene), and incubate overnight at 37°C.
3. Test individual colonies for the presence of the desired insert by colony PCR or plasmid minipreps.
4. Overexpress the GFP::X fusion protein in *E. coli* JM109 cells, under the control of the *tac* promoter (*see* **Note 4**).
   a. Grow cells in 2X YT medium containing 10 μg/mL kanamycin at 30°C.
   b. When the culture reaches an optical density of 0.8–1.0 at 600 nm, add IPTG to a final concentration of 0.1 m*M*, and incubate overnight at 30°C.
   c. Harvest the cells by centrifugation.
5. Purify the fusion protein with N- or C-terminal hexahistidine sequence (6X His-tag) on a metal affinity column, according to manufacture's manual.
6. Using a spectrofluorometer, measure the fluorescence spectra of the purified protein with various concentrations of ligand, and fit the titration curve to the standard equation, $\Delta F = \Delta F_{max}/(1 + K_d/L)^{-1}$, where ΔF is the change in fluorescence emission intensity, K_d is the dissociation constant, and L is the ligand concentration.

If the ligand sensitivity of the GFP::X protein is low, further mutagenesis and screening steps are required.

3.3. Random Mutagenesis and Screening of GFP Biosensors

Random mutagenesis of the insertional fusion gene is performed by using error-prone PCR (*see* **Note 5**). The following screening method is based on simple visual inspection, but other methods can also be applied (*see* **Note 6**).

1. Digest a plasmid DNA containing GFP::X fusion gene, with a restriction enzyme that cleaves outside the target region of mutagenesis.

2. Prepare a PCR mix in a final volume of 100 µL, containing 1 ng of the linearized DNA, 1 µL each of 100 µ*M* mutagenic primers, 10 µL 10X PCR buffer, 10 µL 10X dNTPs, 1 µL 1 *M* 2-mercaptoethanol, 10 µL dimethyl sulfoxide, and 5 U of *Taq* DNA polymerase (*see* **Note 7**). Take the reaction through 25 cycles of amplification, as follows: denaturation at 95°C, 1 min; annealing at 50°C, 1 min; and polymerization at 70°C, 4 min. Purify the sample using a PCR purification kit (*see* **Note 8**).
3. Clone the PCR product into a screening vector containing an appropriate promoter (*see* **Note 9**), transform *E. coli* JM109 cells with the plasmid library, plate on LB medium containing 10 µg/mL kanamycin (assuming the plasmid carries the kanamycin resistance gene), and incubate overnight at 37°C.
4. Select 100 colonies at random, and inoculate the individual colonies in patches (lines of 5–10 mm in length), in identical locations on two LB-kanamycin plates, one of which contains a target ligand. Incubate the plates at 37°C for several hours, then store the plates at 4°C, until the green color fully develops. Under a handheld ultraviolet lamp, pick up colonies with the largest ratio of the fluorescence on the ligand$^+$ plate to that on the ligand$^-$ plate (*see* **Note 10**).
5. Recover plasmids from the selected colonies, and use them as the template DNA for the next cycle of error-prone PCR.
6. Repeat **steps 1–5** several times.
7. Characterize the selected GFP sensors as described in **Subheading 3.2.**

4. Notes

1. The PCR-based method described here is suitable for targeted insertional mutagenesis. If a restriction enzyme site is already present in the GFP gene at an appropriate target site, the classical method using oligo DNA linkers ***(14)***, can also be applied. If a new permissive site for insertion is required, random insertion mutagenesis using the transposon method ***(15)*** is recommended.
2. Instead of the ligation of three DNA fragments described here, an overlap PCR ***(16)*** for connecting the N- and C-terminal fragments of GFP gene can also be used, but this requires extra PCR and purification steps, and relatively long primers.
3. In this step, it is possible to create DNA fragments with various lengths or sequences of spacers at both sides of protein X. The optimum length of spacers may depend on the structures of the individual inserted domains; e.g., a domain having a large distance between the N- and C-termini may need long spacers.
4. Under the control of strong promoters such as T7 promoter, GFP is overexpressed in aggregated form. Under *tac* promoter, wild-type GFP is expressed in the soluble fraction, but some GFP mutant proteins, destabilized by insertion of large polypeptides, are overexpressed in the insoluble form. Such mutant GFPs can be purified under denaturing conditions (8 *M* urea or 6 *M* GuHCl) with a His-tag, and can be efficiently refolded with recovery of fluorescence by dialysis against a buffer containing 50% glycerol.

5. In the protocol first described by Leung et al. ***(17)*** and used here, mutations are biased to either T-to-C or A-to-G transitions. To overcome this limitation, improved protocols have been reported by several researchers ***(18,19)***.
6. To select mutant GFP proteins with a wavelength change in fluorescence upon ligand binding, detection with analytical instruments, such as a fluorescence-activated cell sorter ***(20)*** and microplate reader ***(21)***, are recommended.
7. The wild-type *Taq* DNA polymerase is commercially available from several companies. Replication fidelity varies somewhat, and those with relatively high fidelity are not suitable for error-prone PCR.
8. The reaction buffer for error-prone PCR contains high concentrations of salts, metals, and dNTPs, which co-precipitate under the usual conditions of ethanol precipitation and inhibit the subsequent reaction of some restriction enzymes. Thus, the authors recommend the use of PCR purification kits that do not include an ethanol precipitation step.
9. The expression level of GFP::X protein is important for the following screening step. Not only dark colonies, but also too-bright colonies are unsuitable for visual inspection. When a promoter of appropriate strength is not available, random mutagenesis of the promoter region linked to the GFP::X gene and screening of appropriate colonies may be useful.
10. Since *E. coli* colonies bearing an identical GFP gene often exhibit a range of fluorescence levels, a second screening for confirmation of the ligand sensitivity should be done.

References

1. Hellinga, H. W. and Marvin, J. S. (1998) Protein engineering and the development of generic biosensors. *Trends Biotechnol.* **16,** 183–189.
2. Tsien, R.Y. (1998) The green fluorescent protein. *Annu. Rev. Biochem.* **67,** 509–544.
3. Doi, N. and Yanagawa, H. (1999) Design of generic biosensors based on green fluorescent proteins with allosteric sites by directed evolution. *FEBS Lett.* **453,** 305–307.
4. Doi, N., Itaya, M., Yomo, T., Tokura, S., and Yanagawa, H. (1997) Insertion of foreign random sequences of 120 amino acid residues into an active enzyme. *FEBS Lett.* **402,** 177–180.
5. Doi, N. and Yanagawa, H. (1999) Insertional gene fusion technology. *FEBS Lett.* **457,** 1–4.
6. Küchner, O. and Arnold, F. H. (1997) Directed evolution of enzyme catalysts. *Trends Biotechnol.* **15,** 523–530.
7. Fastrez, J. (1997) In vivo versus in vitro screening or selection for catalytic activity in enzymes and abzymes. *Mol. Biotechnol.* **7,** 37–55.
8. Doi, N. and Yanagawa, H. (1999) STABLE: protein-DNA fusion system for screening of combinatorial protein libraries in vitro. *FEBS Lett.* **457,** 227–230.
9. Sambrook, J., Fritsch, E. F., and Maniatis, T. (1989) *Molecular Cloning: A Laboratory Manual.* Cold Spring Harbor Laboratory, Cold Spring Harbor, NY.

10. Abedi, M. R., Caponigro, G., and Kamb, A. (1998) Green fluorescent protein as a scaffold for intracellular presentation of peptides. *Nucl. Acids Res.* **26,** 623–630.
11. Baird, G. S., Zacharias, D. A., and Tsien, R. Y. (1999) Circular permutation and receptor insertion within green fluorescent proteins. *Proc. Natl. Acad. Sci. USA* **96,** 11,241–11,246.
12. Crameri, A., Whitehorn, E. A., Tate, E., and Stemmer, W. P. C. (1996) Improved green fluorescent protein by molecular evolution using DNA shuffling. *Nat. Biotechnol.* **14,** 315–319.
13. Prijambada, I. D., Yomo, T., Tanaka, F., Kawama, T., Yamamoto, K., Hasegawa, A., et al. (1996) Solubility of artificial proteins with random sequences. *FEBS Lett.* **382,** 21–25.
14. Barany, F. (1985) Single-stranded hexameric linkers: a system for in-phase insertion mutagenesis and protein engineering. *Gene* **37,** 111–123.
15. Hallet, B., Sherratt, D. J., and Hayes, F. (1997) Pentapeptide scanning mutagenesis: random insertion of a variable five amino acid cassette in a target protein. *Nucl. Acids Res.* **25,** 1866–1867.
16. Ling, M. M. and Robinson, B. H. (1997) Approaches to DNA mutagenesis: an overview. *Analyt. Biochem.* **254,** 157–178.
17. Leung, D. W., Chen, E. Y., and Goeddel, D. V. (1989) A method for random mutagenesis of a defined DNA segment using a modified polymerase chain reaction. *Technique* **1,** 11–15.
18. Fromant, M., Blanquet, S., and Plateau, P. (1995) Direct random mutagenesis of gene-sized DNA fragments using polymerase chain reaction. *Analyt. Biochem.* **224,** 347–353.
19. Vartanian, J. P., Henry, M., and Wain-Hobson, S. (1996) Hypermutagenic PCR involving all four transitions and a sizeable proportion of transversions. *Nucl. Acids Res.* **24,** 2627–2631.
20. Cormack, B. P., Valdivia, R. H., and Falkow, S. (1996) FACS-optimized mutants of the green fluorescent protein (GFP). *Gene* **173,** 33–38.
21. Miesenböck, G., De Angelis, D. A., and Rothman, J. E. (1998) Visualizing secretion and synaptic transmission with pH-sensitive green fluorescent proteins. *Nature* **394,** 192–195.

5

Random Insertion of Green Fluorescent Protein into the Regulatory Subunit of Cyclic Adenosine Monophosphate-Dependent Protein Kinase

Pascal J. Baehler, Ricardo M. Biondi, Miguel van Bemmelen, Michel Véron, and Christophe D. Reymond

1. Introduction

Various fusion proteins have been made with green fluorescent proteins (GFPs) (for a recent review, *see* **ref.** ***1***) as a means to localize such proteins within living cells *(2)*. In most cases, GFPs were added at either the C- or N-terminal end of the protein or polypeptide of interest *(3)*. For certain purposes, sueh as fluorescence resonance energy transfer (FRET), GFPs need to be placed at particular locations within the protein *(4)*. Because the crystal structures of most proteins and protein complexes are not currently known, it is not usually possible to predict the optimal position for insertion of the GFP to obtain FRET, or to retain target protein activity. We thus devised a method for random insertion of GFPs within a target protein. The method generates a collection of fusion proteins that can be tested for a desired function.

As a model system, we have selected the regulatory subunit (R) of the cyclic adenosine monophosphate (cAMP)-dependent protein kinase A (PKA). We obtained over 120 clones with GFPs inserted throughout the R-subunit. The GFPs kept their fluorescent properties when inserted at different locations within the R-subunit. Furthermore, the R-subunits were able to bind cAMP and, at least some of them, to interact with the catalytic subunit *(5)*.

The GFP coding region is introduced within the target gene by a series of modifications of the DNA. First, random nicks are introduced into the double-stranded DNA coding for the target protein. Random nicks can be formed using DNase I by standard nick translation methods *(6)*. After nicking, circular mol-

From: *Methods in Molecular Biology, vol. 183: Green Fluorescent Protein: Applications and Protocols*
Edited by: B. W. Hicks © Humana Press Inc., Totowa, NJ

ecules are separated from supercoiled DNA by cesium chloride (CsCl) gradient centrifugation. Generating blunt ends also requires attention, and it can be done either by incubation with DNase I in Mn^{2+} buffer, or by incubation with S1 nuclease *(7)*. We used a combination of S1 nuclease digestion and T4 DNA polymerase action was used. *Sal*I linkers, or other restriction enzyme linkers, are then ligated to the blunt ends. The GFP sequences to be inserted into the target protein were modified with *Sal*I sites by polymerase chain reaction (PCR). A final ligation allows insertion of the GFP sequence within the target gene, for bacterial transformation. This method is versatile, allowing essentially any restriction sites to be utilized. Ligating GFP as a blunt-end fragment into the nicked, linear plasmid was attempted, but proved to be less efficient than the *Sal*I linker method.

Because GFP insertions are random, they can be made into the target gene, into the antibiotic resistance gene, or into portions of the plasmid lacking a promoter. Those insertions made into the antibiotic resistance gene will likely inactivate the lactamase and be removed by antibiotic selection. Screening of fluorescent colonies, using a fluorescence microscope, will allow selection of insertions under the target gene promoter only. Some promoters require specific induction, others, like the one present on pRSETb, are sufficiently leaky to produce a low level of fluorescence in absence of induction. Fluorescence screening ensures selection for GFPs inserted in-frame into the target DNA. Because of the occurrence of small deletions resulting from the nicking procedure, the reading frame may be shifted after the DNA encoding for the GFP, resulting in abherent C-terminal portions of the target protein. These truncated proteins can be detected either by sequencing the plasmid DNA or by immunoblotting using antibodies directed against GFP or the N-terminal part of the target protein. Most fluorescent clones contained GFP insertions within the target gene. We observed a restricted number of insertions outside the coding region for the target protein, which were not further characterized. A collection of ~120 clones was obtained and analyzed further for the site of insertion *(5)*.

The bacterial culture conditions need to be carefully tested, since GFPs require low temperatures and a high level of oxygen to fully fluoresce. To find the best conditions for expression, it is advisable to test fluorescence of whole living bacteria after growth under different conditions, keeping in mind that the highest induction of expression may be deleterious for the bacteria, as indicated by the appearance of lysed bacteria ("watery" colonies on plates). We tested growth on agar plates, and in shaking cultures with or without induction, and at different temperatures. The best conditions varied from clone to clone, but the most reliable condition was to grow the bacteria on agar plates at 22°C, ensuring maximal oxygenation and proper folding of both GFP and R-subunit.

Different lysis and extraction procedures have been used. Here again, obtaining properly folded fluorescent protein may require different preliminary tests. Purification of the target protein fused to GFP can be attained by affinity chromatography based on a molecular tag. We used a 6X His-tag *(8)*, present at the N-terminus of the R-subunit. Efficient recovery of the fused protein was obtained under partially denaturing conditions, namely using urea. Although fluorescence was present, indicating intact GFP, the R subunit was denatured. We thus advise, whenever possible, to purify the fusion proteins under nondenaturing conditions, even when obtaining lower yields. Our R–GFP fusions were finally purified by a combination of Ni-NTA, cAMP-agarose, and HPLC column chromatography.

Devising a simple and rapid functional test for the targeted fusion protein is important, because it will facilitate the screening for proper clones and the protein purification. A challenge is to be capable of screening for function without depending on time-consuming protein purification procedures for each different GFP fusion protein. An alternative might be to attempt to co-transform bacteria or eukaryotic cells with plasmids carrying two interacting proteins of interest, and test their function/interaction in vivo, e.g., one could attempt to detect fluorescence energy transfer of a GFP-tagged PKA regulatory subunit and a blue fluorescent protein-tagged PKA catalytic subunit.

2. Materials

The plasmid, pRSETb-R, was described previously *(9)*. Restriction and modification enzymes used were from Boehringer Mannheim, Promega, Pharmacia, or Gibco. Deoxynucleotide triphosphates (dNTP)s were from Promega. The fluorescence microscope used for screening was a Zeiss Axiovert 25 inverted microscope with a BP-450-490 excitation filter, an FT510 beam splitter, and a BP 515-565 emission filter. Fluorescence spectra were obtained with a Photon Technology International spectrofluorimeter, and data processed using Felix software. The GFP mutants (S65T, W7) used as template for PCR, were kindly provided by Dr. R. Y. Tsien (University of California, San Diego, CA).

2.1. DNA Manipulations

1. DNase I and 1X DNase I buffer (Promega).
2. (50 m*M*) EDTA stop solution (pH 8.0).
3. 1 *M* Tris-HCl (pH 7.6).
4. CsCl.
5. Ethidium bromide (10 mg/mL) stock solution.
6. 0.9% agarose gel.
7. TAE gel buffer: 40 m*M* Tris-acetate (pH 8.3), 1 m*M* EDTA.
8. Isoamyl alcohol.

9. S1 nuclease (Roche, Rotkeuz, Switzerland).
10. S1 nuclease buffer: 50 m*M* sodium acetate (pH 5.7), 200 m*M* NaCl, 1 m*M* $ZnSO_4$, 0.5 % glycerol.
11. Phenol–chloroform.
12. TE: 10 m*M* Tris-HCl (pH 7.6), 1 m*M* EDTA (pH 8.0).
13. T4 DNA polymerase (Roche).
15. dNTPs (10 m*M* each) (Promega). (pH 7.6), 10 m*M* $MgCl_2$, 1 m*M* dithiothreitol.
14. 10X T4 DNA polymerase buffer: 200 m*M* Tris-HCl.
16. 5 m*M* Ammonium acetate in 75% ethanol.
17. T4 DNA ligase.
18. T4 DNA ligase buffer (Roche)
19. 10 m*M* Adenosine triphosphate.

2.2. Bacterial Cell Culture and Screening

1. *Escherichia coli* strain BL21 (DE3) (Stratagene).
2. Luria-Bertoni (LB) Agar selection plates containing ampicillin (100 µg/mL).
3. LB broth.

2.3. Protein Purification

1. LB Agar plates containing ampicillin (100 mg/mL).
2. LB broth containing ampicillin (100 mg/mL).
3. LB broth containing isopropyl-β-D-thiogalactoside (IPTG) (0.5 m*M*).
4. Lysis buffer: 50 m*M* HEPES, pH 7.8, 300 m*M* NaCl, 0.5% NP40, 10 m*M* magnesium acetate, 10% sucrose, and Complete™ protease inhibitor cocktail (Roche).
5. 2-Mercaptoethanol.
6. Ni-agarose (Qiagen).
7. Wash buffer 1: 50 m*M* HEPES, pH 8.0, 100 m*M* NaCl, 10 m*M* imidazole, 2 m*M* 2-mercaptoethanol, 0.1% Tween-20, 1 mg/mL egg-white trypsin inhibitor, 1 m*M* phenylmethylsulfonyl fluoride and 20 µg/mL TLCK.
8. Wash buffer 1 containing 0.3 *M* NaCl.
9. Wash buffer 1 containing 25, 100, or 250 m*M* imidazole.
10. (N^6)-cAMP Sepharose (Sigma).
11. Wash buffer 2: 50 m*M* HEPES, pH 8.0, 100 m*M* NaCl, 2 m*M* 2-mercaptoethanol, and 0.1% Triton X-100.
12. Wash buffer 2 containing 5 m*M* AMP.
13. Wash buffer 2 containing 0.5 *M* NaCl.
14. Wash buffer 2 containing 50 m*M* cyclic guanosine monophosphate (cGMP).
15. 50 m*M* Sodium phosphate (pH 7.2).

3. Methods

3.1. DNA Manipulations

3.1.1. Preparation of Vector DNA

Preliminary assays with small amounts of DNA are absolutely necessary to optimize the conditions for the production of random nicks (*see* **Note 1**).

1. Incubate 25 μg of plasmid DNA with 5 pg of DNase I at 15°C in 1X DNA pol I buffer (Promega) (final volume of 50 μL).
2. Take aliquots after 0, 5, 15, 30, 45, and 60 min, and stop the reaction by adding EDTA up to 25 m*M*. Monitor the progression of nicking by analyzing the aliquots on a 0.9% agarose gel in TAE buffer (electrophoresis for 3–5 h at 50 V).
3. From the gel, choose the incubation time producing ~30–50% circular plasmids with single nicks. Under the above conditions, such plasmids show the largest apparent molecular weight (upper band) on the gel (followed by linear, then supercoiled, molecules).
4. Set up the final reaction by scaling-up the test reaction 25-fold. DNase I is added last, incubated for the predetermined time, and the reaction is stopped by addition of EDTA (25 m*M*).

3.1.2. Isolation of Nicked DNA by CsCl Gradient Centrifugation

In order to eliminate super-coiled plasmids, which may be substrate for S1 nuclease, and/or to remain capable of transforming *E. coli*, nicked circular plasmids need to be purified. In order to achieve sufficient purification, we prefer CsCl gradient separation to gel electrophoresis.

1. Measure the volume of the DNA solution.
2. Add 1 g of solid CsCl/mL and heat the solution to 30°C to facilitate its dissolution.
3. Add ethidium bromide (EtBr), (10 mg/mL stock) up to a final concentration of 740 μg/mL.
4. Transfer the solution to QuickSeal tubes (Beckman). Centrifuge at 45,000 rpm for 16 h in a Beckman Ti65 vertical rotor (or equivalent) at room temperature. After centrifugation, two DNA bands are visible, corresponding to nicked relaxed (upper band) and supercoiled plasmids (lower band).
5. Cut open the tip of the tube and collect relaxed plasmids by inserting a hypodermic needle just below the upper band.
6. Extract ethidium bromide by adding an equal volume of isoamyl alcohol. Mix by briefly vortexing, then centrifuge at 1500*g* for 3 min.
7. Transfer the lower aqueous phase to a new tube and repeat the extraction until the red color is no longer visible in either phase. Recover the DNA by precipitation in 75% ethanol. Centrifuge and discard the supernatant, then resuspend the DNA in 500 μL S1 nuclease buffer.
8. Determine the final concentration of DNA by measuring the absorbance at 260 nm.

3.1.3. Linearization of the Plasmids with S1 Nuclease

After CsCl gradient centrifugation, the purified, nicked, circular plasmids are incubated with S1 nuclease. Conditions must be tested for the appearance of about 30–50% linear plasmids (*see* **Note 2**).

1. Preincubate the solution of DNA for 5 min at 37°C, then add 5000 U of S1 nuclease.

2. After 4 min, transfer 160 µL to a tube containing 15 µL 50 m*M* EDTA (pH 8.0) and 15 µL 1 *M* Tris-HCl (pH 7.6).
3. After 15 and 60 min, take further aliquots (160 µL).
4. Extract DNA with phenol–chloroform and precipitate with 75% ethanol.
5. After centrifugation (12,000*g* for 5 min), drain excess ethanol, and resuspend the DNA in 25 µL TE.
6. Analyze a small aliquot of each sample by agarose gel electrophoresis on a 0.9% gel, using the original plasmid DNA, made linear with an appropriate restriction enzyme, as a marker.
7. Pool the samples containing roughly 50% linear DNA.

3.1.4. Treatment with T4 DNA Polymerase to Produce Blunt Ends

S1 nuclease can generate protruding 5' or 3' ends. In order to obtain blunt ends, DNA needs to be repaired with T4 DNA polymerase.

1. Adjust the volume of the pool to 90 µL with TE.
2. Add 10 µL of 10X T4 DNA polymerase buffer, and 5 U T4 DNA polymerase.
3. Incubate the reaction at room temperature for 1 min (in this step, the 3' termini are removed).
4. Add 10 µL of a solution containing all four dNTPs (10 m*M* each), and incubate 1 h at 37°C.
5. Purify the DNA by extraction with phenol–chloroform (*see* **Note 3**).
6. Precipitate the DNA out of the aqueous phase by adding 100 µL 5 m*M* ammonium acetate and 400 µL ethanol. Incubate 20 min on ice, and centrifuge at 12,000*g* for 10 min at 4°C.
7. Carefully remove the ethanol, and dissolve the DNA in 25 µL TE.

3.1.5. Ligation of Phosphorylated Synthetic SalI Linkers

1. Mix 10 µg linear plasmid DNA with 1 nmol phosphorylated *Sal*I linkers (sequence: GTCGAC), in a final volume of 180 µL TE.
2. Add 20 µL of ligase buffer, 10 m*M* adenosine triphosphate and 2 U T4 DNA ligase.
3. Incubate for 4 h at 16°C.
4. Heat the reaction for 10 min at 68°C, to inactivate the T4 DNA ligase.
5. Add 500 U of *Sal*I, and incubate for 4 h at 37°C.

3.1.6. Preparation of the GFP Sequence by PCR and Ligation

The sequence of S65T or W7 GFP was amplified by 30 cycles of PCR, using the following primers:

a. 5'-CCCGTCGTC**GTCGAC**ATGAGTAAAGGAGAAGAA-3' and
b. 5'-AGTCGG**GTCGAC**TTTGTAATAGTTCATCCATGCC-3'.

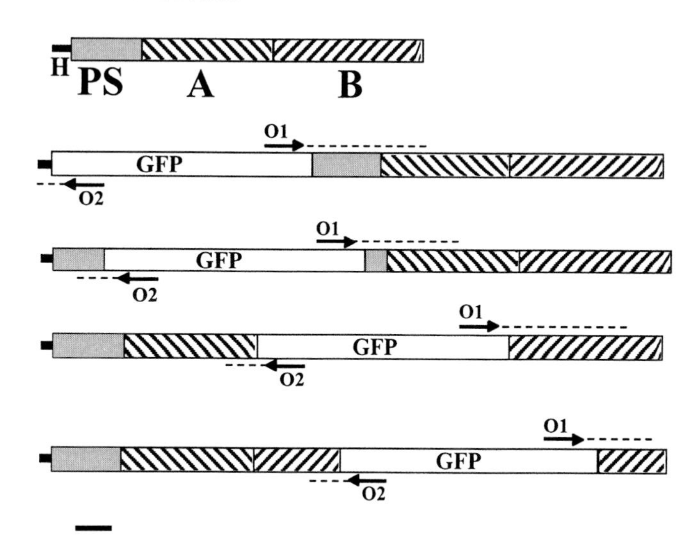

Fig. 1. Four out of 120 isolated clones are compared to the wild type R subunit (upper lane). In order to sequence the site of insertion of the GFP, two primers were designed within the GFP sequence allowing to sequence in both directions (O1 and O2). The primers used were: O1:5'-GCTGCTGGGATTACACA-3', and O2:5'-TAACATCACCATC TAAT-3'. (H, His-tag; PS, pseudosubstrate site; A, B, cAMP binding sites; GFP, green fluorescent protein; dash line, sequenced region; scale bar, 100 bp).

The precise conditions were: 1 min at 95°C, 1 min at 45°C, and 1 min at 72°C, for denaturing, annealing, and elongating steps, respectively.

1. The GFP-encoding PCR fragments were digested with *Sal*I.
2. A 10-fold excess of PCR fragment ends were ligated with the plasmids overnight at 16°C, using T4 DNA ligase in T4 DNA ligase buffer.

3.2. Screening

After transforming *E. coli* strain BL21(DE3) by electroporation, the bacteria were grown on ampicillin-containing LB agar plates at 22°C. Fluorescent colonies were detected on agar plates using an inverted fluorescence microscope. Of $2–5 \times 10^6$ screened colonies, 150 clones with various intensities were picked and their DNA was analyzed by restriction digestion, to determine the site of insertion of the GFP. Random insertion occurred, since the collection of clones showed different sites of insertion, as illustrated by four examples in **Fig. 1**.

3.3. Purification of R–GFP Proteins

3.3.1. Expression of R–GFP in Bacteria

The expression level of the R–GFP fusions varied dramatically from clone to clone. Furthermore, the culture conditions influenced the level of fluorescence. Thus, culture and extraction procedures had to be optimized carefully (*see* **Note 4**).

1. Streak a fluorescent clone on LB-agar plate and select colonies with the highest level of fluorescence. If necessary, repeat the streak, in order to obtain stable expression.
2. Inoculate the selected colony into a 2-L flask containing 500 mL LB with 100 µg/mL ampicillin.
3. Incubate under agitation (250 rpm) at 30°C, up to an OD_{600} of ~0.6.
4. Induce protein expression with 0.5 m*M* IPTG (for *lac* promoters).
5. Maintain the cultures under agitation (250 rpm) for 16 h at 22°C.
6. Harvest the cells by centrifugation for 5 min at 8000*g*.

3.3.2. Cell Lysis

1. Resuspend the cell pellet in 70 mL of chilled (4°C) lysis buffer. Freeze at –80°C for 15 min. Cell suspensions can be stored at –80 °C before further purification steps (*see* **Note 5**).
2. Complete cell lysis, by thawing and adding egg-white lysozyme (250 µg/mL) and DNase I (25 U/mL).
3. Incubate for 30 min at 5°C on a rotatory shaker.
4. Add 2-mercaptoethanol (2 m*M)* to the lysate.
5. Centrifuge for 15 min at 15,000*g* at 4°C.

3.3.3. Ni-NTA Affinity Chromatography

For R-GFP purification, all manipulations were performed at 4°C (*see* **Note 6**).

1. Add imidazole (10 m*M)* to the supernatant containing the fusion protein.
2. Incubate for 45 min on an orbital shaker in the presence of 1.5 mL Ni-NTA agarose beads (Qiagen).
3. Transfer the slurry to a plastic column of 1.5 cm internal diameter.
4. Allow the column to drain, and recirculate the eluant through the column 3×.
5. Wash the beads with 15 mL wash buffer 1.
6. Strip any cAMP bound to the 6X HisGFP-PKA-R fusion protein by equilibrating the column with wash buffer 1 containing 5 m*M* cGMP.
7. Eliminate the cGMP by washing with 10× the bed vol of wash buffer 1 containing 0.3 *M* NaCl.
8. Perform a stepwise elution of bound protein with wash buffer 1 containing 25, 100, or 250 m*M* imidazole (6 bed vol each).

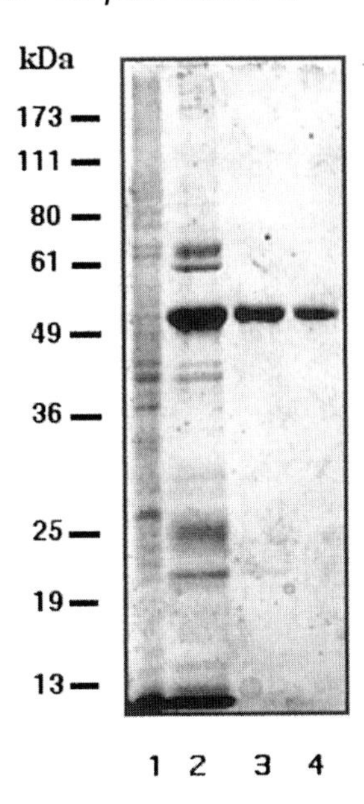

Fig. 2. Coomassie blue staining of proteins separated by SDS-PAGE, after successive steps of purification. Lane 1, crude extract; lane 2, after Ni-NTA chromatography; lane 3, after cAMP–agarose column; lane 4, after HPLC column. Molecular weights were determined from standard mol wt proteins run in parallel.

3.3.4. cAMP-Agarose Affinity Chromatography

The following procedure further purified the R–GFP fusions, and selected for proteins capable of binding to cAMP. A different affinity column must be used for each fusion protein envisaged, based on their specific properties (*see* **Note 7**).

1. Pool fractions from the Ni-NTA column corresponding to the peak of fluorescence.
2. Load the pool into a column containing 0.5 mL (N^6)-cAMP-agarose beads (Sigma) pre-equilibrated with wash buffer 2. Recirculate the eluant through the column three more times.
3. Elute contaminants by several washes with 10× the bed volume of wash buffer 2, containing, successively, 5 m*M* AMP, then 0.5 *M* NaCl.
4. Elute bound R–GFP with three equivalents of bed volume of wash buffer 2 containing 50 m*M* cGMP at room temperature.

3.3.5. High-Performance Size-Exclusion Chromatography

1. Pool fractions from the cAMP-agarose column, corresponding to the peak of fluorescence (**Fig. 2**).

2. Purify at room temperature on a 300 × 7.8 mm BioSep SEC3000 column (Phenomenex, Torrance, CA), fitted with a 75 × 7.8 guard, and connected to a HP1090 Series II HPLC system (Hewlett-Packard).
3. Perform elution at 1 mL/min in 50 m*M* sodium phosphate buffer (pH 7.2). Monitor elution by absorption at 220 and 434 nm.
4. The purity and identity of the isolated protein can be verified by sodium dodecyl sulfate-polyacrylamide gel electrophoresis (SDS-PAGE) followed by Coomassie-blue staining or immunoblotting, using an anti-6X His-tag monoclonal antibody (Qiagen).

3.4. Fluorescence Measurements

The assay was performed at 20°C, using variable amounts of crude extracts or cell suspensions in 50 m*M* HEPES, pH 8.5. Bandwidths were kept constant: excitation bandwidth, 5–9 nm; emission bandwidth, 5–9 nm. Integration time was 0.2–1 s. The GFP S65T fusion proteins were excited at 485 nm, while recording emission over a range from 500–550 nm. GFP/W7 fusion proteins were excited at 434 nm, while recording emission between 450 and 500 nm.

4. Notes

1. When preparing the vector DNA, preliminary assays with small amounts of DNA are absolutely necessary to optimize the conditions for the production of random nicks. DNase I activity may vary from batch to batch, and with time. When using a proper concentration of DNase I, the proportion of circularized DNA increases linearly within ~20 min. The best conditions for obtaining one nick/plasmid are when <50% of the molecules are circularized.
2. Small scale determination of the precise amount of S1 nuclease required for cutting the second strand is important to avoid overprocessing the material. Exposure of nicked plasmid DNA to excessive concentrations of S1 nuclease will result in the formation of large deletions.
3. As an alternative to phenol–chloroform extractions and precipitation, this step can be replaced by a QiaQuick PCR purification procedure (Qiagen).
4. Despite a high level of expression, we found some instability when using *E. coli* BL21(DE3). Thus, expression of the target protein may be preferable in *E. coli* JM109 (DE3) or other bacterial lines designed for protein expression.
5. Cell lysis:
   a. In a first series of experiments, cells were incubated for 1 h on ice in the presence of 1% lysozyme in the presence of protease inhibitors, then lysed by sonication (15X 1 s with 1 s intervals) at 10–13% maximal intensity. Fluorescence spectral properties were modified after sonication in many extracts. Sonication may change GFP fusion protein structure.
   b. A French press may represent a major alternative to lysozyme. However, in our hands, the combination of lysozyme, DNase, and NP40 proved to give the

highest yields of proteins with specific activity. Furthermore, processing large volumes in a French press is time-consuming, and may increase degradation.

c. Some fusion proteins may prove to be difficult to solubilize and tend to stay in the pellet after lysis. In such cases, urea (up to 8 *M)* may be added after lysozyme and DNase I treatment. GFPs are resistant to denaturation at such concentration of urea and fluorescence will not decrease. However, one has to determine the precise urea concentration compatible with the protein biological activity. The urea-containing solution is then added directly onto the Ni-NTA agarose and purification is conducted as described above.

6. In Ni-NTA affinity chromatography, the optimal imidazole concentration must be determined for each fusion protein. The goal is to prevent unspecific binding of *E. coli* proteins to the column, while binding as much fusion protein as possible. Steps to eliminate bound cAMP are specific for the PKA R-subunit and can be omitted for the purification of other fusion proteins.
7. Imidazole may interfere with the binding of the fusion protein to the cAMP–agarose affinity column. A Sepharose G50 chromatography may be inserted before the cAMP–agarose separation. Fluorescence, which usually appears in the excluded volume, can be used to track the elution of the fusion protein from the Sepharose column.
8. A potential problem is that, despite the detection of fluorescence, the fused protein may be nonfunctional. The GFP may then be inserted into an active site, or possibly destroy the conformation of the fused protein. However, recent work indicates that fluorescence correlates with the correct folding of the fused protein domains when expressed in *E. coli* ***(10)***. There is thus a good chance that many random insertions selected for their green fluoresent property would turn out to be functional fusion proteins; This could be achieved by random insertions at loops or between domains.

References

1. Tsien, R. Y. (1998) The green fluorescent protein. *Annu. Rev. Biochem.* **67,** 509–544.
2. Tsien, R. Y. and Miyawaki, A. (1998) Seeing the machinery of live cells. *Science* **280,** 1954–1955.
3. Chalfie, M., Tu, Y., Eusskirohen, G., et al. (1994) Green fluorescent protein as a marker for gene expression. *Science* **263,** 802–805.
4. Lankiewicz, L., Malicka, J., Wiczk, W., et al. (1997) Fluorescence resonance energy transfer in studies of inter-chromophoric distances in biomolecules. *Acta. Biochim. Pol.* **44,** 477–489.
5. Biondi, R. M., Baehler, P. J., Reymond, C., et al. (1998) Random insertion of GFP into the cAMP-dependent protein kinase regulatory subunit from *Dictyostelium discoideum. Nucleic Acids Res.* **26,** 4946–4952.
6. Kelly, R. B., Cozzarelli, N. R., Deutschen, M., et al. (1970) Enzymatic synthesis of deoxyribonucleic acid. XXXII. Replication of duplex deoxyribonucleic acid by polymerase at a single strand break. *J. Biol. Chem.* **245,** 39–45.

7. Heffron, F., So, M., McCarthy, B., et al. (1978) In vitro mutagenesis of a circular DNA molecule by using synthetic restriction sites. *Proc. Natl. Acad. Sci. USA* **75,** 6012–6016.
8. Hengen, P. (1995) Purification of His-Tag fusion proteins from *Escherichia coli. Trends Biochem. Sci.* **20,** 285–286.
9. Etchebehere, L. C., van Bemmelen, M. X., Anjard, C., et al. (1997) The catalytic subunit of Dictyostelium cAMP-dependent protein kinase — role of the N-terminal domain and of the C-terminal residues in catalytic activity and stability. *Eur. J. Biochem.* **248,** 820–826.
10. Waldo, G. S., Standish, B. M., Berendzen, J., et al. (1999) Rapid protein-folding assay using green fluorescent protein. *Nat. Biotechnol* **17,** 691–695.

6

Circular mRNA Encoding for Monomeric and Polymeric Green Fluorescent Protein

Rhonda Perriman

1. Introduction

Many proteins with unusual structural properties are comprised of multiple repeating amino acid sequences, and are often fractious to expression in recombinant systems. To facilitate recombinant production of such proteins for structural and engineering studies, the author has developed a method for producing messenger RNAs on circular RNA templates. This circularization process is derived from a rearranged group I intron, from which circular RNA is produced through the splicing activity of autocatalytic group I RNA elements (**Fig. 1**; *1,2*). Because the only cofactors required for splicing of the group I intron are magnesium and guanosine, the process can take place in a variety of organisms, making it amenable to a wide variety of protein expression systems *(1–4)*.

This chapter details the design and construction of circular mRNAs containing the open reading frame (ORF) encoding for green fluorescent protein (GFP). Included on the circular GFP mRNA constructs are translation initiation sequences designed to recruit either prokaryotic or eukaryotic ribosomes. By removing in-frame stop codons, the author has also designed and tested circular, infinite mRNAs encoding GFP. The mRNAs produce extremely long protein chains of polyGFP, demonstrating that both prokaryotic and eukaryotic ribosomes can internally initiate, and repeatedly transit, a circular mRNA *(3,5)*. The author has also analyzed fluorescence spectra from *E. coli* expressing the monomeric GFP or polyGFP from circular mRNA, and find that only the monomeric forms of GFP are fluorescent. The application of circular mRNA technology may provide a unique means of producing very long repeating sequence proteins (e.g., silks, mollusk shell framework) *(6–8)*, opening the way for

From: *Methods in Molecular Biology, vol. 183: Green Fluorescent Protein: Applications and Protocols*
Edited by: B. W. Hicks © Humana Press Inc., Totowa, NJ

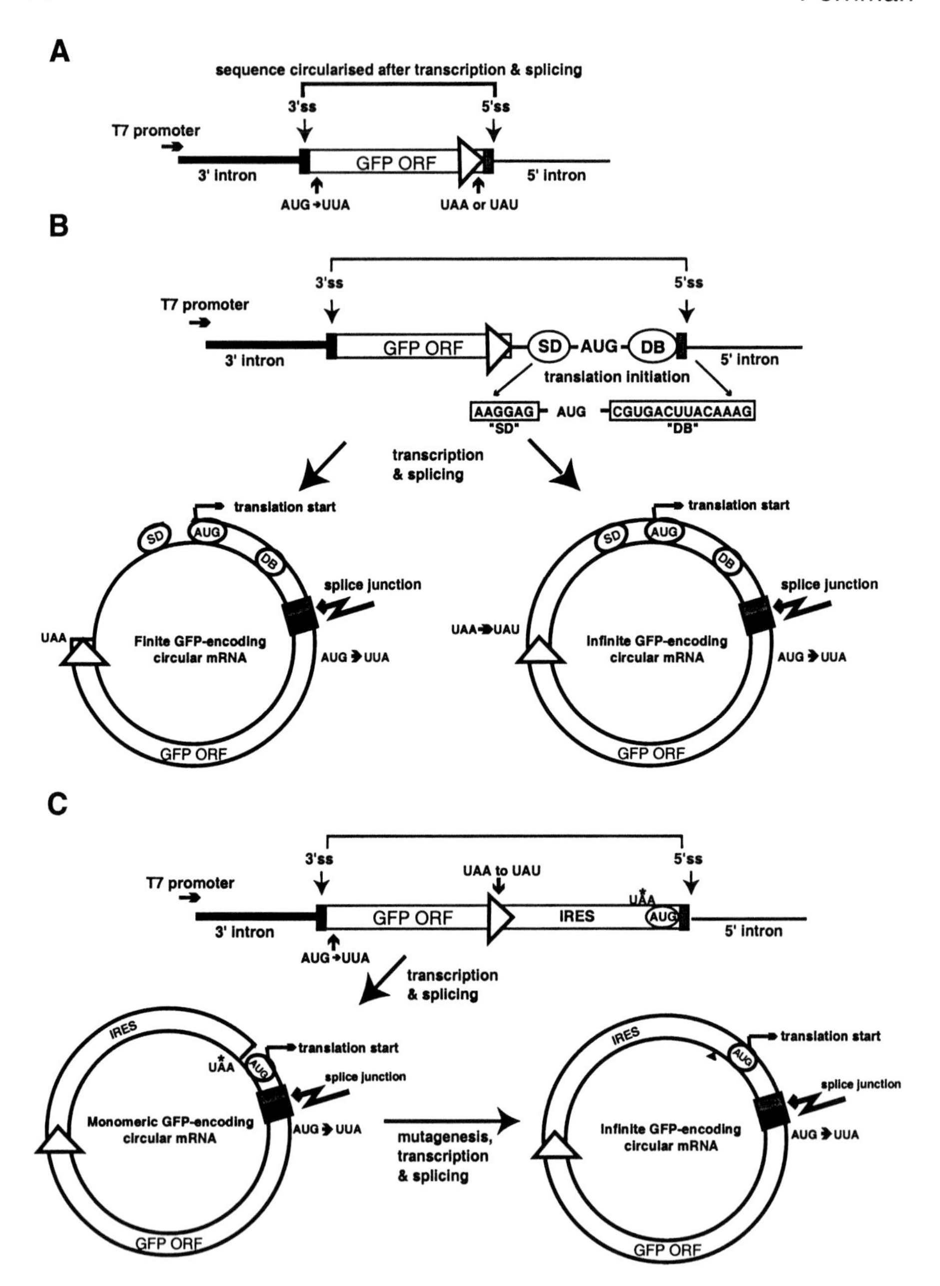

Fig. 1. Design features of plasmids containing rearranged group I intron elements for circular GFP mRNA expression of monomeric GFP or polyGFP in **(B)** *E. coli* or **(C)** rabbit reticulocyte lysates. Transcription and splicing results in circularization of the bracketed sequence, between the 3' (3'ss) and 5' (5'ss) splice sites shown in each figure. **(A)** Relevant region of circular GFP mRNA plasmid containing the GFP ORF.

development of proteinaceous materials with novel properties (e.g., *see* **ref. 9** for review). The methods described below can easily be adapted to the production of any desired protein sequence.

2. Materials

2.1. Making Plasmids for Production of Circular GFP-mRNA

1. Multicopy plasmid vectors (the author uses pBluescript from Stratagene) and restriction enzymes from common vendors were used in buffers as directed by manufacturers.
2. *Escherichia coli* strains: CJ236 (*ung-*, *dut-*) and a standard *rec-* laboratory strain suitable for maintaining and amplifying plasmids (e.g., XL1-Blue from Stratagene).
3. Luria-Bertoni (LB) broth (1 L): 10 g Bacto-tryptone, 10 g NaCl, 5 g yeast extract, pH to 7.6. LB-ampicillin.
4. LB broth containing 100 µg/mL ampicillin.

(Fig. 1. *continued*) Site of T7 RNA polymerase promoter sequence used for in vitro and in vivo (*E. coli*) RNA production is indicated. The 3' and 5' group I intron sequences are shown. AUG to UUA is the mutation introduced by the GFP-AUG oligonucleotide, to remove the initiating GFP-AUG (*see* **Subheading 3.1.**). UAA to UAU is the mutation introduced by the GFP-stop oligonucleotide to remove the GFP-stop codon, thus allowing creation of infinite circular ORFs (see **Subheading 3.1.**). **(B)** Circular GFP mRNA plasmid containing the GFP ORF and *E. coli* protein expression cassette. Enlarged, boxed nucleotides are the SD and DB motifs of the translation initiation sequence. Circular species show either the monomeric or polyGFP mRNA created after transcription and splicing. Translated portion of circular mRNAs is shown as double circle. The monomeric GFP-encoding mRNA, created after transcription and splicing, encodes a ~30 kDa GFP species. The polyGFP-encoding mRNA has UAU in place of UAA termination codon, and is devoid of stop codons in the GFP reading frame. The initiating AUG (translation start) and fused 5'ss/3'ss (jagged arrow-splice junction) are shown on the circular mRNA species for both monomeric and polymeric constructs. Other abbreviations are: GFP, green fluorescent protein ORF; SD, Shine-Dalgarno sequence; AUG, initiating codon; DB, downstream box. **(C)** Same as **(B)**, except circular GFP mRNA plasmid containing the GFP ORF and IRES for mammalian protein expression cassette. Linear species shows relevant portions of circular GFP mRNA plasmids including both GFP-AUG and GFP-stop mutations (*see* **Subheading 3.1.**). The circular species show each GFP-encoding circular mRNA after transcription and splicing. The monomeric GFP-encoding mRNA contains a single UAA termination codon, as indicated by *, and encodes ~50 kDa protein. The polyGFP-encoding species has a two base insertion at this position (indicated by triangle), and is devoid of stop codons in the GFP reading frame. "IRES" is the internal ribosome entry sequence required for ribosome recruitment (*see* **Subheading 3.1.4.**).

5. LB agar: LB broth containing 15 g/L agar.
6. LB-amp agar: LB agar containing 100 µg/mL ampicillin.
7. M13KO7 helper phage (Pharmacia).
8. Kanamycin (10 µg/mL in LB-amp).
9. 20% Polyethylene glycol/2.5 *M* NaCl.
10. TES: 50 m*M* Tris-HCl, pH 7.5, 5 m*M* EDTA, 0.5% sodium dodecyl sulfate (SDS).
11. Phenol–chloroform–isoamyl alcohol (24:24:1 v/v/v).
12. 3 *M* Sodium acetate, pH 6.0.
13. 70% Ethanol.
14. 1% Agarose gel containing 1 µg/mL ethidium bromide, 10X TBE (1 L): 108 g Tris, 55 g boric acid, 9.3 g EDTA.
15. 5X Annealing buffer: 200 m*M* Tris-HCl, pH 7.5, 100 m*M* $MgCl_2$, 250 m*M* NaCl, containing 1 m*M* each of deoxyribonucleoside triphosphate (dNTP)s, 1 m*M* dithiothreitol (DTT), 1 m*M* adenosine triphosphate (ATP), 400 U T4 DNA ligase (New England Biolabs), and 3 U Sequenase™ (U.S. Biochemicals, or other DNA polymerase; *see* **Note 1**).
16. 10X Filling buffer: 100 m*M* Tris-HCl, pH 7.5, 50 m*M* $MgCl_2$, 75 m*M* DTT.
17. dNTP mix containing 0.5 m*M* each deoxynucleotide.
18. Klenow DNA polymerase.
19. 0.5 *M* EDTA, pH 7.0.
20. 10X Dephosphorylation buffer: 1 *M* NaCl, 500 m*M* Tris-HCl, pH 7.9, 100 m*M* $MgCl_2$, 10 m*M* DTT.
21. Calf intestinal phosphatase.
22. 10X Ligation buffer: 500 m*M* Tris-HCl, pH 7.5, 100 m*M* $MgCl_2$, 100 m*M* DTT, 10 m*M* ATP, 0.25mg/mL bovine serum albumin (BSA).
23. Deoxyoligonucleotides: These can be obtained from many commercial sources, or, if available, an in-house facility (*see* **Note 2** for design tips). Oligonucleotides required for circular mRNA constructs in **Subheading 3.1.**
    a. GFP-AUG: alters the initiating AUG of the GFP ORF 5'-GGAGATATAA CCTTAAGTAA AGGAGA-3'.
    b. GFP-stop: alters the GFP stop codon 5'-AACTATACAA ATATTGAGCT CTCATGA-3'.
    c. *E. coli*-Shine-Dalgarno sequence (SD): adds ribosome recruitment signals to the circular mRNA plasmid for expression in *E. coli* 5'ATTGACCTGAGATC GCTTTTTGCTTTGTAAGTCACGTTAGAGCTAGCCATCTTGTGTCTC CTTGTGCAGACCTCTCGAGCTCCAT-3'.
    d. Internal ribosome entry sequence (IRES)-stop: adds two nucleotides to the circular RNA plasmid for polyGFP expression in rabbit reticulocyte lysates 5'-GGGACTAAGGCGGAAT TCTCGAGCTCCATG-3'.

2.2. Testing for Circular mRNA

All reagents for RNA work should be used for this purpose only to minimize contamination from RNase. If desired, H_2O used to prepare these solutions can be treated with diethyl pyrocarbonate (DEPC), by adding 1 mL DEPC/1 L water. Mix and stand, then incubate overnight at 55°C to inactivate DEPC (this is important, because active DEPC can chemically modify ribonucleic acids).

1. rNTPs (10 m*M* each A, C, and G, 1 m*M* U).
2. 5X Transcription buffer: 200 m*M* Tris-HCl, pH 7.5, 30 m*M* $MgCl_2$, 10 m*M* spermidine, 50 m*M* NaCl.
3. RNasin (Promega).
4. α-^{32}P-uridine triphosphate (UTP) (3000 Ci/mmol, Amersham).
5. T7 RNA polymerase.
6. RQ1 DNase I (Promega).
7. 4.5% polyacrylamide gel containing 7 *M* urea.
8. Autoradiography film (Kodak).
9. Formamide dye mix: 100% formamide, 1% bromophenol blue, 1% xylene cyanol.

2.3. Testing Protein Expression in E. coli

1. *E. coli* strain BL21-DE3 (or other T7-expressing *E. coli* strain; *see* **Note 3**).
2. Isopropyl-β-D-thiogalactopyranoside (IPTG).
3. 50 m*M* HEPES, pH 7.5.
4. 8 *M* Urea, 0.1 *M* NaH_2PO_4, 10 m*M* Tris-HCl, pH 6.0.
5. Bradford assay kit (Bio-Rad).
6. SDS-PAGE solutions: 30 polyacrylamide: 0.8 *bis*-acrylamide solution (a premade 30% stock can be purchased from Bio-Rad); 4X stacking buffer (100 mL): 6.06 g Tris, 4 mL 10% SDS (pH 6.8); 4X separating buffer (100 mL): 18.17 g Tris, 4 mL 10% SDS (pH 8.8); 10X protein running buffer (1 L): 30 g Tris, 144 g glycine, 10 mL 10% SDS; protein running dye: 2 mL glycerol, 2 mL 10% SDS, 0.25 mg bromophenol blue, 2.5 mL 4X stacking buffer, 0.5 mL β-mercaptoethanol.
7. Western blotting materials: nitrocellulose membrane; BSA; polyclonal GFP antibody (Clontech); immunoglobulin G-goat-antirabbit secondary antibody; 5-bromo-4-chloro-3-indolyl phosphate (BCIP), dissolved at 25 mg/500 µL in dimethyl formamide; nitroblue tetrazolium (dissolved at 25 mg/350 µL in dimethyl formamide + 150 µL water); Western transfer buffer (4 L): 9.6 g Tris, 45.2 g glycine, 800 mL methanol, to 4 L with H_2O; TBS-T (500 mL): 0.45 g NaCl, 50 mL, 1 *M* Tris-HCl, pH 7.5, 500 µL Tween; Western substrate buffer (10 mL): 250 µL, 4 *M* NaCl, 50 mL 1 *M* $MgCl_2$, 1 *M* Tris-HCl, pH 9.5.
8. Autoradiography materials: ^{35}S-methionine (Amersham); Flexi™ rabbit reticulocyte lysate in vitro translation kit (Promega); 10 mg/mL RNase A solution; film (Kodak).

3. Methods

3.1. Making Plasmids for Production of Circular GFP-mRNA

A general-purpose plasmid, containing self-splicing introns, which produces circular mRNA, must be created (**Fig. 1**; **refs.** ***1,3***). The ORF encoding GFP ***(10)***, is inserted, as an end-filled *BstB*1-*Sac*I fragment, into an end-filled *Nco*I-

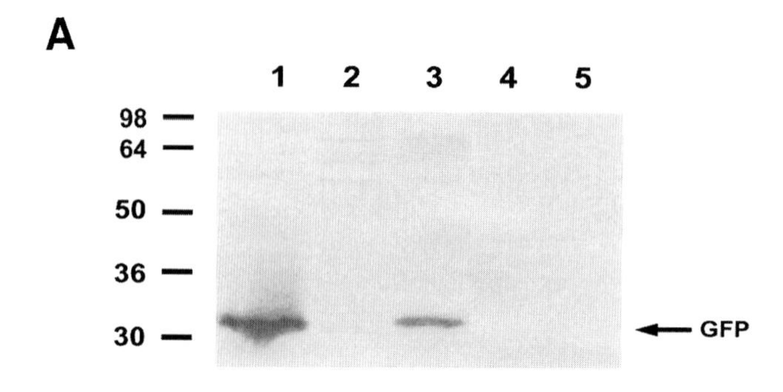

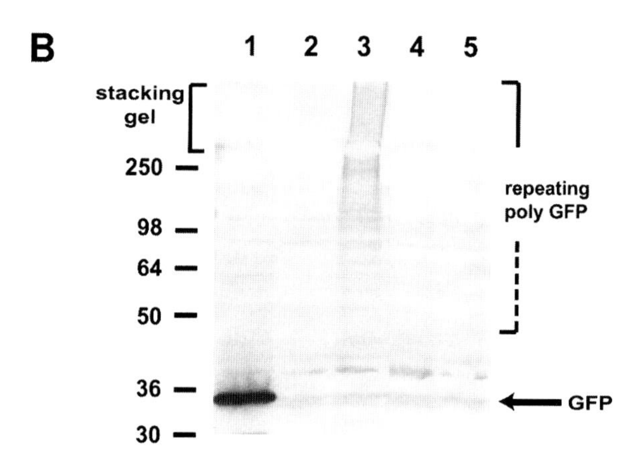

Fig. 2. Translation of GFP from circular mRNAs in *E. coli*. Western blot for GFP expression from strains expressing various monomeric **(A)** and polymeric **(B)** GFP-encoding circular mRNAs, and linear control mRNA. **(A)** Lane 1, linear GFP mRNA control; lane 2, vector control; lane 3, monomeric GFP-encoding circular mRNA; lane 4, monomeric GFP-encoding circular mRNA without translation initiation sequence; lane 5, monomeric GFP-encoding lacking the 5' half group I intron element. Arrowed "GFP" is protein species. Protein size markers are indicated. **(B)** As for **(A)**, except that infinite-encoding circular mRNAs are being expressed.

*Sac*I-digested circular mRNA plasmid (**Fig. 1A**). This plasmid can be used after subsequent engineering to allow expression of either monomeric or poly GFP in either eukaryotic or prokaryotic systems (*see* **Note 4**). The translation initiation sequences are inserted in-frame with the GFP ORF (**Fig. 1**). For the prokaryotes, this produces a circular mRNA transcript containing 795 nucleotides, from which a ~30-kDa monomeric GFP (**Fig. 1B** and **Fig. 2**) or polyGFP (the GFP stop codon is mutated to code for isoleucine) can be expressed. For

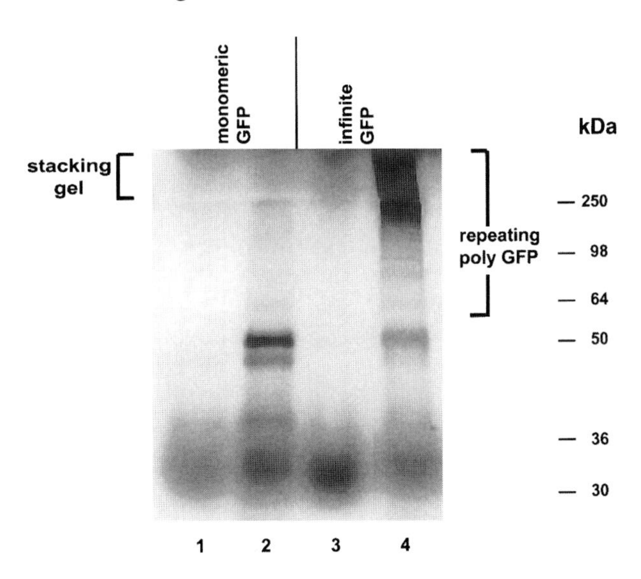

Fig. 3. Translation of GFP from circular mRNAs in rabbit reticulocyte lysates. Autoradiograph of in vitro translation products from monomeric (lanes 1 and 2) and polymeric (lanes 3 and 4) GFP-encoding circular RNA plasmids. Lane 1, monomeric GFP-encoding RNA lacking the 5' half intron; lane 2, monomeric GFP-encoding circular mRNA; lane 3, as for lane 1 except the polyGFP-encoding construct was translated; lane 4, polyGFP-encoding circular mRNA (as is the case in *E. coli*, a heterogeneous collection of proteins is seen, ranging in size from ~50 to >300 kDa).

eukaryotic expression using rabbit reticulocyte lysate, the DNA construct produces a circular mRNA transcript containing 1311 nucleotides encoding for a ~50-kDa monomeric GFP, or a polyGFP (after mutation of the GFP stop codon) (**Fig. 1C** and **Fig. 3**).

3.1.1. Site-Directed Mutagenesis of GFP Start and Stop Codons

Prior to addition of ribosome entry signals, site directed mutagenesis is used to create two sequence alterations in the circular GFP mRNA plasmid. The first mutation (GFP-AUG) changes the existing initiating GFP-AUG, and is necessary to ensure that no translation of the GFP ORF occurs from unspliced linear RNAs. The author changed the AUG to UUA, which encodes for leucine. This construct is used to make monomeric GFP in *E. coli*, and it acts as a template for the second mutation. The second mutation (GFP-stop) alters the GFP stop codon, to code for isoleucine (UAA to AUU).

1. Transform CJ236 (*ung*$^-$, *dut*$^-$) *E. coli*-competent cells with cylase-GFP plasmid. Select transformed colonies by spreading dilutions of transformed cells onto LB-amp agar plates after incubating overnight at 37°C.

2. Pick a single colony, and inoculate 2 mL LB-amp in a sterile test tube. Shake at 37°C for >6 h.
3. Transfer 60 μL to a fresh 3-mL aliquot of LB-amp liquid in a new tube, then add 2×10^7 pfu M13KO7 helper phage, and shake at 37°C for 2 h. Add kanamycin to 10 μg/mL, and continue shaking overnight at 37°C.
4. Pellet the cells, and remove 2 mL supernatant. Add 0.54 mL 20% polyethylene glycol/2.5 *M* NaCl solution to the supernatant. Incubate for 15 min at room temperature, then spin at 10,000*g* for 10 min to pellet phage particles.
5. Resuspend pellet in 100 μL TES. Add an equal volume of phenol–chloroform–isoamyl alcohol, then vortex to mix. Spin at 10,000*g* for 5 min, to separate phases. Transfer top phase to a new tube and repeat the phenol–chloroform–isoamyl alcohol extraction.
6. Add 0.1 vol 3 *M* NaAc (pH 6.0) and 2.5 vol ethanol, to precipitate single-stranded DNA. Spin at 10,000*g* for 10 min. Remove the supernatant, and wash the pellet with 70% ethanol (be careful not to disrupt pellet), and dry remaining solvent under vacuum. Resuspend pellet in 30 μL water. About 3 μL of this single-stranded DNA solution should be enough to use as template for site-directed mutagenesis (as a guide, 3 μL should be easily visible by ethidum bromide staining, after electrophoresis through a 1% agarose gel).
7. Mix 3 μL single-stranded DNA with 50 ng phosphorylated oligonucleotide containing mutagenic sequence designed to alter GFP-AUG or GFP-stop (*see* **Note 2**) in a total volume of 10 μL. Incubate at 65°C for 10 min.
8. Add 10 μL 5X annealing buffer containing 1 m*M* each dNTP, 1 m*M* DTT, 1 m*M* ATP, 400 U (New England Biolabs units; *see* Materials) T4 DNA ligase, and 3 U DNA polymerase (*see* **Note 1**). Incubate at 37°C for 1 h.
9. Transform XL1-Blue (or equivalent rec^- lab strain) *E. coli* with 0.1 vol mutagenesis reaction from **step 8**. Select for transformed bacteria on LB-amp agar after overnight incubation at 37°C.
10. Isolate plasmid DNA by alkaline lysis from six or more cultures grown in liquid LB-amp, and analyze for mutant sequences by restriction digest and DNA sequencing (*see* **Note 5**).

3.1.2. Mutagenesis of Plasmids for Monomeric GFP Expression in E. coli

1. A sequence designed for translation initiation in *E. coli* is inserted, by site-directed mutagenesis (*see* **Subheading 3.1.1.**), into the circular mRNA construct from **Subheading 3.1.1.** in which only the GFP-AUG is mutated. The oligonucleotide for this is called "*E. coli*-SD" (*see* **Subheading 2.1.** and **Note 1**) and contains an SD ***(10–13)***, an AUG codon for initiation, and a downstream box (DB) (*8*; **Fig. 1B**). Although this sequence does recruit ribosomes in vivo (*3*; **Fig. 3**), it is possible that alterations could enhance its activity (*see* **Note 4**).
2. The *E. coli* translation sequence is inserted in-frame and downstream of the GFP ORF in the cyclic RNA transcription unit (**Fig. 1A,B**).

3. To create this construct, follow the site directed mutagenesis procedure outlined in **Subheading 3.1.1.** substituting the *E. coli*-SD oligonucleotide for the GFP-AUG or GFP-stop oligonucleotides (**Fig. 1B**).

3.1.3. Mutagenesis of Plasmids for PolyGFP Expression in E. coli

1. For poly GFP translation in *E. coli*, the author inserted the *E. coli*-SD oligonucleotide containing the sequence used in **Subheading 3.1.2.**, but into plasmid from **Subheading 3.1.1.**, in which both the GFP-AUG and the GFP-stop had been mutated.
2. Follow site-directed mutagenesis protocols in **Subheading 3.1.1.**, remembering to verify the exact sequence of this construct, very carefully, by dideoxynucleotide sequencing. For success of the polyGFP translation system, it is critical that each nucleotide within the circularized mRNA is accounted for (*see* **Note 5**).

3.1.4. Plasmids for Monomeric GFP Translation from Rabbit Reticulocyte Lysates

For translation in rabbit reticulocyte lysates, the IRES from the picornavirus, encephelomyocarditis virus *(14)*, is introduced downstream of the GFP ORF in the DNA construct from **Subheading 3.1.1.**, in which both the GFP-AUG and the GFP-stop are mutated. The IRES fragment is inserted, by subcloning, into the circular GFP mRNA plasmid (**Fig. 1C**). The author used an IRES fragment that had been modified *(5)*, to remove all potential in-frame internal stop codons, so that polyGFP mRNA constructs could eventually be created (*see* **Subheading 3.1.5.**).

1. Digest 1–5 μg picornavirus IRES-containing plasmid *(5)* with restriction enzymes *Xho*I and *Bgl*II, using standard protocols and instructions from the manufacturers. Likewise, digest 1–5 μg circular GFP mRNA plasmid with restriction enzymes *Eco*RI and *Nco*I.
2. Digestion with these restriction enzymes leaves incompatible ends, so both the IRES fragment and the circular GFP mRNA vector are repaired, using Klenow DNA polymerase to generate blunt ends. For a 20-μL Klenow reaction, mix restricted DNA, 1 μL of 0.5 m*M* dNTPs, 2 μL 10X filling buffer, 1–5 U Klenow, and water to 20 μL. Incubate for 15 min at 30°C, and stop the reaction by adding 1 μL 0.5 *M* EDTA.
3. **Steps 3** and **4** for digested GFP DNA only. Increase volume to 50 μL, extract with phenol–chloroform–isoamyl alcohol, precipiate with ethanol, and isolate DNA as in **Subheading 3.1.1.** Resuspend the DNA at 50–100 ng/μL in water.
4. To prevent self-ligation, add 2 μL of 10X dephosphorylation buffer, 0.5 U calf intestinal phosphates/pmol DNA, and water to 20 μL. Incubate for 60 min at 37°C and terminate by addition of 1 μL 0.5 *M* EDTA. Repeat **step 3** above.
5. **Steps 5** and **6** for IRES DNA fragment only. To prevent religation into the original vector, the author uses a "freeze-squeeze" procedure to isolate the IRES-contain-

ing fragment. Run digested DNA on a 1% agarose gel containing 1 μg/mL ethidium bromide in 1X TBE. Visualize DNA using an ultraviolet (UV) light-box.

6. Poke a small hole in a 0.5-mL tube and plug with small ball of tissue (e.g., Kimwipe); rest this tube in a 1.5-mL tube. Excise the gel containing the IRES fragment from the gel and place it in the 0.5-mL tube. Freeze on dry ice for ~5 min. Spin the 0.5-mL tube (within 1.5-mL tube) at 10,000*g* for 10 min, and collect DNA-containing supernatant now in the 1.5-mL tube. Repeat ethanol precipitation, and resuspend the isolated fragment in water at 100–200 ng/μL.
7. For ligation of vector and fragment, the author uses a molar ratio of ~1 vector: 2–5 fragment. Mix 50–100 ng GFP vector, 200–500 ng IRES fragment, 1 μL 10X ligation buffer, 400 U New England Biolabs T4 DNA ligase (*see* Materials), and add water to 10 μL. Incubate 4–12 h at 16°C.
8. Isolate the ligated DNA by ethanol precipitation, and resuspend the pellet in 10 μL water. Transform *E. coli* XL1-Blue with 2–5 μL or the resuspended DNA. Analyze for the IRES insert, by restriction digest and/or sequencing (*see* **Note 5**). This plasmid is now ready for monomeric protein expression in rabbit reticulocyte lysate.

3.1.5. Plasmids for PolyGFP Translation in Rabbit Reticulocyte Lysates

1. An additional round of site-directed mutagenesis is required to create circular polyGFP mRNA encoding plasmids designed for translation in rabbit reticulocyte lysates.
2. Despite using the circular GFP mRNA plasmid lacking the GFP-stop, insertion of the IRES creates an in-frame stop codon at exactly one complete revolution of the 1311 nucleotide circle (*see* **Figs. 1C** and **4**).
3. A two nucleotide insertion by site directed mutagenesis using IRES-stop oligonucleotide is required to create the circular polyGFP-ORF. Follow the site directed mutagenesis protocol using the cyclic GFP plasmid containing the IRES (**Fig. 1C**) as the template for mutagenesis.

3.2. Testing Circularization of In Vitro RNA Transcripts

Prior to introduction into *E. coli* or rabbit reticulocyte lysates, the circularization efficiency for each of the constructs can be tested by making in vitro RNA transcripts using T7 RNA polymerase and analyzing RNA products on denaturing polyacrylamide gels. All the requirements for the circular mRNA production are within a transcription reaction.

1. Linearize 1–2 μg plasmid template DNA with appropriate restriction enzyme and resuspend in 20 μL sterile RNase-free water.
2. Assemble the transcription mix in the following order: 5 μL rNTP mix (10 m*M* rATP, rCTP, rGTP + 1 m*M* rUTP), 10 μL 5X transcription buffer, 5 μL 0.1 m*M* DTT, 6 μL dH_2O, 1 μL RNasin, 2 μL α-^{32}P-UTP (3000 Ci/mmol), 1–2 μg linear DNA in 20 μL water, 1 μL T7 RNA polymerase.
3. Incubate at 37°C for 1 h. This reaction should be set up at room temperature (not ice) to avoid precipitation of rNTPs or linear DNA template, which can occur in

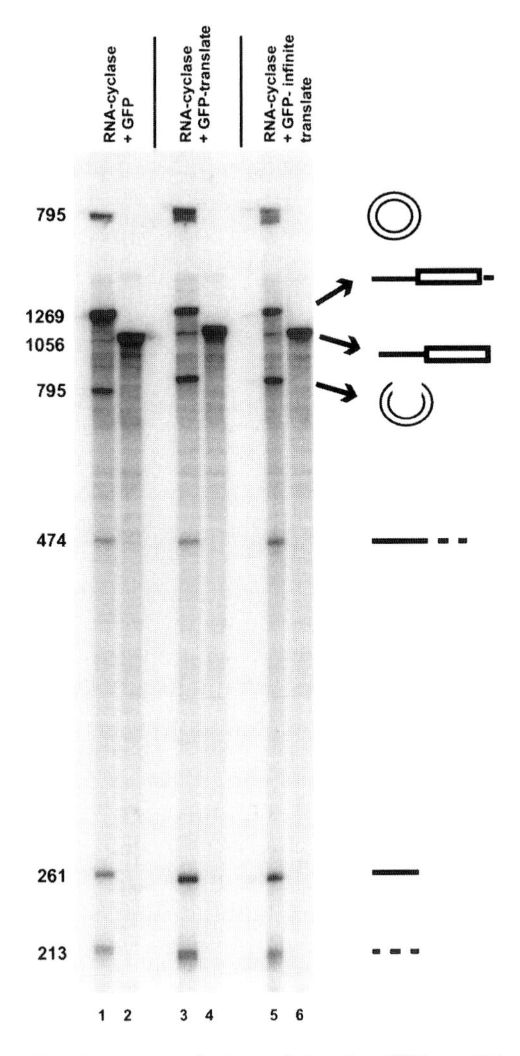

Fig. 4. Example of in vitro transcription of circular GFP mRNA from plasmids run on 4% PAG containing GFP (lanes 1 and 2), monomeric GFP + *E. coli* initiation cassette (lanes 3 and 4), polyGFP + *E. coli* initiation cassette (lanes 5 and 6). Lanes 2, 4, and 6 are negative controls, in which the circular GFP mRNA plasmids have been linearized to exclude the 5' intron sequence from the transcribed RNA. These RNA species represent the 3' intron +internal circularizing GFP ORF, which is also the first step of the group I splicing reaction. Lanes 1, 3, and 5 are complete circular GFP mRNA plasmids, and show products of the group I splicing reaction. Product designations, and their approximate length, are indicated to the right and left of the panel, respectively, and, from top to bottom, are: circularized RNA; unspliced precursor; 3' intron + internal circularizing GFP ORF; nicked circular RNA; ligated 5' and 3' introns; 5' intron; 3' intron. Note the difference in migration of the nicked circular species and the closed circular species.

the presence of spermidine (found in the transcription buffer). Add 1 µL of RQ1 DNase I to remove DNA template from the mixture. Terminate DNase I treatment by addition of 50 µL formamide dye mix.

4. Load 1–2 µL of the transcription mix onto a 40-cm, 4.5% denaturing (7 *M* urea) polyacrylamide gel. Run until bromophenol blue has reached the bottom of the gel. Dry the gel and analyze RNA products by exposing using either autoradiographic film or a Phosphoimager screen. Circularized RNA can be detected as a species distinct from other self-splicing RNA products in a variety of ways (e.g., **Fig. 4**; **Note 6**).

3.3. Testing for Protein Expression

3.3.1. Testing for Expression by Electrophoretic Analysis

3.3.1.1. Protein Extraction from *E. coli*

1. Transform *E. coli* strain BL21(DE3), or another *E. coli* strain expressing T7 RNA polymerase and designed for protein expression, with plasmids containing monomeric or poly GFP, and select on LB-AMP agar plates (*see* **Note 7**).
2. Grow 1-mL cultures of a single colony from each transformant in LB-AMP overnight at 37°C with shaking.
3. Seed 5 mL LB-AMP to A_{600} of ~0.05, with overnight cultures.
4. Grow at 37°C, with shaking, to A_{600} ~0.4. Add IPTG (1 m*M*) and grow cultures for another 4 h.
5. Pellet the cells, and resuspend them in 400 µL of 50 m*M* HEPES, pH 7.5. Lyse cells by 2–3 15-s bursts using a sonicator. Clarify by centrifugation at 10,000*g* for 10 min at 4°C. Decant, and save the supernatant. Protein extracts can be stored at –70°C.
6. Sonication destroys the repeating polyGFP multimers from the infinite encoding constructs (data not shown). Protein from these can be isolated as follows. Pellet cells and resuspend in 8 *M* urea–0.1 *M* NaH_2PO_4–10 m*M* Tris, pH 6.0. Freeze on dry ice, then heat to 90°C for 5 min. Clarify by centrifugation at 10,000*g* for 10 min at 4°C. Aliquots can be stored at –70°C.

3.3.1.2. Detection of Monomeric and PolyGFP by Western Blots

1. Determine protein concentrations of extracts with Bradford assay kit per manufacturer's specifications.
2. For detection of polyGFP, equivalent total protein amounts (as determined in **step 1**) are separated on 8% separating–4% stacking polyacrylamide–SDS gels. For detection of monomeric GFP, 12.5% separating–4% stacking polyacrylamide–SDS gels are used. The author finds that 5–10 µL total cell lysate is sufficient for Western analysis and visualization of both monomeric GFP and polyGFP products (**Fig. 2A,B**).
3. Proteins are electrophoretically transferred to nitrocellulose membrane at 4°C, 50 V for ≥ 2 h in Western transfer buffer.
4. The nitrocellulose membrane is blocked by incubating in 3% BSA in TBS-T for 30 min at room temperature. Polyclonal GFP antibody is added at 1:1000 dilu-

tion in TBS-T–3% BSA, and incubated for ~2 h at room temperature. The membrane is washed in TBS-T 3 × 15 min, then incubated in TBS-T–3% BSA containing IgG-goat-antirabbit secondary antibody at 1:1000 dilution for 1 h at room temperature. The membrane is then washed in TBS-T 3 × 15 min, and GFP products visualized via conjugated alkaline phosphatase, in a solution of 17 µL BCIP and 33 µL nitroblue tetrazolium in 5 mL substrate buffer.

3.3.1.3. Testing for Protein Expression in Rabbit Reticulocyte Lysates by Autoradiography

To test for protein expression from circular mRNA constructs encoding monomeric GFP or polyGFP from the mammalian expression cassette (**Fig. 1**), mRNAs are transcribed in vitro, then incubated in a rabbit reticulocyte lysate translation system (*see* **Materials** and **Note 8**) using ^{35}S-methionine for autoradiographic detection of proteins (**Fig. 4**).

1. In vitro transcription can be done as described in **Subheading 3.2.**, **steps 1** and **2**, except radiolabeled α-^{32}P-UTP is omitted. Instead, substitute 10 m*M* rUTP for 1 m*M* rUTP. Include RQ1 DNase treatment, to remove template DNA.
2. Add an equal volume (50 µL) of phenol–chloroform–isoamyl alcohol; mix, and spin at 10,000*g* for 5 min. Transfer top phase to new tube, and precipitate the RNA with ethanol. Resuspend the precipitated RNA in RNase-free water at ~1–2 µg/µL (usually requires ~10 µL of water).
3. The author typically uses 2–3 µg in vitro transcribed RNA in a 25-µL in vitro translation reaction, and follows a proportionally scaled-down version of the standard 50-µL protocol found in the technical bulletin supplied with rabbit reticulocyte lysates from Promega (with exceptions listed in **Note 8**). The author uses ^{35}S-methionine as the radiolabeled amino acid. Generally, one-tenth of this reaction (2–3 µL), is sufficient for electrophoresis in 10% separating–4% stacking SDS-polyacrylamide gel (both products can be analyzed using 10% polyacrylamide as the separating gel).
4. Following electrophoresis, the gel is dried and used to expose film for autoradiography or, if available, phosphoimager analysis (**Fig. 3**). In vitro translation products of GFP are visualized via incorporated ^{35}S-methionine.

3.3.2. Detection of GFP by Fluorimetry

GFP expression from the circular mRNAs in *E. coli* can also be analyzed by looking for emission of green light, upon photoexcitation ***(11)***. This can be done by observing *E. coli* colonies on a UV lightbox after growth on LB agar containing IPTG, or, for more sensitive and quantitative detection, by fluorimetry. The author can detect fluorescence from *E. coli* expressing the monomeric circular GFP mRNA constructs using either method, but The author is unable to detect any fluorescence from the polyGFP expressing constructs (*see* **Note 3**).

1. Fluorimetry, if available, is the method of choice for detecting GFP expression from circular mRNAs. The author uses a Perkin-Elmer (LS50B) luminescence

spectrometer to do an emission scan with a 5 nm slit width at 240 nm/min. Excitation is fixed at 397 nm and emission is scanned between 350 and 550 nm (peak emission for the GFP derivative is 508 nm).

2. Cultures of GFP-expressing *E. coli* (2–3 mL) are grown overnight at 37°C. The cells are pelleted and resuspended in 1 mL of water to analyze. A four-sided glass cuvet, with a 10-mm path length works well in these assays (*see* **Note 3**).

4. Notes

1. Several other DNA polymerases work well in this protocol. The author has used T7 DNA polymerase (from which Sequenase™ is a derivative), T4 DNA polymerase, and Klenow. Buffer requirements may vary, so be sure to check if using a different enzyme.
2. Deoxyoligonucleotides for site-directed mutagenesis using the method described in this chapter should be designed such that the mutant nucleotides are central (e.g., a "band-aid" effect). For point mutations, such as those required to remove start or stop codons in the circular GFP mRNA constructs, potential hybridization 5' and 3' of the mutagenic region should be at least 10–12 bases, to ensure efficient annealing. For larger sequence alterations, such as the insertion of the *E. coli* expression cassette, The author recommends extending this potential hybridization to at least 17 nucleotides. Also, synthetic deoxyoligonucleotides do not have phosphates at their 5' termini. For any ligation of the mutagenic strand, these oligonucleotides must be phosphorylated, using standard molecular biology protocols involving T4 DNA kinase.

   The author carried out sequential mutagenesis reactions (i.e., first, the author altered the GFP-AUG) then this was used as a template in a second mutagenic reaction using the GFP-stop oligonucleotide), but there is no reason why both could not be attempted simultaneously, by including both oligonucleotides in the annealing step. If this is done, one needs to acquire the GFP-AUG as a single mutation as well as in combination with the GFP-stop.
3. Prokaryotic in vivo translation of the monomeric GFP from the circular mRNA constructs is relatively inefficient, compared with the identical linear constructs *(3)*. Thus, reduced fluorescence makes visualization by UV illumination with a light-box difficult. To be certain that one is not observing autofluorescence (which can be remarkably deceiving), it is essential to compare with non-GFP expressing *E. coli*. The author could convincingly observe fluorescence from the monomeric GFP constructs in *E. coli* using a UV light-box, and verified this by fluorimetry. In contrast, the author was unable to distinguish any fluorescence above background using the UV light-box method or fluorimetry when looking at the polyGFP mRNA construct *(3)*. Presumably, this means that multimeric linked units of GFP are unable to fluoresce. The author has not analyzed why this is the case, but the polyGFP may be inefficiently posttranslationally modified *(17)*, and/or not able to fold into the correct conformation to allow formation of the cyclic tripeptide chromophore. Possibly the addition of a linker region between the repeating GFP units would allow enough space for correct processing, fold-

ing, and fluorescence. Additionally, because a heterogeneous collection of proteins is observed, ranging in size from ~50 kDa (the size of the monomer) to >300 kDa, possibly much of the polyGFP is made from incomplete GFP units that might interfere with posttranslational processing.

4. Although the author has not tested other sequences, previous data ***(3)*** show that both the SD and DB sequence motifs, in the translation initiation sequence of the circular mRNAs, are necessary to produce maximal protein levels, as measured in this experimental system. The author finds that the SD sequence contributes more than the DB motif ***(3)***. Since very little is known about whether there is a role for 5' ends of mRNA contributing to translation in *E. coli*, possibly the SD and DB used in this study may not be best for exclusively internal initiation. Other RNA sequence motifs may enhance direct ribosome recruitment, and thus improve circular mRNA translation. Data also show that the apparent effects of removing either the SD or the DB motif is minimized during expression from the polyGFP ORF, compared to the monomeric GFP ORF ***(3)***, which indicates that initiation is the rate-limiting step in translation in *E. coli* from circular mRNAs, at least in the context of sequences contained in our circular GFP mRNA constructs.
5. The author recommends always sequencing mutagenic clones, particularly in the region containing the mutation(s) because site-directed mutagenesis can sometimes introduce sequenced changes other than those intended. In addition to verifying them by restriction digest. The author also strongly recommends sequencing when fragments and/or vector have been end-filled. This is particularly critical for the circular GFP mRNA constructs, in which each nucleotide must be accounted for, to ensure the reading frame remains correct.
6. In vitro transcribed RNAs from the circular GFP mRNA plasmids can be analyzed for circularized products in several ways. Analysis of circular GFP mRNA expression products on high- and low-% denaturing acrylamide gels leads to a shift in mobility of nonlinear (i.e., circular) molecules, whereby circles run proportionally faster than their linear counterparts on a low % gel, but proportionally slower on a high-% gel ***(12)***. Only the migration of nonlinear, circularized RNA species are affected in this way. Two dimensional denaturing gel electrophoresis ***(1)*** is another way of visualizing circular RNAs, which is an extension of the first, but here a diagonal of linear RNA molecules is produced. The circular molecules appear above the diagonal, because of the reduction in mobility of nonlinear species in higher-% polyacrylamide gels.
7. There are several commercially available strains of *E. coli* expressing T7 RNA polymerase from a chromosomal gene copy. The author uses BL21(DE3) from Novagen, in which T7 is under control of a regulated *lacUV5* promoter that can be induced using IPTG. The Novagen catalog (for 2000) and web site ***(18)*** contain excellent overviews on protein expression and some detail regarding the advantages and disadvantages of each strain variant. Because the author has not tested strains other than BL21(DE3), comment cannot be made on their suitability for expressing protein from our circular GFP mRNA constructs. If desired, circularized mRNAs can also be analyzed by purifying total RNA from IPTG-induced *E. coli* and doing Northern hybridization ***(3)***.

8. The author found the following modifications necessary to optimize in vitro translation products from our circular GFP mRNA plasmids. The author titrated addition of 25 m*M* magnesium acetate and found 0.5/25 µL reaction optimal for monomeric and polyGFP translation from IRES circular GFP mRNA constructs. Similarly, the author found 1.2/25 µL reaction of 2.5 m*M* potassium chloride increases translation efficiency. DTT was not added. An incubation time of 60 min is sufficient. Following incubation, and prior to loading on polyacrylamide gel, The author finds it necessary to add 0.5 µL 10 mg/mL RNase A to a 3-µL aliquot of the reaction, to digest aminoacyl tRNAs, which produce background bands. The GFP produced in the rabbit reticulocyte lysate system does not fluoresce in our hands. The author was unable to obtain any GFP translation products using the coupled transcription–translation kits from Promega.

Acknowledgments

This work was done in the laboratory of Professor Manuel Ares, under partial support by grant GM40478 from the National Institutes of Health. Rhonda Perriman was also supported by an American Cancer Society Postdoctoral fellowship. The author thanks Roland Nagel for review and Manny Aves for stimulating discussion, support, and critical review of this manuscript.

References

1. Ford, E. and Ares, M. (1994) Synthesis of circular RNA in bacteria and yeast using RNA cyclase ribozymes derived from a group I intron of phage T4. *Proc. Natl. Acad. Sci. USA* **91,** 3117–3121.
2. Puttaraju, M. and Been, M. D. (1996) Circular ribozymes generated in *Escherichia coli* using group I self-splicing permuted intron-exon sequences. *J. Biol. Chem.* **271,** 26,081–26,087.
3. Perriman, R. and Ares, M. (1998) Circular mRNA can direct translation of extremely long repeating sequence proteins *in vivo*. *RNA* **4,** 1047–1054.
4. Long, M. B. and Sullenger, B. A. (1999) Evaluating group I intron catalytic efficiency in mammalian cells. *Mol. Cell Biol.* **19,** 6479–6487.
5. Chen, C. Y. and Sarnow, P. (1995) Initiation of protein synthesis by the eukaryotic translational apparatus on circular RNAs. *Science* **268,** 415–417.
6. Prince, J. T., McGrath, K. P., DiGirolamo, C. M., and Kaplan, D. L. (1995) Construction cloning and expression of synthetic genes encoding spider dragline silk. *Biochemistry* **34,** 10,879–10,885.
7. Oshimi, Y. and Suzuki, Y. (1977). Cloning of the silk fibroin gene and its flanking sequences. *Proc. Natl. Acad. Sci. USA* **74,** 5363–5367.
8. Sudo, S., Fujikawa, T., Nagakura, T., et al. (1997) Structures of mollusc shell framework proteins. *Nature* **387,** 563–564.
9. Heslot, H. (1998) Artificial fibrous proteins: a review. *Biochemie* **80,** 19–31.
10. Shine, J. and Dalgarno, L. (1974) The 3'-terminal sequence of *Escherichia coli* 16S ribosomal RNA: complementarity to nonsense triplets and ribosome binding sites. *Proc. Natl. Acad. Sci. USA* **71,** 1342–1346.

11. Steitz, J. A. and Jakes, K. (1975) How ribosomes select initiator regions in mRNA: base pair formation between the 3' terminus of 16S rRNA and the mRNA during initiation of protein synthesis in *Escherichia coli. Proc. Natl. Acad. Sci. USA* **72,** 4734–4738.
12. Gold, L. (1988) Posttranscriptional regulatory mechanisms in *Escherichia coli. Annu. Rev. Biochem.* **57,** 199–233.
13. Sprengart, M. L., Fuchs, E., and Porter, A. G. (1996) The downstream box: an efficient and independent translation initiation signal in *Escherichia coli. EMBO J.* **15,** 665–674.
14. Hellen, C. U. and Wimmer, E. (1995) Translation of encephalomyocarditis virus RNA by internal ribosomal entry. *Curr. Top. Microbiol. Immunol.* **203,** 31–63.
15. Chalfie, M., Tu, Y., Euskirchen, G., Ward, W. W., and Prasher, D. C. (1994) Green fluorescent protein as a marker for gene expression. *Science* **263,** 802–805.
16. Puttaraju, M. and Been, M. D. (1992) Group I permuted intron-exon (PIE) sequences self-splice to produce circular exons. *Nucl. Acids Res.* **20,** 5357–5364.
17. Heim R., Prasher D. C., and Tsien R. Y. 1994 Wavelength mutations and post-translational autoxidation of green fluorescent protein. *Proc. Natl. Acad. Sci. USA* **91,** 12,501–12,504.
18. http://www.novagen.com; URL verified Oct. 10, 2000.

II

Detection and Imaging of Green Fluorescent Protein

7

Fluorescence Lifetime Imaging (FLIM) of Green Fluorescent Fusion Proteins in Living Cells

Ammasi Periasamy, Masilamani Elangovan, Elizabeth Elliott, and David L. Brautigan

1. Introduction

1.1. The Need and Instrumentation for Fluorescence Lifetime Imaging

Fluorescence microscopy is an exceptional tool for looking inside cells and tissues. Recent advances in fluorescence microscopy, including improved optics, sensitive fluorescent dyes, and high-sensitivity cameras, coupled with technological advances in computers and sophisticated software now permit quantitative measurement and noninvasive acquisition of spectroscopic information from a single living cell *(1–4)*. Because of the specificity inherent in current fluorescence labeling techniques, and the sensitivity in fluorescence microscopic techniques, it is possible to detect very small amounts of proteins with very high sensitivity and precision. The enormous advantage of fluorescence is that most fluorophores used in biology are excited by absorption, and emit light by fluorescence, all within 100 ns. This time-scale coincides with the time-scale of molecular interactions in biological systems under physiological conditions. As a consequence, changes in an active biological system may be followed, theoretically by monitoring the fluorescence lifetime, i.e., the duration of the excited state. To calculate the lifetime, multiple images are collected within ns (a considerable technical challenge) then the decay of fluorescent intensity vs time fit to one or more exponentials. The lifetime of a fluorophore can be used to monitor functional and structural aspects of living specimens.

Instrumental methods for measuring fluorescence lifetimes are divided into two major categories: frequency-domain *(5)* and time-domain *(5–12)*. Fre-

From: *Methods in Molecular Biology, vol. 183: Green Fluorescent Protein: Applications and Protocols*
Edited by: B. W. Hicks © Humana Press Inc., Totowa, NJ

quency-domain fluorometers excite the fluorescence with light, which is sinusoidal and modulated at radio frequencies (for nanosecond decays), then measure the phase shift and amplitude attenuation of the fluorescence emission, relative to the phase and amplitude of the exciting light. Thus, each lifetime value will cause a specific phase shift and attenuation at a given frequency. For single lifetime samples, the lifetime may be calculated directly from either the phase shift or magnitude of the attenuation (or both, since both are available from a single measurement). For multiple lifetimes, many measurements are required over a range of excitation frequencies. In time-domain methods, pulsed light is used as the excitation source, and fluorescence lifetimes are measured from the fluorescence signal directly, or by using single photon counting. The lifetime could be determined easily, using the two-gate-window images (**Fig. 1**) without major computational processing *(8)*.

Fluorescence lifetime imaging (FLIM) microscopy is an important advance in fluorescence light microscopy. Existing fluorescence microscopy techniques do not allow imaging of dynamic molecular interactions between cellular components on a precise spatial and temporal scale. This imaging modality visualizes or monitors spatial and temporal information of environmental changes in living specimens. Recent technological advances in high-speed lasers, as well as sensitive, high-speed, gated image detection devices and image-processing techniques, plus highly specific fluorescent probes, such as green fluorescent proteins, have facilitated the development of FLIM. This chapter describes the design and operation of a FLIM microscope that uses a picosecond-gated, multichannel plate-image intensifier, providing two-dimensional (2-D) time-resolved images of biological specimens. The technique is demonstrated in adherent tissue culture cells that have been transiently transfected to express green, cyan, or yellow fluorescent proteins fused a domain of DNA gyrase that can be dimerized by addition of the double-headed compound coumermycin A. Specimens were placed on an infinity-corrected Nikon epifluorescence microscope, coupled to a Coherent tunable, femtosecond, Ti-sapphire, pulsed laser and a frequency doubler, to provide 440 nm excitation light (**Fig. 2**). After synchronizing the high-speed, gated-image intensifier to the excitation laser pulses, time-resolved nanosecond images of fluorescent emission were acquired. These images were processed pixel-by-pixel for single exponential decay to obtain images based on fluorescence lifetimes (**Figs. 3** and **4**). The FLIM method is expected to have many applications in studying the dynamic behavior of proteins in living cells.

1.2. Theoretical Principles of Lifetime Imaging

When a fluorophore absorbs light, one of the electrons goes from its ground electronic state (S0) to an excited vibrational level within a higher electronic

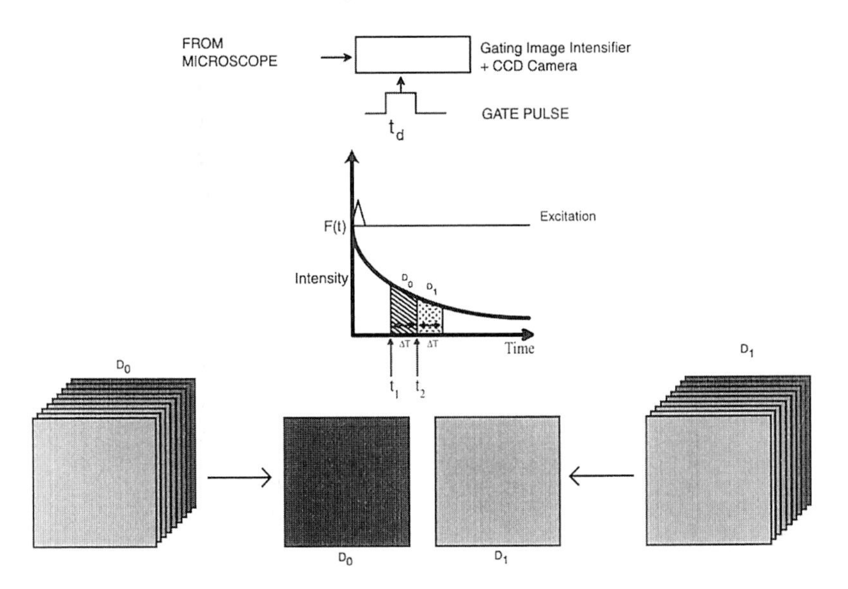

Fig. 1. Illustration of gating procedure to acquire time-resolved FLIM image acquisition for single exponential decay. For double exponential decays, four-gated images should be acquired, then processed using equations, as described in the literature ***(11)***. (Also on CD-ROM.)

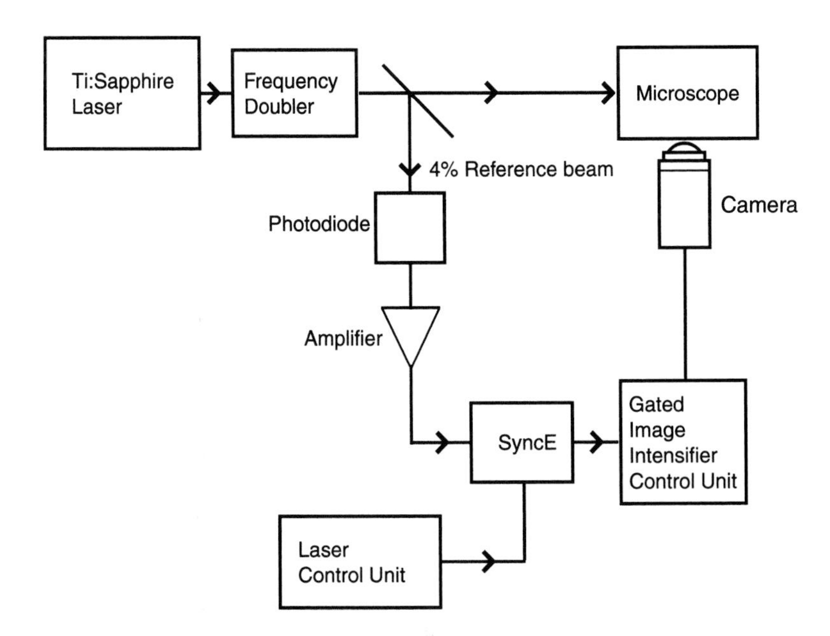

Fig. 2. Schematic illustration of the fluorescence lifetime imaging (FLIM) system. SyncE-synchronization electronics. (Also on CD-ROM.)

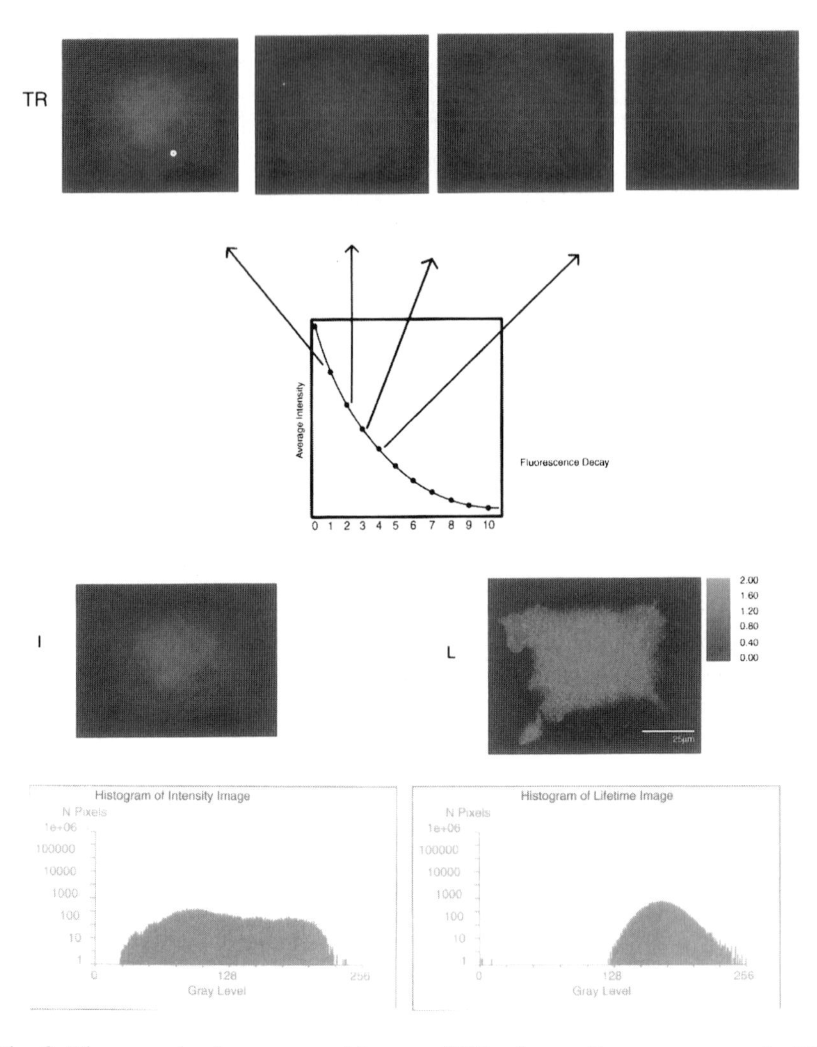

Fig. 3. Time-resolved nanosecond images (**TR**) of cyan fluorescent protein (CFP) tagged to the GyrB protein. These preliminary images were acquired using lifetime imaging microscopy equipment by synchronizing the LaVision PicoStar gating camera to Coherent Ti:sapphire laser. COS7 cells expressing CFP-GyrB were placed on a Nikon TE 300 epifluorescence microscope stage and the laser was tuned to 880 nm and doubled to 440 nm to excite the CFP fluorophore molecules tagged to the proteins. A 60× water-immersion lens was used to acquire the TR and the lifetime image was processed on a pixel-by-pixel basis using the single exponential decay **Eq. 5** (2 and 3). The histogram of the lifetime image (**L**) is very narrow, compared to the intensity image (**I**). The intensity image shows some spatial heterogeneity of CFP distribution, compared to the lifetime image. The average lifetime of CFP, in the absence of acceptor (YFP), is ~1.7 ns. (For optimal, color representation please see accompanying CD-ROM.)

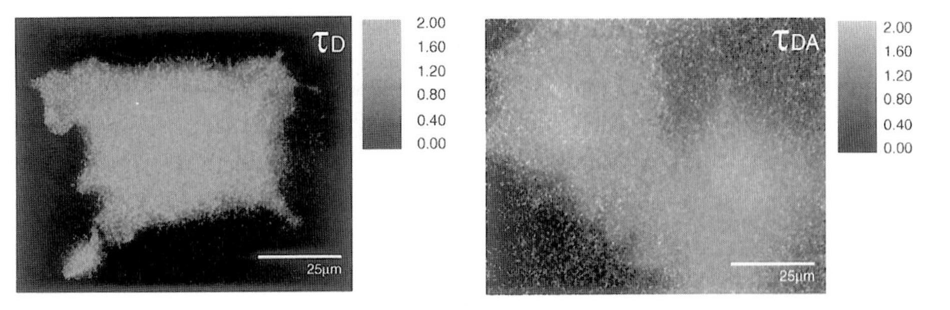

Fig. 4. FLIM–FRET data for CFP-YFP-GyrB fusion protein dimerization in COS7 living cells, by adding 100 μ*M* coumermycin. Time-resolved donor images were acquired in the presence and absence of the acceptor (not shown). The single exponential decay of the donor-alone lifetime (τ_D, cyan color) and the donor in the presence of acceptor (τ_{DA}, yellow color) were processed pixel-by-pixel using the decay equations (2 and 3). The occurrence of protein dimerization was shown by quantitating the lifetimes on two different conditions ($\tau_D = 1.7$ ns; $\tau_{DA} = 1.4$ ns). The efficiency (*E*) and the distance (*r*) between two fluorophores (CFP and YFP) in the fusion proteins were calculated ($E = 18\%$; $r = 6.8$ nm). Moreover, note the distribution of lifetimes of the protein complex (τ_{DA}), and that kind of spatial and temporal resolution cannot be obtained in the steady state FRET imaging. Note that different cells were used to quantitate the donor lifetime, in the presence and the absence of acceptor (YFP). (For optimal, color representation please see accompanying CD-ROM.)

energy level, the excited state (S1). Within picoseconds the electron decays through bond vibrations and collisions with neighboring molecules to the lowest of the multiple vibrational levels within the excited electronic state (S1) without emitting a photon. The excited electron can then decay over a few nanoseconds to one of the excited vibrational levels within the ground electronic state by emission of a photon (fluorescence), whose wavelength is longer than the excitation wavelength. Molecules can also undergo conversion to the triplet state, and decay over microseconds to milliseconds to a ground state by emission of a photon (phosphorescence) (**Fig. 5A**).

The fluorescence lifetime (τ) is defined as the average time that a molecule remains in an excited state prior to returning to the ground state. For single exponential decay of fluorescence, the fluorescence intensity, $I(t)$, as a function of time, t, after a brief pulse of excitation light, is described by **Eq. 1**:

$$I(t) = I_0 \exp(-t/\tau) \tag{1}$$

where I_0 is the initial intensity immediately after the excitation pulse. In practice, the fluorescence lifetime is defined as the time in which the fluorescence intensity decays to 1/e of the intensity, immediately following excitation (**Fig. 5B**). Excited-state lifetime measurements are independent of excitation

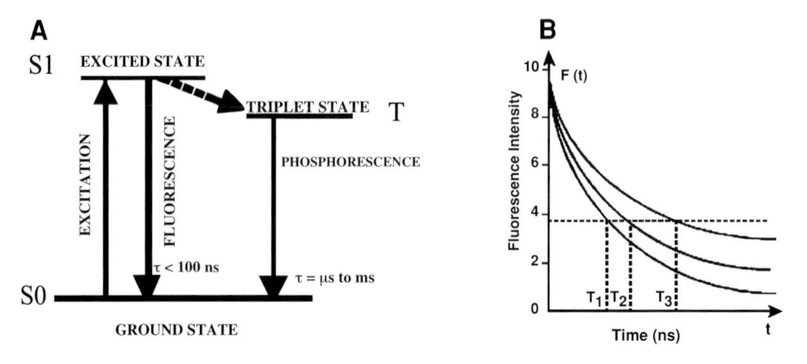

Fig. 5. **(A)** Energy-level diagram showing the lifetime range of fluorescence and phosphorescence. **(B)** Illustration of multiexponential fluorescence lifetime decay. (Also on CD-ROM.)

light intensity, probe concentrations, and light scattering, but highly dependent on the local environment of the fluorophore.

Conventional fluorescence microscopy gives measurements of intensity that provide images to reveal primarily the distribution and amount of stain in the cell. In contrast, the time-resolved lifetime fluorescence microscopic technique potentially allows monitoring of interactions between cellular components at very high temporal resolution. To date, most measurements of fluorescence lifetimes have been performed in solution or cell suspensions, although fluorescence lifetime measurements through a microscope have been reported, using photon counting ***(12)***. Unfortunately, these photon-counting instruments do not allow visualization (either 2-D or 3-D) of fluorescent lifetimes in two dimensions. FLIM was developed to overcome this drawback and still provide the ability to use the power of fluorescence lifetime measurements ***(7–9)***. A fluorophore in a microscopic sample may exist, e.g., in two environmentally distinct regions and have a similar fluorescence intensity distribution in both regions but different fluorescence lifetimes. Measurements of fluorescence intensity alone would not reveal any difference between two or more regions, but imaging of the fluorescence lifetime can reveal such regional differences. Shown in the example **Fig. 4**, dimerization caused a change in lifetime but no change in steady state fluorescence.

2. Materials

2.1. Laser, Microscope, and Camera System

1. A schematic of the FLIM microscope system we use is shown in **Fig. 2**. Discussed below are various components involved in setting up the lifetime imaging system (*see* **Note 1**).

a. A tunable Verdi-pumped Ti:sapphire laser system, providing 150-femtosecond excitation pulses, at repetition rate of up to 76 MHz (Coherent, Santa Clara, CA) is tuned to 880 nm.
b. The frequency is doubled using Coherent frequency doubler to 440 nm to excite the cyan fluorescent protein (CFP).
c. This pulsed blue laser line was coupled to a Nikon inverted epifluorescent microscope, through beam-steering optics.
d. A 60×-water immersion, 1.2 numerical aperture (NA) objective lens was used for the experiments. For laser coupling to a microscope, one should consider a nonphase objective lens, to minimize any light scattering.
e. The output side port of the microscope was coupled to a high-speed, gated image intensifier, to acquire the time-resolved fluorescent images.
f. The high-speed gated camera system (PicoStar HR; LaVision, Goettingen, Germany) consists of camera head (CCD and image intensifier) and its control units (TTL-I/O synchronization board and A/D converter). This intensified gated/modulated camera gate pulse driver has a bandwidth of 1 GHz, and it has internal pulse forming circuitry to provide gate widths less than 300 ps, at trigger rates from single shot to 110 MHz. The camera features an internal microcontroller with a front panel LCD display and keypad for all functions. The intensifier and the CCD are provided in a remote housing, with a flexible connection to the respective control units. The output of this intensifier was coupled, by fiber optics or lens, to a 16-bit CCD chip at 12.5 MHz readout (640 × 480 pixels). After synchronizing the gated camera to the excitation laser pulses, time-resolved (TR) nanosecond images were acquired, as shown in **Fig. 3**.

2.2. DNA Vector Preparation

1. The pECFP and pEYFP vectors (Clontech) for use as polymerase chain reaction (PCR) templates.
2. A vector containing the coding region of the B subunit of bacterial DNA gyrase (GyrB) (kindly provided by Dr. Michael Farrar, Merck).
3. The pRK7 vector similar to Clontech's pcDNA3 vector, having a CMV promoter with excellent mammalian-cell expression (kindly provided by Dr. Ian Macara, University of Virginia).
4. PCR primers: 5' primer to introduce an *Xba*I (underlined) site into the ECFP and EYFP inserts: 5'-GCTCTAGAGCCACCATGGTGAGCAAG-3', and a 3' primer to add three Gly residues, a *Bam*HI (bold) site, and a *Eco*RI (underlined) site to the ECFP and EYFP inserts, 5'-CGGAATTCTCA**GGATCC**GCCGCCCCCT TGTACAGCTCGTCCAT-3'.
   5' primer to introduce a *Bam*HI (underlined) site into the GyrB insert: 5'-GCGG ATCCAGCAATTCTTATGACTCC-3', and a 3' primer to introduce an *Eco*RI (underlined) site after the GyrB stop codon: 3'-AAGGTGATACTTCCGACT CTTAAGCG-5'.
5. T4 DNA ligase and buffer (New England Biolabs).

6. Restriction enzymes and buffers: *Xba*I (Gibco-BRL), *Bam*HI and *Eco*RI (New England Biolabs).
7. Agarose gel (Gibco-BRL) and electrophoresis apparatus (Fisher).
8. *E. coli* strain XL-1 Blue (Stratagene).
9. Plasmid DNA agarose gel purification kit and plasmid miniprep kit (Qiagen).

2.2.1. Cell Culture and Slide Preparation

1. COS7 cells and medium: High glucose Dulbecco's modified Eagle's medium (DMEM) (Gibco-BRL) supplemented with 10% newborn calf serum.
2. Fibronectin-coated cover glasses.
3. Fugene 6 transfection reagent (Roche)
4. 10 m*M* coumermycin A_1 (Sigma) in dimethyl sulfoxide stock solution.
5. 4% Paraformaldehyde in phosphate-buffered saline (PBS).
6. 50 m*M* ammonium chloride in PBS.

3. Methods

The cloning of the jellyfish GFP, and expression of GFP fusion proteins in a wide variety of cell types, has proven that this fluorescent protein can be a versatile marker for both gene expression and protein localization in living cells or tissue ***(13–15)***. This fluorescent protein does not require any cofactors, substrates, or additional gene products. GFP retains its fluorescent properties when fused to other proteins, allowing fluorescence microscopy to be used to visualize dynamic changes in protein localization in living cells. Recently, several mutant variants of the GFP, with emission in blue, green, cyan, and yellow spectrum have become available ***(4)***. The cyan fluorescent protein (CFP) version is mutated at six residues (F64L, S65T, Y66W, N146I, M153T, and V163A), and the yellow fluorescent protein (YFP) version, is mutated at four residues (S65G, V68L, S72A, and T203Y). The expression of genetic vectors encoding protein fusions with these mutant forms of GFP provides a general method for labeling proteins in cells or tissue. Characterizing these GFPs spectroscopically is important, so that it is easier to quantitate the expression of signal in many applications, such as protein interactions under physiological conditions using the fluorescence resonance energy transfer (FRET) method ***(13)***.

3.1. CFP Plasmid Construction and Cell Transfection

3.1.1. DNA Vector Preparation Protocol

1. Perform PCR on the pECFP and pEYFP plasmids, using the two primers listed in Materials to engineer in the *Xba*I and *Bam*HI/*Eco*RI sites. Similarly, perform PCR on the GyrB plasmid, using the two primers listed above, to engineer in the *Bam*HI and *Eco*RI sites.

2. Digest the ECFP and EYFP PCR products with *Xba*I and *Eco*RI, the GyrB PCR product with *Bam*HI and *Eco*RI.
3. Digest the pRK7 vector with *Xba*I and *Eco*RI. Separately ligate the ECFP and EYFP fragments into the pRK7 vector using T4 DNA ligase. Purify the product plasmid by agarose gel electrophoresis. Digest the pRK7-ECFP and pRK7-EYFP vectors with *Bam*HI and *Eco*RI, and, as above, ligate with the GyrB fragment.
4. Transform *Escherichia coli* (strain XL-1 Blue) with the pRK7-ECFP-GyrB and pRK7-EYFP-GyrB plasmids.
5. For eukaryotic cell transfection, perform a plasmid DNA miniprep and isolate the plasmid DNA using a Qiagen miniprep kit following the manufacturer's directions.

3.1.2. Cell Culture and Slide Preparation

1. Transiently transfect COS7 cells (plated onto fibronectin-coated cover glasses in 35-mm dishes) with either 0.5 μg pRK7-ECFP-GyrB, 1 μg pRK7-EYFP-GyrB, or a mixture of both vectors (at the same ratio) using Fugene 6 transfection reagent (according to manufacturer's directions). Incubate cells in 10% NCS-DMEM for at least 12 h at 37°C.
2. To induce dimerization of the GyrB fusion proteins for FRET, treat the cells with 100 μ*M* coumermycin, an antibiotic composed of two GyrB binding arms (CFP:GyrB-C-GyrB:YFP). Add the coumermycin to fresh, warmed, serum-free DMEM, before adding it to the culture dish; incubate the cells for 20 min at 37°C. Keep a duplicate set of cultures untreated for controls.
3. Immediately rinse the cover glasses with PBS and fix in 4% paraformaldehyde in PBS for 15 min at room temperature. Rinse twice in PBS, and then quench the paraformaldehyde with 50 m*M* ammonium chloride for 10 min. Rinse twice again with PBS before mounting and sealing the cover glasses onto glass slides. Slides may be stored at 4°C.

3.2. Time-Resolved Image Acquisition and Processing

A rapid lifetime determination method, which uses two windows in the single exponential decay curve was used to calculate the lifetime (**Figs. 1** and **3**). This method was employed to calculate τ values for CFPs, because it reduces the time involved in lifetime determinations, and provides results whose accuracy is comparable to that of conventional curve-fitting methods *(9)*. As shown in **Fig. 1**, the fluorescence decay is detected at two different delay times, t_1 and t_2, with the gate width, ΔT. The gated fluorescence signals (D_0 and D_1) can be described as for single exponential decay *(9)*. The lifetime, τ, and pre-exponential factor, A, can be extracted from **Eqs. 2** and **3**:

$$\tau = (t_2 - t_1) / \ln (D_0/D_1) \quad (2)$$

$$A = (D_0/\tau) \exp (-t_1/\tau) [1\text{-}\exp (\Delta T/\tau)] \quad (3)$$

Thus, the fluorescence lifetimes and pre-exponential factors can be calculated directly from four parameters (D_0, D_1, t_1, and t_2), without fitting a large number of data points, as is required by conventional least-squares methods (*see* **Note 3**).

1. For determining τ of CFP and YFP, transiently transfected and fixed COS7 cells treated with 100 μM coumermycin are placed on the microscope stage (*see* **Note 2**).
2. The average power of the laser at the specimen plane (1 mW at 76 MHz) needs to be carefully selected to reduce photodamage and photobleaching of the specimen (*see* **Note 4**).
3. Synchronize the camera gating to the excitation of the specimen by the laser pulse generator.
4. Following delivery of an excitation pulse, the camera is turned on for a very brief interval (ΔT), at some delay time (t_1) after the pulse, and the emitted fluorescence signal from the camera is acquired on the cooled CCD.
5. The image derived from the second gate is collected in the same way after changing the delay time (t_2, $t_2 > t_1$), then the gating intensifier is turned on.

The fluorescence lifetime 2-D image was obtained with single exponential decay (**Figs. 3** and **4**) by processing the two time-shifted gate images on a pixel-by-pixel basis (using **Eqs. 2** and **3**), then displaying the image on the color monitor *(9)*. The algorithm is simple, easy to implement, and suitable for large 2-D data processing (*see* **Notes 5** and **6**).

4. Notes

1. By adjusting the time-delay unit and displaying the respective signal on the oscilloscope, one can accomplish the synchronization of the signals between the excitation laser pulses to the gating electronics of the gated image intensifier. The signal from the specimen should be stronger to obtain a lifetime image with good signal-to-noise (S/N) ratio, e.g., a well-labeled cell could take ~500 ms accumulation time to obtain a decent S/N image. Commercial units are available, but do not have a temporal resolution for data like the unit described in this chapter.
2. Depending on experimental condition, the live specimen could be maintained on the microscope stage. It requires only a few seconds to acquire the lifetime data.
3. The specimen image without labeling should be used to eliminate the autofluorescence lifetime. Also, the system should be calibrated with known standard fluorophore molecules before implementing this technology for the biological specimen.
4. The excitation intensity should be chosen in such a way that the change in gray-level intensity should be <10 gy-level intensity units, for ~1 min continuous data collection mode. As mentioned previously, the photobleaching is negligible since the collection of the lifetime data require only a few seconds.
5. Even though the lifetime imaging mode provides high temporal and spatial resolution, the double exponential decay processing option helps to discriminate the binding and nonbinding condition of proteins, calcium, etc. The number of

gates one could acquire is controlled by the strength of the signal levels from the biological specimen, and the repetition rates of the laser pulse and the gating camera. So far, we have derived the decay equations for double exponential decays, which required four gated images, whether the cells are singly, doubly, or triply labeled. The lifetime measurements report change in the environmental (dynamic activity) conditions of the biological system.

6. One could obtain a 3-D (*x*, *y*, lifetime) distribution of the lifetime of live or fixed biological specimen. It is also possible for one to collect images from different optical sections of the specimen, but it would require more deconvolution processing of the acquired images, since the described system uses a wide-field fluorescence microscopy system. One could also use confocal or multiphoton lifetime imaging system to acquire lifetime images at different optical sections, but that would require a spot scanning photomultiplier tube as a detector and different hardware and electronics configurations.

Acknowledgments

This work was supported by grants from W. M. Keek Foundation and NCI CA 40042.

References

1. Inoue, S. and Spring, K. R. (1997) *Video Microscopy: The Fundamentals*, 2nd ed., Plenum, New York.
2. Herman, B. and Lemasters, J. J. (1993) *Optical Microscopy: Emerging Methods and Applications.* Academic, New York.
3. Periasamy, A. and Herman, B. (1994) Computerized fluorescence microscopic vision in the biomedical sciences. *J. Comput. Asst. Microsc.* **6,** 1–26.
4. Periasamy, A. (2001) *Methods in Cellular Imaging.* Oxford University Press, New York.
5. Lakowicz, J. R. (1983) *Principles of Fluorescence Spectroscopy.* Plenum, New York.
6. O'Connor, D. V. and Phillips, D. (1984) *Time Correlated Single Photon Counting.* Academic, New York.
7. Periasamy, A., Wang, X. F., Wodnicki, P., Gordon, G. W., Kwon, S., Diliberto, P. A., and Herman, B. (1995) High-speed fluorescence microscopy: lifetime imaging in the biomedical sciences. *J. Microsc. Soc. Amer.* **1,** 13–23.
8. Periasamy, A., Wodnicki, P., Wang, X. F., Kwon, S., Gordon, G. W., and Herman, B. (1996) Time-resolved fluorescence lifetime imaging microscopy using a picosecond pulsed tunable dye laser system. *Rev. Sci. Instrum.* **67,** 3722–3731.
9. Periasamy, A., Siadat-Pajouh, M., Wodnicki, P., Wang, X. F., and Herman, B. (1995) Time-gated fluorescence microscopy in clinical imaging. *Microsc. Analyt.* (March), 33–35.
10. Ballew, R. M. and Demas, J. N. (1989) An error analysis of the rapid lifetime determination method for the evaluation of single exponential decays. *Analyt. Chem.* **61,** 30–33.

11. Sharman, K. K., Asworth, H., Demas, J. N., Show, N. H., and Periasamy, A. (1999) Error analysis of the rapid lifetime determination (RLD) method for double exponential decays: evaluating different window systems. *Analyt. Chem.* **71,** 947–952.
12. Morgan, C. G., Mitchell, A. C., and Murray, J. G. (1992) In situ fluorescence analysis using nanosecond decay time imaging. *Trends Analyt. Chem.* **11,** 32–35.
13. Periasamy, A. and Day, R. N. (1999) Visualizing protein interactions in living cells using digitized GFP imaging and FRET microscopy. *Meth. Cell Biol.* **58,** 293–314.
14. Sullivan, K. and Kay, S. (1999) Green Fluorescent Protein. *Meth. Cell Biol.* **58,**
15. Cubitt, A. B., Heim, R., Adams, S. R., Boyd, A. E., Gross, L. A., and Tsien, R. Y. (1999) Understanding, improving and using green fluorescent proteins. *Trends Biochem. Sci.* **20,** 448–455.

8

Fluorescence Resonance Energy Transfer (FRET) Applications Using Green Fluorescent Protein

Energy Transfer to the Endogenous Chromophores of Phycobilisome Light-Harvesting Complexes

Jasper J. van Thor and Klaas J. Hellingwerf

1. Introduction

Fluorescence resonance energy transfer (FRET) is a technique that can be used to estimate intermolecular distances between pigment molecules, which is an approach first proposed by Stryer *(1)*. The theoretical basis for FRET was originally put forward by Förster *(2,3)*, and is related to "Fermi's golden rule" for electronic interactions. This chapter explains how and when FRET can be used, and what the physical basis is of the energy transfer events. The point is stressed that, in most cases, FRET cannot be used to directly measure intermolecular distances, but only to estimate them. The reason for these uncertainties are made clear. First, the mechanisms of fluorescence and fluorescence energy transfer are briefly introduced.

Consider a pigment molecule that has absorbed a photon with energy hν matching the energy difference between a ground-state energy level and an energy level of the first excited state. Absorption of a photon with an energy matching the first transition is usually called the $S_0 - S_1$ transition, and corresponds to the absorption band with the longest wavelength (the lowest frequency) of the pigment. Usually absorption bands are broad because of the thermal population of substates with discrete energy levels of both the ground state and the excited state.

When the excited state of a molecule is populated, there are several possible fates for this high-energy state. It may decay back to the ground state without emission, thereby converting the electronic resonance energy into vibrational,

From: *Methods in Molecular Biology, vol. 183: Green Fluorescent Protein: Applications and Protocols*
Edited by: B. W. Hicks © Humana Press Inc., Totowa, NJ

rotational and/or translational energy (heat), which is also referred to as an "internal conversion" process. For pigments utilizing internal conversion as a major deactivation channel, the excited state typically lives no longer than a few picoseconds. For other pigments, photoproducts are formed with altered electronic ground-state conformations. A well-known example of this is photoisomerization of the 11-*cis*-retinal chromophore to all-*trans*-retinal in visual rhodopsin. In this case, the excited state has enough energy to cross a barrier, leading to a new ground state with an isomerized double-bond configuration. In the case of green fluorescent protein (GFP), neither the thermal decay to the ground state, nor the formation of a photoisomerized ground state product, is competitive with the radiative decay process. The isomerization process is inhibited by tight packing of the chromophore (fluorophore) within the protein interior. With a certain low probability, the excited (singlet) state can also be converted to a triplet state, which is referred to as "intersystem crossing." Triplet states have a lower energy and can be very long-lived (microseconds to hours), because the triplet → singlet transition is "forbidden." Those molecules that have an intrinsically high rate of triplet-formation become phosphorescent because of radiative decay from the triplet to the ground state.

When none of these processes occur with a high probability, the pigment is fluorescent. In this case, the excited state decays radiatively with a lifetime, τ, typically of nanoseconds (ns). However, within the excited state, there is an ultrafast relaxation process that produces the lowest energy level accessible to the excited state. This relaxation process is accompanied by some energy loss, which is the reason that fluorescence emission is red-shifted compared to the absorbed light; its size is expressed as the "Stokes shift." As a rule, FRET donors need to be fluorescent.

Several possible mechanisms exist for nonradiative transfer of excited state resonance energy between pigment molecules. A first distinction is made between transfer via exchange of electrons and via Coulombic (electronic) interactions. Three different descriptions of energy transfer can be formulated, two of which result from Coulombic interactions.

First, the exchange (Dexter) mechanism is operative only when the excited state is long-lived transfer via Coulombic interactions does not occur, because transitions are forbidden, and the distance between the pigments is very small. The transfer of excited-state energy in this case is accomplished via actually exchanging a donor-electron, from an excited-state orbital, with the unoccupied excited-state orbital of an acceptor, and simultaneously exchanging in return an electron from a ground-state orbital. In most cases, however, contributions from this exchange mechanism can be neglected.

Second, FRET occurs when the donor excited state is sufficiently long-lived (ns); the energy of the emission of the donor is matched by an optical transition of the acceptor pigment, and the distance between the donor and acceptor molecules is short (in the nm range). Even though this energy transfer is radiationless, the emission wavelength of the fluorescent donor is considered, since the amount of transferred energy is the same as the energy that is otherwise released as fluorescence in the absence of an acceptor. The efficiency of this form of energy transfer is extremely sensitive to the intermolecular distance. The rate of energy transfer by the FRET mechanism is a function of the inverse of the sixth power of the distance (*see* below). FRET is also known as "Förster energy transfer," and is also referred to as weak coupling.

Third, strong excitonic coupling also occurs via Coulombic interactions, but only when two pigment molecules are very close (typically about an Å). Strong coupling can be interpreted in terms of mixed molecular electronic orbitals; the excited-state orbital is in essence shared by a dimer of pigment molecules. A well-known case of strongly coupled chromophores is the bacteriochlorophyll *a* dimer, forming the primary donor of the bacterial reaction center.

Except for strong excitonic coupling, all the abovementioned mechanisms result in depopulation of the excited state, and can therefore contribute to the overall rate (k_{total}) of deactivation:

$$\tau_{total} = \frac{1}{k_{total}} = \frac{1}{k_{fluorescence} + k_{thermal\text{–}deactivation} + k_{triplet} + k_{photo\text{–}product} + k_{exchange} + k_{FRET}} \tag{1}$$

However, a scenario with significant contributions by all terms summed in **Eq. 1** is very improbable. Depending on the physical state of a pigment, the rate of some reactions is negligible, thus defining its specific optical properties. In the absence of energy transfer, for fluorescent molecules, **Eq. 1** is reduced to:

$$\tau_{total} = \frac{1}{k_{total}} = \frac{1}{k_{fluorescence} + k_{thermal\text{–}deactivation} + k_{triplet} + k_{photo\text{–}product}} \tag{2}$$

or:

$$\tau_{total} = \frac{1}{k_{total}} = \frac{1}{k_{fluorescence} + k_{other}} \tag{3}$$

Now, a so-called radiative rate constant ($k_{radiative}$) is introduced, which in fact equals the rate of fluorescence decay ($k_{fluorescence}$) in these expressions.

$$\tau_{total} = \frac{1}{k_{total}} = \frac{1}{k_{radiative} + k_{other}} \tag{4}$$

$$\phi_{fluorescence} = \frac{k_{radiative}}{k_{total}} \tag{5}$$

Note that the fluorescence lifetime that is determined experimentally equals τ_{total}, and not $\tau_{fluorescence}$. The latter parameter is an intrinsic property of the pigment molecule that can only be derived when both τ_{total} (or $\tau_{observed}$) and the fluorescence quantum yield, $\Phi_{fluorescence}$ are known. The radiative rate is the rate of fluorescence decay that would be determined experimentally, when the fluorescence quantum yield equals 1.

The radiative rate constant of any pigment molecule can be calculated, using theoretical grounds, from its absorption spectrum, since the latter is directly related to the transition dipole strength. Usually, it is determined experimentally by measuring both the rate of fluorescence decay and the absolute quantum yield of fluorescence. For wild-type GFP, both parameters are known: excitation at 395 nm, at pH 8.0 $\Phi_{fluorescence} = 0.79$, and $\tau_{observed} = 3.3$ ns, so $k_{total} = 1/3.3 \times 10^{-9}\ s^{-1} = 3.03 \times 10^{8}$ s ***(4,5)***. Therefore $k_{radiative} = (0.79)3.03 \times 10^{8}\ s^{-1} = 2.39 \times 10^{8}\ s^{-1}$. It is thus derived from **Eqs. 4** and **5** that, under the conditions mentioned, the total rate of deactivation of the excited state of GFP, via mechanisms other than fluorescence, equals $6.36 \times 10^{7}\ s^{-1}$, which is obviously much smaller than the rate of radiative decay. It is also known that these values do not include contributions from energy transfer ($k_{exchange} + k_{FRET} = 0\ s^{-1}$).

Now consider conditions in which a molecule of GFP is in close proximity to a pigment molecule that is able to absorb light with a wavelength of 508 nm (which equals $\lambda_{emission}$ of GFP). In this case, $k_{exchange}$ would be a negligible contribution compared to the other deactivation processes. In general, with FRET applications based on GFP, the rate of the exchange mechanism ($k_{exchange}$) can neither compete with the rate of fluorescence decay ($k_{fluorescence}$), nor with the rate of resonance energy transfer (k_{FRET}). Therefore, $k_{exchange}$ is generally neglected in cases in which Coulombic interaction can lead to energy transfer. Strong excitonic interactions can also be neglected in all cases involving GFP, since donor and acceptor molecules must be almost in physical contact for this interaction to occur. Clearly, this cannot happen, because the β-barrel shape of the GFP shields the fluorophore from such contacts. For applications of GFP (and many other fluorophores), the rate of energy transfer is therefore always approximated by k_{FRET} only. When energy transfer occurs from GFP to an acceptor molecule, an additional pathway of deactivation is introduced. Equation 1 can be used to calculate k_{FRET}, with the following simplification:

$$k_{total} = k_{radiative} + k_{other} + k_{FRET} \tag{6}$$

which amounts to:

$$(k_{total})_{FRET} = (k_{total})_{no\ FRET} + k_{FRET} \tag{7}$$

and therefore:

$$\frac{(\phi_{fluorescence})_{FRET}}{(\phi_{fluorescence})_{no\ FRET}} = \frac{(\tau_{total})_{FRET}}{(\tau_{total})_{no\ FRET}} = \frac{\dfrac{1}{(k_{radiative}+k_{other}+k_{FRET})}}{\dfrac{1}{(k_{radiative})(k_{other})}}$$

$$= \frac{(k_{total})_{no\ FRET}}{(\tau_{total})_{no\ FRET}+(k_{FRET})} \qquad (8)$$

Relative changes of fluorescence quantum yield can be obtained experimentally by using a fluorescence spectrophotometer, a device present in many biological labs. In order to determine k_{FRET}, conditions must be defined within the same sample, where GFP does and does not participate in energy transfer, respectively. In that case, relative changes of the fluorescence quantum yield can be used to solve **Eq. 8**, assuming that $(k_{total})_{no\ FRET}$ can be taken from the literature. Whenever possible, the absorption of GFP should be quantified in the sample, so that the absolute fluorescence quantum yield can then be determined using a standard dilution series of purified recombinant GFP. However, this may not always be possible, depending on the nature of the sample. In the example discussed below, the abundance of the GFP fusion protein was not high enough in the sample, which also contained high concentrations of phycobilisome light-harvesting complexes, to quantify the absorption of GFP. Another possibility is to determine $(k_{total})_{FRET}$ directly, by determining the fluorescence lifetime. This, however, is a complex experiment and requires specific expertise and equipment.

For obvious reasons, GFP has found many applications for FRET-experiments. In this chapter, experimental requirements for FRET applications are considered, and illustrated with an application involving energy transfer from GFP to the endogenous chromophores of light-harvesting antennae. Since a number of GFP mutants are presently available with altered spectroscopic properties (*see* Chapter 1 of this volume, and, for a review, *see* **ref. 6**), many endogenous chromophores can now also be selected for FRET studies, including chlorophylls, carotenoids, flavins, photoreceptor molecules, and others.

One possible application is the in vivo detection and quantification of redox intermediates by FRET. For example, certain members of the flavoproteins (such as ferredoxin:NADP$^+$ reductase; [FNR]) tend to accumulate semiquinone states (i.e., single-electron reduced) during electron transport turnover. Such semiquinone states are known to have increased absorption, compared to the fully oxidized and reduced states, usually ~500 nm. The formation of such states could be detected from the increase of FRET efficiency in a fusion protein consisting of wild-type GFP and the particular flavoprotein. It should be taken into account, however, that the fluorescence intensity of all phenolate anion mutants of GFP shows a distinct dependence on the pH. Therefore, for in vivo applications, wild-type GFP is recommended, since its pH sensitivity is much lower.

2. Materials

1. A scanning spectrophotometer. Spectrophotometers of high quality are produced by Aminco (the DW 2000), Cary, Hewlett Packard, Pharmacia LKB, and Shimadzu.
2. A scanning fluorescence spectrophotometer. Suitable fluorimeters are produced by Aminco-Bowman, Hewlett Packard, Perkin Elmer, Shimadzu, and Spex. For fluorescence anisotropy measurements, the fluorimeter must be equipped with polarizers. The Aminco-Bowman Series 2 Luminescence Spectrometer was used for the experiments described here.
3. Purified recombinant GFP, which can either be obtained commercially (from Clontech) or prepared by described methods *(7)*.
4. A solution of sodium fluorescein in 0.1 *M* NaOH (λ_{max} = 490 nm), which can be used as a fluorescence quantum yield standard as well as a control for fluorescence depolarization (*see* **Subheading 3.** and **Note 6**).
5. Software for data manipulation and calculation. The software package, Origin 6.0, can be used for all the operations described in this chapter. Operations on sets of data with a different x-axis scaling (such as absorption and fluorescence spectra measured with different spectral resolution) can be performed with Origin 6.0, but also with software programs for spectroscopic data analysis, such as Grams. Packages that can perform integrations of curves, besides Origin 6.0, are, among others, Tablecurve, Grams, and Peakfit.
6. Control phycobilisomes and recombinant phycobilisomes isolated from a *Synechocystis* PCC 6803 mutant expressing a fusion protein containing the N-terminal phycobilisome-binding domain of FNR fused to GFPuv *(8)*.

3. Methods

3.1. Calculation of FRET Parameters

Many different versions of the formulas describing weak interaction can be found in the literature. Here, the relevant formulas are tabulated, together with references. In addition, suggestions are provided about how to proceed practically with these calculations.

The Förster formulas can be derived from a quantum mechanical description of electron transfer, known as "Fermi's golden rule" (for formal notation and explanation of these formulas, *see* **ref. *9***). The Förster equation describing excitation energy transfer is an approximation, because the electronic coupling-term of Fermi's golden rule (containing the quantum-mechanical overlap wavefunctions for the ground and excited states) is approximated by accounting only for the dipole–dipole interactions. This approximation is usually considered to be very appropriate ***(10,11)***. For a comprehensive review of the subject, *see* **ref. *11***).

1. The most complete form of the Förster equation, describing the rate of energy transfer from donor (D) to acceptor (A) ($k_{DA} = k_{FRET}$) is:

$$k_{DA} = \frac{9 \times 10^{45} (\ln 10) \kappa^2 k^D_{radiative}}{128\ \pi^5 n^4 N R^6_{DA}} \int f_D(\nu)\, \varepsilon_A(\nu) \nu^{-4} d\nu \qquad (9)$$

in which: $k_{DA} = k_{FRET}$ (s^{-1}); κ = pigment geometric factor; $k^D{}_{radiative}$ = radiative decay rate of the FRET donor (s^{-1}) (*see* above); n = refractive index; N = Avagadro's number (number of molecules/mol) (mol^{-1}); R_{DA} = actual distance between donor and acceptor pigment (nm) (for explanation of further symbols, *see* below).

2. However, it is more practical to use two separate equations. **Eq. 10** describes the energy transfer rate, k_{DA}, as a function of the distance between the pigments. **Eq. 12** contains the spectral parameters and the orientation factor, κ^2 for the donor and acceptor pigment:

$$k_{DA} = k^D_{radiative} \left(\frac{R_0}{R_{DA}}\right)^6 \qquad (10)$$

in which: $k^D{}_{radiative}$ = radiative decay rate of the donor (s^{-1}); R_0 = Förster critical radius (nm), the distance at which FRET efficiency E_{FRET} equals 0.5.; R_{DA} = distance between donor and acceptor (nm).

Equation 10 is often presented in a different form, which directly relates R_{DA} to the FRET efficiency E_{FRET} (E_{FRET} = 0.5, when $k_{FRET} = k_{radiative}$):

$$E_{FRET} = \left(1 + \left(\frac{R_{DA}}{R_0}\right)^6\right)^{-1} \qquad (11)$$

3. A separate expression allows calculation of the Förster critical distance (or radius), R_0.

$$R_0^6 = 8.79 \times 10^{17} \kappa^2 n^{-4} \int f_D(\nu) \varepsilon_A(\nu) \nu^{-4} d\nu \qquad (12)$$

in which R_0 = Förster critical radius (nm); κ = geometric factor determined by the orientation of donor and acceptor pigments (for averaged orientation, use $\kappa^2 = 2/3$); n = refractive index (for proteins, in most cases, a value of 1.5 is used); $f_D(\nu)$ = normalized fluorescence emission spectrum of donor (*see* **Subheading 3.3.**); $\varepsilon_A(\nu)$ = absorption spectrum of acceptor (*see* **Subheading 3.3.**); ν = frequency of radiation (cm^{-1}).

3.2. Obtaining a Value for k^2

1. κ^2 reflects the orientation of the emission dipole moment of the donor, with reference to the transition dipole moment of the acceptor. Consider two vectors representing the orientation of both dipole moments. Then:

$$\kappa = (\cos \alpha - 3 \cos \beta \cos \gamma) \qquad (13)$$

in which α is the angle between the dipoles, and β and γ are the angles between the two dipoles and the vector connecting their origins. The value for κ^2 may vary between 0 and 4. For randomly oriented chromophores, a value of $\kappa^2 = 2/3$ is used. Methods are available that allow determination of κ^2 directly. However, they require very specific sample conditions and equipment, and are therefore not discussed here. When κ^2 is not known, it introduces a large uncertainty into the results.

2. As is shown in the example in **Subheading 3.5.**, for GFP, the rate of rotational diffusion is significantly slower than the rate of fluorescence decay, which means that the orientation factor, κ^2, should be known, in order to be able to determine the rate of energy transfer. Rotational diffusion of pigments in solution, which are not protein-bound, is usually faster than the rate of fluorescence decay. For such cases, the average value of $\kappa^2 = 2/3$ applies.

3.3. Obtaining a Value for the Spectral Overlap Integral

1. The fluorescence emission spectrum of the donor $F_D(\lambda)$ must first be measured using a fluorimeter. When a solution of recombinant GFP at pH 8.0 and an OD_{395} of 0.05, is analyzed, a strong signal will be detected when the sample is excited at 395 nm, with detection at 508 nm (*see* **Note 1**).
2. $F_D(\lambda)$ is then be plotted on a frequency scale, to obtain $F_D(\nu)$. At this point, the spectrum has an arbitrary emission scale (i.e., dimensionless) as a function of frequency, in wave numbers (cm^{-1}). Normalize this spectrum so that its integral (the area below the curve) equals 1 (cm^{-1}). The normalized spectrum is denoted $f_D(\nu)$. The formula for this operation is:

$$f_D(\nu) = \frac{F_D(\nu)d\nu}{\int F_D(\nu)d\nu} \qquad (14)$$

3. Next, the absorption spectrum of the acceptor pigment in pure form is measured. The absorption spectrum should be scaled to the molar extinction coefficient (to be taken from the literature), with the proper factor, in order to obtain $\varepsilon_A(\lambda)$. Plot the spectrum on the frequency scale $\varepsilon_A(\nu)$, so that it has dimensions cm^{-1} (on the x-axis) and M^{-1} cm^{-1} (on the y-axis).
4. The third term in the spectral overlap integral equals ν^{-4}. Note that this contribution is independent of the fluorescence and absorption properties of the pigments. The result is that at lower frequencies (longer wavelengths) the size of the spectral overlap integral increases, and therefore also the rate of energy transfer.
5. The spectral overlap integral can now be determined graphically. For this, $f_D(\nu)$, $\varepsilon_A(\nu)$, and ν^{-4} are multiplied, and the integral of the resulting spectral overlap image is determined. The spectral overlap integral has dimensions of M^{-1} cm^3.
6. Consider the implications of this result. The rate of energy transfer is a function of the overlap integral, which contains the spectrum of the donor emission and the acceptor absorption. This suggests that both spectra should fit together as

well as possible, for energy transfer to proceed efficiently. This is often an important criterion when FRET couples are selected. However, it is the oscillator strength (i.e., the extinction coefficient) that primarily determines the size of the spectral overlap integral. In the example shown below, the donor emission and the acceptor absorption do not match optimally, but the spectral overlap integrals, nevertheless, have significant values because the molar extinction coefficient of the acceptors used is appreciable. Furthermore, as can be seen from **Eq. 10**, the FRET efficiency is highly dependent on the donor-acceptor distance. In addition, it is necessary to calculate whether energy can also be transferred in the reverse direction. This can occur when both pigments are fluorescent, provided that the Stokes shifts are not very large.

3.4. Experimental Determination of k_{FRET}

1. There are several methods that can be used to obtain a value for k_{FRET} directly. Much depends on the nature of the sample, and the equipment available (*see* **Note 2**, containing **Eq. 15**). Assuming that only a fluorimeter and a spectrophotometer are available, either an absolute or a relative value for the fluorescence quantum yield of the donor (under FRET conditions) can be determined. The first possibility is obviously more desirable, but requires that the absorption of the donor can be detected (*see* **Note 3**).
2. In this case, the fluorescence intensity of the donor (under FRET conditions) should be compared with the fluorescence intensity of a dilution series of the donor with OD values relevant for the sample (recombinant GFP, in this case, for which a value for $\Phi_{fluorescence}$ is taken from the literature) (*see* **Notes 2** and **4**). **Equation 8** can then be used to calculate k_{FRET}.
3. The fluorescence quantum yield can also be determined by comparison with a standard. For GFP, a solution of sodium fluorescein in 0.1 *M* NaOH (λ_{max} = 490 nm) can be used, for which $\Phi_{fluorescence} = 0.93$. For determination of $\Phi_{fluorescence}$ of GFP, fluorescein and GFP samples with identical OD values (small differences can of course still be corrected for) at the wavelength of excitation should be measured (e.g., $OD_{485} \sim 0.01$).
4. The ratio of the integrals of the emission spectra then provide the ratio of the fluorescence quantum yields. When comparing the emission from GFP and fluorescein, no correction is necessary for the wavelength dependence of the photomultiplier. However, when the emission maxima are further separated, the spectra need to be corrected for the instrument response. Note that for wild-type GFP, $\Phi_{fluorescence}$(485 nm excitation)/$\Phi_{fluorescence}$(395 nm excitation) = 0.82 *(7)*.
5. Using **Eqs. 8**, **11**, and **12**, the donor–acceptor distance can then be calculated for different values of κ^2.

The example used in this chapter deals with energy transfer from GFP to antenna pigments. However, many FRET applications, using GFP, deal with FRET between different mutants of GFP, since endogenous acceptor pigments are not always available. The first described GFP-based FRET couple was a blue fluorescent protein plus an S65T mutant (contains a phenolate anion chromophore). For this pair, R_0 is on the order of 40–43 Å *(6)*. The most popular FRET pair at present is the combination of the cyan and yellow GFP mutants, with R_0 values on the order of 49–52 Å *(6)*. These R_0 values are given for the case in which $\kappa^2 = 1$, and therefore Eq. 16 should be used to calculate k_{FRET} (*see* **Note 5** containing **Eq. 16**). Using **Eqs. 11** and **12**, a spectroscopic ruler is readily constructed for the cyan and yellow fluorescent protein FRET couple (**Fig. 1**).

3.5. Obtaining the Rate Constant of Rotational Diffusion via Fluorescence Anisotropy Measurements

Consider a solution of GFP that is excited with 395 nm light, which is filtered through a polarizer. In that case, the direction of the electromagnetic wave that is used to excite the GFP solution is linearly oriented, and therefore only GFP molecules are excited that happen to have their transition dipole moment oriented in the same direction. This selective excitation is (most appropriately) called "photoselection." If the excited molecules are stationary then the fluorescence emission will be oriented as well. In fact, when the emission dipole moment has the same orientation as the absorption transition dipole moment, then the fluorescence intensity will be highest, if the direction of the polarizer in front of the detector is identical to the one of the polarizer used for excitation. In contrast, when GFP is allowed to diffuse freely in solution, the selective orientation of the emitted light will be (partially) lost. Under conditions when there is no FRET, the excited state of GFP (when excited at 395 nm) has a lifetime of 3.3 ns, whereas the rotational correlation lifetime ($\tau_{rotational} = 1/k_{rotational}$) of GFP under standard sample conditions (pH 8.0, 20°C, low osmolarity) equals approx 15 ns *(8,12,13)*. When these time constants are compared, it can be understood that at least some of the orientation of the emitted light will be retained. Some fluorimeters are equipped with polarizers, allowing measurement of the steady-state fluorescence polarization.

1. Define the fluorescence intensity, I_{hh}, by measuring with both the excitation and the detection polarizer in horizontal (h) position. Likewise, define other intensities as I_{hv} (excitation: horizontal, detection: vertical), and so on.
2. The degree of polarization of (emitted) light can be expressed in two alternative ways: fluorescence polarization and fluorescence anisotropy. Here, only the expressions for fluorescence anisotropy (A) is treated, which is defined according to:

$$A = \frac{I_{\mathrm{vv}} - I_{\mathrm{vh}}}{I_{\mathrm{vv}} + 2\,I_{\mathrm{vh}}} \qquad (17)$$

3. The theoretical maximum for the anisotropy value is 0.4; the minimum is –0.2 (*see also* **Eq. 21**). When the absorption and emission dipoles are parallel, the anisotropy will be 0.4, when there is no rotational depolarization.

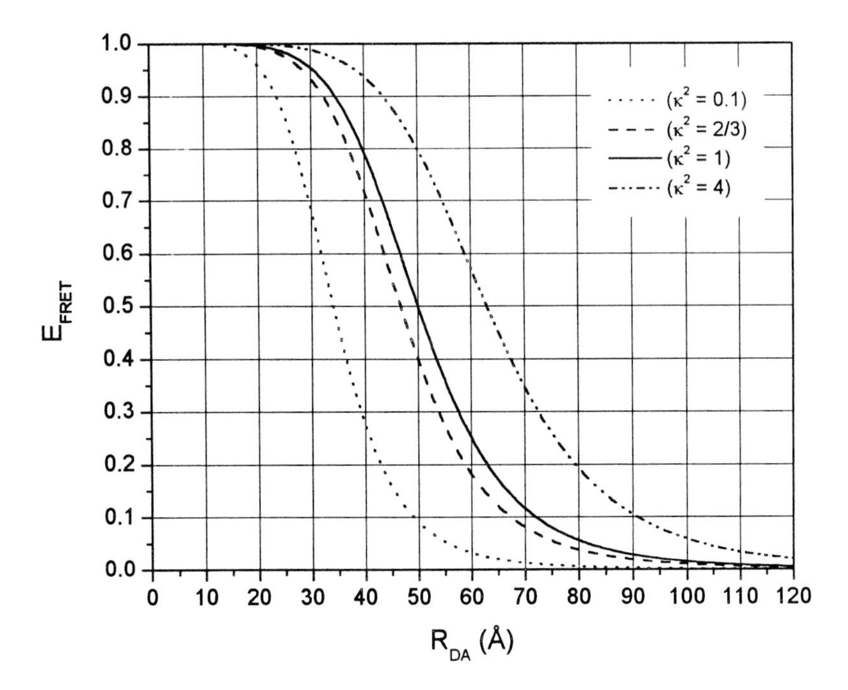

Fig. 1. Spectroscopic ruler for the cyan and yellow fluorescent protein FRET couple. The efficiency of energy transfer E_{FRET} between the chromophores of the cyan fluorescent protein (ECFP) and the yellow fluorescent protein (EYFP) is plotted for the donor–acceptor distance R_{DA} according to **Eqs. 11** and **12** for values of 0.1, 2/3, 1 and 4 for κ^2.

4. Experimentally, erroneous values for A are often obtained, if no careful correction is made for the instrument response of the fluorescence detector, because a large difference in excitation intensity occurs when the excitation polarizer is switched from the vertical to the horizontal position. This results in a nonproportional output of the photomultiplier. Therefore, a G-factor is also obtained (an instrument-correction factor), and the correct anisotropy is calculated according to:

$$A = \frac{I_{vv} - GI_{vh}}{I_{vv} + 2GI_{vh}} \quad (18) \quad \text{with} \quad G = \frac{I_{hv}}{I_{hh}} \quad (19)$$

5. The rotational diffusion rate constant can be obtained via anisotropy measurements by two methods. The first method requires simultaneous time-resolved measurement of the fluorescence and its anisotropy. As a function of time, both the fluorescence intensity (in the case of GFP, with a decay-time constant of 3.3 ns), and the fluorescence anisotropy will decrease. The decay of the latter results from rotational diffusion *(13)*.

6. The $k_{rotational}$ can also be obtained from a simple steady-state experiment. Such measurements do not require specialized equipment, such as a set-up for laser-flash-induced, time-resolved fluorescence spectroscopy. $k_{rotational}$ can be derived from the steady-state fluorescence anisotropy directly, using the Perrin equation:

$$\frac{1}{A} = \frac{1}{A_0}\left(1 + \frac{\tau_{fluorescence}}{\tau_{rotational}}\right) \quad (20)$$

   in which A = steady-state anisotropy; A_0 = intrinsic anisotropy (directly after excitation); $\tau_{fluorescence}$ = observed fluorescence lifetime (in fact, τ_{total} ;= 3.3 ns for GFP); $\tau_{rotational}$ = rotational correlation time constant (= $1/k_{rotational}$) (ns).
7. Note that, under FRET conditions, the steady-state anisotropy value increases because of a decrease of the observed fluorescence lifetime (τ_{total}). When the rate of energy transfer is known, then τ_{total} is also known, and $\tau_{rotational}$ can subsequently be determined from the steady state anisotropy measurement.
8. The intrinsic anisotropy value (A_0) depends on the angle φ between the absorption and emission transition dipoles according to:

$$A_0 = \frac{1}{5}(3\cos^2\varphi - 1) \quad (21)$$

   In the case of GFP, A_0 was experimentally determined to be ~0.38, which is very close to the theoretical maximum ***(8,13)***.
9. A value for $k_{rotational}$, calculated using the Perrin equation, is only meaningful when the rotational motion is isotropic (i.e., equal in all directions). Probably, GFP does not rotate isotropically, when bound to large molecules. When the fluorophore does not diffuse isotropically, $k_{rotational}$ is underestimated when calculated according to the Perrin equation. In addition, such calculations cannot distinguish between a homogeneous and a heterogeneous distribution of rotational correlation rate constants. In spite of these restrictions, this method does have the advantage that a value for $k_{rotational}$ can be obtained straightforwardly, using standard equipment (*see* **Note 6**).

3.6. FRET Between the Fluorophore of GFP and the Endogenous Phycocyanobilin Pigments of Recombinant Phycobilisome Light Harvesting Complexes

In cyanobacteria, GFP has recently been used for a FRET experiment, in order to characterize the function of a domain present in an electron transport protein, (FNR), which causes this protein to bind to the phycobilisome light-harvesting complexes of *Synechocystis* PCC 6803 ***(8)***. Phycobilisomes are very large, water-soluble aggregates that contain linear tetrapyrrole chromophores as the light-harvesting pigments for the oxygenic photosynthetic apparatus in these bacteria. The phycobilisomes can readily be purified in intact form, using cell-free extracts, prepared in a high-ionic-strength buffer, and sucrose-density centrifugation.

The phycobilisomes, when isolated, require specific buffer and temperature conditions, in order to maintain their proper energy-transfer characteristics ***(14)***. Upon dilution into a low-ionic-strength buffer, e.g., 50 m*M* phosphate buffer, pH 8.0, these complexes dissociate within minutes ***(14,15)***. With wild-type phycobilisome complexes from *Synechocystis* PCC 6803, the dissociation of phycocyanin from allophycocyanin, apparent from the increase of the intensity of fluorescence emission at 650 nm, proceeds at this time-scale, as well.

A mutant was constructed that no longer expressed the wild-type form of the FNR protein with the (N-terminally placed) phycobilisome-binding domain, and, as a result, the purified phycobilisomes of this mutant no longer contain FNR. In this mutant background, a fusion protein was expressed, composed of the phycobilisome-binding domain of FNR at the N-terminus translationally fused to GFPuv. Recombinant phycobilisome complexes from this mutant contained this GFP-fusion protein and no FNR.

1. These complexes show a detectable, but relatively weak, fluorescence emission maximum at 510 nm, with a corresponding excitation maximum at 395 nm; absorption of GFP could not be detected.
2. The emission is absent in control preparations of phycobilisomes. In these recombinant complexes, the emission of GFP is well separated from the emission of phycocyanin and allophycocyanin, the two major chromoproteins of the phycobilisomes.
3. The fluorescence anisotropy of the phycobilisome-bound GFP is higher than the corresponding value determined for free, monomeric, recombinant GFP (**Table 1**), indicating that its excited state is shorter-lived and/or the rotational motion is decreased, compared to free monomeric GFP. The measured fluorescence anisotropy for recombinant GFPuv (**Table 1**) agreed well with the measured anisotropy of the GFP variant S65T, i.e., 0.325 ***(12)***. Most likely, the rotational correlation time-constant for the phycobilisome-bound GFP fusion protein is larger than for free monomeric GFP.
4. For the phycobilisome-bound GFP, $\tau_{rotational}$ is calculated to be approx 50 ns. A minimal value of 13 ns is estimated, within the obtained accuracy of the measurements performed, with intact recombinant phycobilisomes (**Table 1**). This could indicate that the GFP-fusion protein, when bound to the phycobilisomes, moves by anisotropic rather than by isotropic rotational diffusion.
5. Note that, for intact phycobilisome complexes, $\tau_{rotational}$ is large, probably a few hundred nanoseconds.
6. As mentioned previously buffer conditions are critical to phycobilisome stability. Upon transfer of the recombinant complexes, containing the GFP fusion protein, into 10 m*M* phosphate buffer, the fluorescence quantum yield of the emission at 508 nm increased, until a maximum was reached 20 min later (**Fig. 2**). A decrease of the fluorescence anisotropy was measured in parallel, upon dissociation of the complex, until a value was reached comparable to that of recombinant monomeric GFP (**Table 1**). Most likely, energy transfer from GFP to phycobilisome-associ-

Table 1
Fluorescence and Rotational Diffusion Characteristics of Monomeric Recombinant GFP, and the Phycobilisome-Bound GFP Fusion Protein

	A	τ_f (ns)	$\tau_{rotational}$ (ns)
30 kDa recombinant GFP	0.325 ± 0.002	3.3	$13.8 > \tau_{rotational} > 14.8$
39 kDa fusion-GFP (PBS-bound)	0.40 ± 0.02	0.7	$\tau_{rotational} >> 13$
39 kDa fusion-GFP (dissociated)	0.33 ± 0.02	3.3	$11 > \tau_{rotational} > 23$

The fluorescence anisotropy (A) was determined for the free, monomeric recombinant form of GFPuv, the phycobilisome (PBS) bound form in high-ionic strength buffer, and the dissociated form, 20 minutes after dilution into low-ionic strength buffer. The rotational correlation times were calculated using the Perrin equation ($\tau_{f} = \tau_{total}$).

ated chromophores was completely disrupted within 20 min after dilution, and rotational diffusion of the GFP fusion protein had gained an isotropic mode, comparable to monomeric GFPuv. Quantification of the fluorescence intensity 20 min after dilution revealed a ratio of 0.25 GFP fusion protein for each phycobilisome complex.

7. These conclusions are based on the assumption that excitation energy transfer from GFP to phycocyanobilin chromophores proceeds with rates that are high enough to compete with the fluorescence decay rate. These rates are a function of the spectral overlap integral that contains the integral of the donor emission spectrum and the acceptor absorption spectrum *(2)*. The emission maximum of GFP is at 510 nm whereas the β-155 chromophore of phycocyanin, that absorbs maximally at 590 nm *(16)*, is the most likely acceptor.
8. Förster overlap integrals were calculated for all the individual phycocyanin chromophores, using their deconvoluted absorption spectra *(16)*, and the absorption spectrum of the allophycocyanin chromophore, α-84, assuming an extinction coefficient of 235 mM^{-1} cm^{-1} at 650 nm for the αβ monomer *(17)*. The calculated spectral overlap integrals for GFP, and all possible acceptor chromophores in the phycobilisomes, are given in **Table 2**.
9. Excitation energy transfer from phycocyanobilin chromophores to GFP was neglected. In intact phycobilisomes, excitation energy is transferred to the terminal emitters with very high efficiency, and emission is in the red region of the spectrum (for a review, *see* **ref. *18***), in which the chromophore of GFP does not absorb.
10. From the observed changes in fluorescence quantum yield in the presence of the acceptor, and the known fluorescence lifetime in absence of the acceptor, the total rate of energy transfer from GFP to the acceptors (**Fig. 2**; time 0) was calculated to be 1.16×10^9 s^{-1}.
11. Since the orientations of the chromophores are unknown, values for the squared orientation factor κ^2 of 1/3, 2/3, 1 1/3, and 4 were used to calculate the Förster radii, reflecting unfavorable, average, good, and optimal orientation of the transition dipoles, respectively.

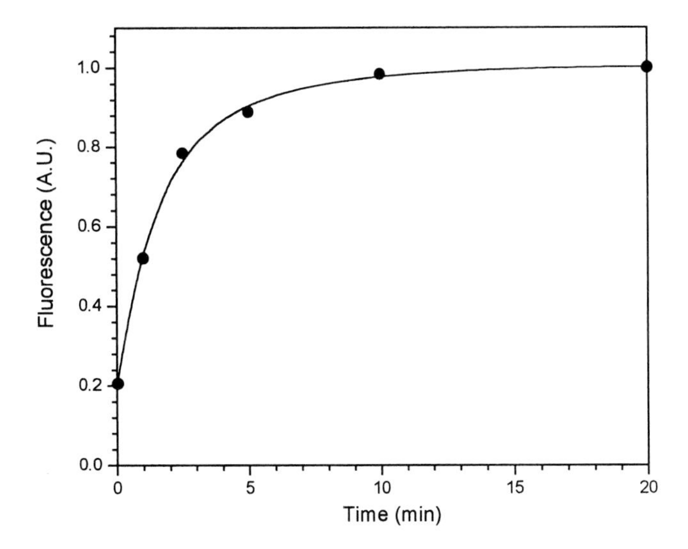

Fig. 2. Relative fluorescence intensity of the GFP-fusion protein during dissociation of the recombinant phycobilisome complexes. At t = 0, intact recombinant phycobilisomes containing the GFP-fusion protein (for explanation: *see* text) were transferred to low-ionic strength conditions, in order to initiate dissociation. The fluorescence intensity of the GFP fusion protein was corrected for changes in the scattering of the sample at the relevant wavelengths.

12. **Table 2** presents the R_0 and R_{DA} values for all acceptor chromophores. The R_{DA} values were calculated, assigning the total energy transfer rate $1.16 \times 10^9\ s^{-1}$ to each individual chromophore, exclusively. Judging from these values, it is possible that energy transfer proceeds from GFP to more than one acceptor chromophore simultaneously. This would result in an over-all increase of the Förster radii. Alternative scenarios that take these considerations into account are discussed elsewhere *(8)*. From these calculations, it is concluded that the estimated distance between GFP and the phycocyanobilin acceptor chromophore(s), in the recombinant complexes, is 3 nm, minimally, and 7 nm, maximally. We conclude that these results are relevant to the localization of FNR with respect to the phycobilisome-bound chromophores. Note that, in this case, the uncertainty of the calculated end-result is introduced not only because of the unknown orientation factor, κ^2, but also because the true identity of the acceptor(s) is not known.

4. Notes

1. The measurement of good-quality fluorescence spectra requires proper instrument settings. For fluorimeters, the slit widths and the photomultiplier voltage are important parameters, both affecting the signal amplitude, the signal:noise ratio, and the spectral resolution. For measurements with sufficiently high con-

Table 2
Förster Radii and Donor-Acceptor Radii Assuming Exclusive Energy Transfer to the Separate Acceptors

Energy transfer rate for GFP k_{ET}^{DA} (s^{-1})	Acceptor Chromophore	Spectral overlap integral (cm^3 M^{-1})	R_0 (Å)				R_{DA} (Å)			
			$\kappa^2 =$ 1/3	$\kappa^2 =$ 2/3	$\kappa^2 =$ 4/3	$\kappa^2 =$ 4	$\kappa^2 =$ 1/3	$\kappa^2 =$ 2/3	$\kappa^2 =$ 4/3	$\kappa^2 =$ 4
$1.16.10^9$	PC-β-155	$2.08.10^{-13}$	46	52	58	70	36	40	45	54
$1.16.10^9$	PC-α-84	$1.09.10^{-13}$	42	47	52	63	32	36	40	49
$1.16.10^9$	PC-β-84	$5.1.10^{-14}$	37	41	46	55	28	32	36	43
$1.16.10^9$	APC-α-84	$1.95.10^{-13}$	46	51	58	69	35	40	45	53

The total energy transfer rate determined experimentally was assigned to the respective phycocyanobilin chromophores separately, assuming exclusive energy transfer to these acceptors. A refractive index for phycobilisomes of $n = 1.567$ ***(18)*** was used for the calculations, for several values of κ^2.

centrations of GFP (i.e., ≥100 n*M*), try narrow slits widths first (1–4 nm for both excitation and emission). The photomultiplier voltage should not be set too high, because this will result in topping of the emission and excitation spectra. In addition, photomultipliers operated at high voltage are known to lose their linear characteristics because of saturation. On the other hand, when the voltage is set too low, the signal:noise ratio will deteriorate.

2. An important experimental consideration for FRET measurements is that the absorption of the samples should not exceed 0.01–0.1 at the relevant wavelengths, because the following relation holds:

$$10^{-\mathrm{OD}} = T \tag{15}$$

in which T is the fraction of light intensity transmitted at a given optical density, OD). Therefore, with a path length of 1 cm, the transmission at OD = 0.01 equals 0.98, and at 0.1, it equals 0.79. When the sample is too concentrated, emitted fluorescence can be reabsorbed, resulting in trivial energy transfer that can mistakenly be interpreted as radiationless energy transfer. This also applies to a solution of one single fluorophore, since self-absorption will result in an overall decrease of the apparent excited-state life-time, and therefore in a decrease of the calculated fluorescence quantum yield.

3. When the absorption of the donor is not detectable in the sample, a trick should be devised to modulate the efficiency of energy transfer. One way to achieve this is to disrupt the protein–protein interactions involved. For example, a specific protease cleavage site can be engineered into a GFP-fusion protein, so that the fluorescence can be measured in the same sample under subsequent conditions of FRET and no FRET, respectively. Note that GFP itself is known to be resistant to proteases such as chymotrypsin, subtilisin, and enterokinase. Alternatively, the absorption of the acceptor chromophore can be modulated, e.g., by reduction/oxidation (GFP is relatively resistant to reducing and oxidizing agents, but proper controls should be performed).

4. An additional control, which should be performed when wild-type GFP is used, is to determine the extent (if any) of photoconversion between the neutral and the anionic form of GFP, which occurs as a result of excitation at 280 nm and 395 nm. The conversion between these species can be detected by comparing the excitation amplitudes at 395 and 480 nm, respectively, before and after the FRET measurements *(7)*.

5. Divergent forms of the Förster equations and parameters are sometimes encountered in the literature. Some texts provide versions of the Förster formulas that include $\Phi^D{}_{fluorescence}$. Usually, the excited state decay of the donor is given by "$k^D{}_{radiative}$" or "$1/\tau^D{}_{radiative}$." When the fluorescence quantum yield is included, the excited-state decay meant in such cases is $k^D{}_{total}$, so that $k^D{}_{radiative}$ is replaced by the product of $k^D{}_{total}$ and $\Phi^D{}_{fluorescence}$ in **Eqs. 9**, **10**, and **16**. The formal definition of radiative decay assumes a contribution of radiative deactivation only, and no contribution from thermal deactivation (recall that $k^D{}_{total}$ includes both).

Occasionally, R_0 values are given for certain donor–acceptor pairs *(6)*. This single parameter contains all spectral information for a specific donor–acceptor pair, and consequently the rate of energy transfer (k_{FRET}) can be readily calculated by applying **Eq. 10**. However, unless otherwise stated, these values are usually calculated without taking the orientation factor, κ^2, into account. In fact, many of these R_0 values are calculated using a value of 1 for κ^2, reflecting a more optimal orientation than average. In order to calculate k_{FRET}, **Eq. 10** should then be replaced by:

$$k_{DA} = \kappa^2 \, k^{D}_{radioative} \left(\frac{R_0}{R_{DA}}\right)^6 \quad (16)$$

6. Obviously, when two polarizers are introduced into the light path, the signal will decrease significantly. As an example, under sample conditions, when a photomultiplier voltage of 600 V produces sufficient signal intensity in the absence of polarizers, the voltage will probably have to be increased to 1000 or 1200 V when the polarizers are introduced into the light beams. In order to be certain that the high voltage does not cause nonlinear behavior, it is necessary to obtain a G-factor value for each separate measurement. In addition, it is advisable to measure (with one photomultiplier voltage setting) the anisotropy of a dilution series of GFP, in the concentration range that is relevant for the sample to be tested. For GFP at pH 8.0, and 20°C, an anisotropy value of about 0.325 should be obtained *(8)*. A good control experiment for detection of rotational depolarization is measurement of the steady-state anisotropy of a 100 n*M* solution of sodium fluorescein in 0.1 *M* NaOH. The anisotropy of this sample should be ~0.01 (A_0 = 0.4; $\tau_{fluorescence}$ = 4.0 ns), and provide a calculated $\tau_{rotational}$ of ~100 ps. A final consideration is that the steady state anisotropy should always be measured with excitation and detection wavelengths set at the excitation and emission maxima.

References

1. Stryer, L. (1978) Fluorescence energy transfer as a spectroscopic ruler. *Annu. Rev. Biochem.* **47,** 819–846.
2. Förster, T. (1948) Zwischenmoleculare energiewanderung und fluoreszenz. *Ann. Physik. (Leipzig)* **2,** 55–75.
3. Förster, T. (1959) Transfer mechanisms of electronic excitation. *Faraday Soc. Disc.* **27,** 1–17.
4. Patterson, G. H., Knobel, S. M., Sharif, W. D., Kain, S. R., and Piston, D. W. (1997) Use of the green fluorescent protein and its mutants in quantitative fluorescence microscopy. *Biophys. J.* **73,** 2782–2790.
5. Perozzo, M. A., Ward, K. B., Thompson, R. B., and Ward, W. W. (1988) X-ray diffraction and time-resolved fluorescence analyses of *Aequorea* green fluorescent protein crystals. *J. Biol. Chem.* **263,** 7713–7716.
6. Tsien, R. Y. (1998) The green fluorescent protein. *Annu. Rev. Biochem.* **67,** 509–544.

7. van Thor, J. J., Pierik, A. J., Nugteren-Roodzant, I., Xie, A., and Hellingwerf, K. J. (1998) Characterization of the photoconversion of green fluorescent protein with FTIR spectroscopy. *Biochemistry* **37,** 16,915–16,921.
8. van Thor, J. J., Gruters, O. W., Matthijs, H. C., and Hellingwerf, K. J. (1999) Localization and function of ferredoxin:NADP(+) reductase bound to the phycobilisomes of *Synechocystis*. *The EMBO J.* **18,** 4128–1436.
9. Moser, C. C., Keske, J. M., Warncke, K., Farid, R. S., and Dutton, P. L. (1992) Nature of biological electron transfer. *Nature* **355,** 796–802.
10. Knox, R. S. (1975) Excitation energy transfer and migration: theoretical considerations in *Bioenergetics of Photosynthesis*. (Govindjee, ed.) Academic, New York, pp. 183–221.
11. van Amerongen, H., Valkunas, L., and van Grondelle, R. (2000) in *Photosynthetic excitons*. World, Singapore.
12. Swaminathan, R., Hoang, C. P., and Verkman, A. S. (1997) Photobleaching recovery and anisotropy decay of green fluorescent protein GFP-S65T in solution and cells: cytoplasmic viscosity probed by green fluorescent protein translational and rotational diffusion. *Biophys. J.* **72,** 1900–1907
13. Striker, G., Subramaniam, V., Seidel, C. A. M., and Volkmer, A. (1999) Photochromicity and fluorescence lifetimes of green fluorescent protein. *J. Phys. Chem.* **102,** 8612–8617.
14. Gantt, E., Lipschultz, C. A., Grabowski, J., and Zimmerman, B. K. (1979) Phycobilisomes from blue-green and red algae. Isolation criteria and dissociation characteristics. *Plant physiol.* **63,** 615–620.
15. Maxon, P., Sauer, K., Zhou, J., Bryant, D. A., and Glazer, A. N. (1989) Spectroscopic studies of cyanobacterial phycobilisomes lacking core polypeptides. *Biochim. Biophys. Acta* **977,** 40–51.
16. Demidov, A. A. and Mimuro, M. (1995) Deconvolution of C-phycocyanin β-84 and β-155 chromophore absorption and fluorescence spectra of cyanobacterium *Mastigoclaudus laminosus*. *Biophys. J.* **68,** 1500–1506.
17. Bryant, D. A., Guglielmi, G., Tandeau de Marsac, N., Castets, A., and Cohen-Bazire, G. (1979) The structure of cyanobacterial phycobilisomes: a model. *Arch. Microbiol.* **123,** 113–127.
18. van Thor, J. J., Mullineaux, C. W., Matthijs, H. C. P., and Hellingwerf, K. J. (1998) Light harvesting and state transitions in cyanobacteria. *Botanica Acta* **111,** 430–443.
19. Grabowski, J. and Gantt, E. (1978) Photophysical properties of phycobiliproteins from phycobilisomes: fluorescence lifetimes, quantum yields and polarization spectra. *Photochem. Photobiol.* **28,** 39-45.

9

Bioluminescence Resonance Energy Transfer Assays for Protein–Protein Interactions in Living Cells

Yao Xu, Carl Hirschie Johnson, and David Piston

1. Introduction

We recently developed a bioluminescence resonance energy transfer (BRET) system for assaying protein–protein interactions *(1,2)*, which has been used successfully for studying the interaction of circadian clock proteins isolated from cyanobacteria, and tested in *Escherichia coli* cells *(1)*, and the dimerization of human β_2-adrenergic receptors in mammalian cells *(3)*. BRET results from the nonradiative energy transfer between a donor and an acceptor. It is related to fluorescence resonance energy transfer (FRET) *(4,5)*, except that, in FRET, there are two fluorophores, one that absorbs exogenous excitation (the donor) and passes the energy to the other fluorophore (the acceptor); in BRET, the donor is a luciferase that generates its own luminescence emission in the presence of a substrate, and can pass the energy to an acceptor fluorophore. For either BRET or FRET to work, the donor's emission spectrum must overlap the acceptor's absorption spectrum, their transition dipoles must be in an appropriate orientation, and the donor and acceptor must be in close proximity (usually within 30–80 Å of each other, depending on the degree of spectral overlap) *(6)*. During a BRET assay for molecular interactions, the first criterion is fixed for a given donor–acceptor combination, but the relative orientation and distance between the donor and acceptor will change depending on the strength of the interaction. Although its use as a protein interaction assay is novel, BRET is a natural phenomenon. In fact, resonance energy transfer between the calcium ion-dependent photoprotein aequorin and green fluorescent protein (GFP) is a natural example of BRET.

From: *Methods in Molecular Biology, vol. 183: Green Fluorescent Protein: Applications and Protocols*
Edited by: B. W. Hicks © Humana Press Inc., Totowa, NJ

The cloning and development of GFP mutants with varied spectral properties made it possible to use them as FRET pairs to detect protein–protein proximity in real time in living cells, by coexpressing a donor GFP variant (shorter-wavelength excitation) fused to one candidate protein and an acceptor GFP variant (longer-wavelength excitation) fused to the other candidate protein ***(7–13)***.

One advantage of FRET-based interaction assays is that the GFP-fusion proteins can be targeted with signal sequences to specific compartments so that protein interactions can be monitored within cellular compartments in vivo in the native organism (if transformable). In practice, however, excitation of the donor fluorophore gives rise to several problems that can limit the usefulness of the FRET approach: photobleaching of the donor fluorophore, autofluorescence of the cells/tissues, damage to the cells/tissues by the excitation light, stimulation of the tissue if it is photoresponsive (e.g., retina), and direct excitation of the acceptor fluorophore that is independent of the resonance transfer. In BRET, these problems are avoided because there is no requirement for an excitation light. The donor fluorophore of the FRET pair is replaced by a luciferase, in which bioluminescence from the luciferase in the presence of a substrate excites the acceptor fluorophore through the same resonance energy transfer mechanisms as FRET.

Another advantage of BRET compared to FRET, is that it is easier to quantify the relative expression levels of a luciferase donor and a fluorophore acceptor (as in BRET) because the two molecules can be measured independently (luminescence intensity for the donor, fluorescence intensity for the acceptor); if both donor and acceptor are fluorophores whose spectra overlap (as in FRET), it is difficult to measure the intensity of one fluorophore without getting some contamination in the spectrum from the fluorescence of the second fluorophore.

The major liability of the BRET technique is that the intensity of luminescence generated by the luciferase donor is dim, and requires specialized equipment to measure. Fortunately, the commercial sector is beginning to address this problem (*see* **Note 1**). Therefore, BRET offers the advantages of FRET without having its major disadvantages, which accrue from the use of the excitation light. We anticipate that BRET will prove valuable for investigating protein interactions within native cells, especially for integral membrane proteins or proteins targeted to specific organelles ***(1,2)***.

In BRET experiments to test for protein–protein interactions in *E. coli*, we have used two different luciferase donors: *Renilla* luciferase (RLUC) or *Gaussia* luciferase (GLUC). Both of these luciferases have an emission peak at 480 nm, and use coelenterazine as a substrate. Coelenterazine is a hydrophobic and cell-permeable molecule. As an acceptor, we have used a red-shifted GFP mutant, an enhanced yellow fluorescent protein (EYFP), which has a broad excitation spectrum, with a peak at 513 nm (but is still efficiently excited at much shorter wavelengths, such as 480 nm) and an emission peak of 527 nm. RLUC and GLUC were chosen as the donor luciferases, because their emission spectra are similar to the cyan mutant of *Aequorea* GFP (λ_{max} ~480 nm), which has been shown to exhibit FRET with EYFP ***(9,10)***. The excitation peak of EYFP does not perfectly match the emission peak of RLUC or GLUC, but the emission spectrum of RLUC or GLUC is sufficiently broad to provide good excitation of EYFP. Each of these donor and acceptor molecules were genetically fused to putative interacting proteins, in this case, to circadian clock proteins (KaiA and KaiB) from cyanobacteria ***(1,14)***. We found that when the LUC and EYFP fusion proteins were brought into proximity by Kai protein interaction, part of the energy from LUC emission is transferred to the EYFP, which emits yellow light, and results in a bimodal emission spectrum (**Fig. 1**). Thus, the intermolecular BRET phenomenon can be used to monitor the protein–protein interactions, both in vivo and in vitro, by quantifying the emission ratio at 530:480 nm.

2. Materials

2.1. Plasmids, Vectors, and Strains

1. pRL-null (Promega, Madison, WI) contains the RLUC coding region.
2. pcDNA3 GLUC (Prolume) harbors a gene encoding GLUC.
3. pEYFP (Clontech, Palo Alto, CA) encodes an EYFP, which is a red-shifted mutant of the *Aequorea victoria* GFP.
4. pEGFP-NI (Clontech) encodes EGFP, and includes a kanamycin-resistance gene.
5. p44N carries the circadian clock genes *kai*A, *kai*B, and *kai*C from the cyanobacterium *Synechococcus* sp. strain, PCC 7942 ***(14)***.
6. pRSET expression vector (Invitrogen, Carlsbad, CA) has an ampicillin (AMP)-resistant selection marker.
7. *E. coli* strain BL21 (DE3) is a host strain (Novagen, Madison, WI) for expressing genes under the control of a T7 promoter, and *E. coli* DH5α is a cloning host strain.

2.2. Reagents, Buffers, and Media

1. Luria-Bertani (LB) medium: 10 g/L Bacto-tryptone, 5 g/L Bacto-yeast extract, 10 g/L NaCl (pH 7.0).
2. In vivo BRET assay buffer: (M9 medium salts) 12.8 g/L $Na_2HPO_4 \cdot 7H_2O$, 3 g/L KH_2PO_4, 0.5 g/L NaCl, 1.0 g/L NH_4Cl.
3. Coelenterazine (BioSynth, Naperville, IL): 100 μ*M* stock solution (*see* **Note 2**).
4. In vitro BRET assay buffer: 50 m*M* KCl, 50 m*M* NaCl, 2.5 m*M* $MgCl_2$, 2 m*M* EDTA, 5 m*M* dithiothreitol, 0.2% Nonidet P-40, 100 μg/mL phenylmethylsulfonyl fluoride, 2 μg/mL leupeptin, 2 μg/mL aprotinin, 20 m*M* HEPES, pH 8.0.
5. Isopropyl β-D-thiogalactoside (IPTG): 1.0 *M* stock solution.

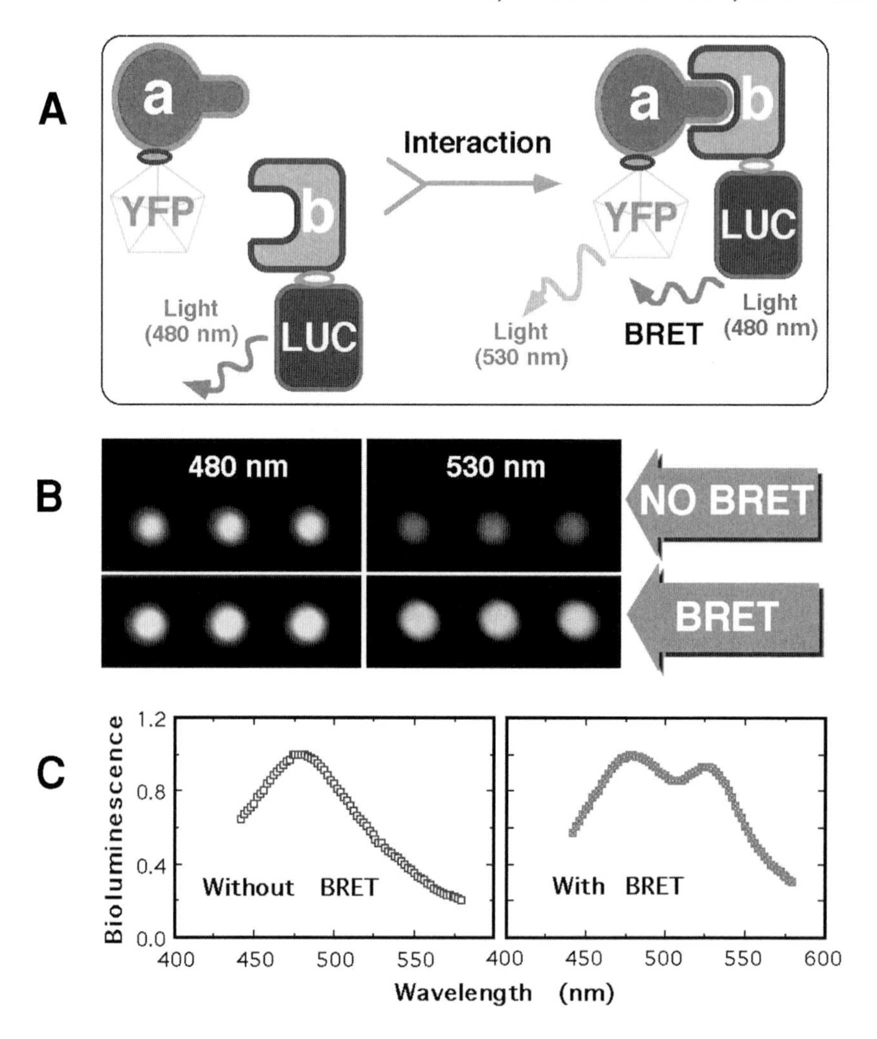

Fig. 1. Bioluminescence resonance energy transfer (BRET) as a tool for monitoring protein-protein interaction. **(A)** Schematic diagram of BRET principle. One protein of interest (b) is genetically fused to the donor luciferase, LUC, from either *Renilla* or *Gaussia*, and the other candidate protein (a) is fused to the acceptor fluorophore, YFP. If the two proteins interact strongly enough to bring LUC and YFP into close proximity, a longer wavelength emission can be generated by BRET. **(B)** Visual imaging of BRET signals with a charge-coupled device camera, through filters transmitting light of 480 nm or 530 nm. *E. coli* cultures co-expressing fusion proteins RLUC•KaiB and EYFP•KaiB) exhibit BRET, but cultures co-expressing RLUC•KaiB and EYFP•KaiA have no BRET (method as described in **Subheading 3.4.**). **(C)** Comparison of luminescence emission spectra, using a fluorescence spectrophotometer between the BRET- and non-BRET *E. coli* strains (method as described in **Subheading 3.3.**). (For optimal, color representation please see accompanying CD-ROM.)

2.3. Apparatus

1. SPEX Fluorolog spectrofluorometer with a 250 W xenon arc lamp (Spex, Edison, NJ) (*see* **Note 1**).
2. FB12 Luminometer (Zylux, Maryville, TN) (*see* **Note 1**).
3. Charge-coupled device camera (TE/CCD512BKS, Princeton Instruments, Trenton, NJ).
4. 480- and 530-nm interference filters (Ealing Electron-Optics, Holliston, MA).

3. Methods

The BRET technique for monitoring protein-protein interaction involves several steps: choosing expression vectors that are appropriate for the organism to be tested, creation of gene fusions of RLUC/GLUC and YFP to proteins of interest (*see* **Note 3**), co-transformation and co-expression of gene fusions in living cells (*see* **Notes 4** and **5**), and BRET assays in vivo or in vitro. Here, we use the interaction of the cyanobacterial circadian clock protein, KaiB, in *E. coli*, as an example to illustrate the BRET technique for protein–protein interaction.

3.1. Construction of BRET Gene Fusions

The following constructs were made to express the gene fusions under the control of the T7 promoter in *E. coli*.

1. Positive-control BRET construct pT7/RLUC•EYFP for expression of fusion protein RLUC•EYFP (*see* **Note 6**).
2. Negative-control BRET constructs, pT7/RLUC, pT7/GLUC, or pT7/EYFP, for expression of RLUC, GLUC, or EYFP alone, and pT7/EYFP•KaiA, for expression of fusion protein, EYFP•KaiA (*see* **Notes 7** and **8**).
3. BRET donor construct, pT7/RLUC•KaiB, with an ampicillin-resistance selection marker for expression of BRET donor fusion protein, RLUC•KaiB (*see* **Note 8**).
4. BRET acceptor construct, pT7/EYFP•KaiB, with a kanamycin-resistant selection marker for expression of BRET acceptor fusion protein, EYFP•KaiB (*see* **Note 8**).

3.2. Co-Transformation and Co-Expression of BRET Gene Fusion

1. Grow *E. coli* strain BL21 (DE3) to use as the expression host (*see* **Note 9**).
2. Transform 20–100 μL of competent host cells, with 100 ng LUC donor or its control plasmids carrying kanamycin resistance gene, by standard procedures *(15)*.
3. Confirm RLUC or GLUC expression in transformed kanamycin-resistant colonies, by luminescence activity assay: Transfer 0.5 mL of an overnight culture, derived from a single transformed or untransformed colony, into an Eppendorf tube; add coelenterazine to a final concentration of 1 μ*M*; votex for 1 s, then read the relative luminescence units immediately, using the FB12 Luminometer (*see* **Note 10**).

4. Make competent cells from kanamycin-resistant strains expressing LUC donor, by standard methods *(15)*.
5. Co-transform 20–100 µL competent LUC donor strains with 100 ng BRET YFP acceptor or its control plasmids carrying the ampicillin resistance gene, and select the double transformants on LB plates containing both 100 µg/mL ampicillin and 50 µg/mL kanamycin. For YFP construct transformation alone, use ampicillin only to select colonies.
6. Examine YFP expression in transformed colonies by fluorescence assay: Spin 1.5 mL of an overnight *E. coli* culture for 20 s, using a microcentrifuge at high speed; wash once, and resuspend in 1.5 mL in vivo BRET assay buffer (*see* **Note 11**); measure fluorescence emission spectrum, scanned from 505 to 580 nm, on SPEX Fluorolog spectrofluorometer with a 250 W xenon arc lamp, with excitation at 470 nm (*see* **Note 12**).

3.3. BRET Assays

3.3.1. Bioluminescence Spectral Emission Acquisition In Vivo

1. Inoculate a single colony from a freshly streaked plate of an *E. coli* strain expressing a BRET construct, and incubate overnight in LB medium containing 50 µg/mL kanamycin and/or 100 µg/mL carbenicillin (for ampicillin-resistant strains) at 37°C, with shaking (*see* **Note 13**).
2. Collect 1.5–3.0 mL cells, centrifuge, wash once, and resuspend in 1.5 mL in vivo BRET assay buffer (*see* **Note 11**).
3. Transfer the washed cells into a 4-mL fluorescence cuvet, and put it into the SPEX Fluorolog spectrofluorometer, leaving the xenon arc lamp off (to eliminate the possibility of inadvertent fluorescence excitation).
4. Gently bubble air into the cell suspension in the cuvet (*see* **Note 14**).
5. Add fresh coelenterazine to the cell suspension, to a final concentration of 1–5 µ*M*.
6. Measure the luminescence emission spectrum between wavelengths 440 and 580 nm, in 2-nm steps, with 2–10 s integrations for each step.
7. Determine the extent of BRET by evaluating the emission spectra, especially the emergence and magnitude of the second peak at ~530 nm in the luminescence spectra in the experimental samples, compared with the positive and negative control samples (**Fig. 1C**).

3.3.2. BRET Imaging In Vivo

1. Prepare 1.5 mL overnight-grown *E. coli* cultures, as described previously.
2. Wash cells twice, and resuspend the cell pellet in 300 µL *in vivo* BRET assay buffer containing 3 µ*M* fresh-prepared coelenterazine.
3. Immediately place 5-µL aliquots of the suspension in individual wells of a Nunc Microwell plate (*see* **Note 14**).
4. Divide each group of samples (i.e., by strain) into two sets of wells in the plate.
5. Place a 480-nm (± 5 nm) interference filter on top of one set of wells, and a 530-nm (± 5 nm) filter on top of the other set of wells.

6. Capture the luminescence emission images of both sets of wells simultaneously with a cooled-charge-coupled device camera.
7. Compare the imaging signals, as shown in **Fig. 1B** (*see* **Note 15**).

3.3.3. Calculation of the Magnitude of BRET

1. Obtain BRET signals of *E. coli* cultures, as described in **Subheadings 3.3.1.** and/or **3.3.2.**
2. Quantify the luminescence intensities at 480 and 530 nm.
3. Calculate the magnitude of BRET (*see* **Note 16**).
4. Compare the BRET magnitude among the experimental, positive-control, and negative-control samples, to determine if there has been resonance energy transfer. The relative magnitude of BRET can be used to infer whether protein–protein interaction has occurred.

3.3.4. BRET Assay In Vitro

1. Inoculate single colonies of a BRET LUC donor strain, a BRET YFP acceptor strain, and a positive-control BRET strain expressing the RLUC•EYFP fusion protein, into each of three 500-mL Erlenmeyer flasks containing 100 mL LB medium, with either 50 μg/mL kanamycin or 100 μg/mL carbenicillin, and incubate overnight at 37°C, with shaking.
2. Wash cells once with fresh LB medium, then resuspend in the same medium containing 1 m*M* IPTG.
3. Induce expression of gene fusion, with shaking at 37°C for 3 h.
4. Place the flasks on ice for 5 min, then harvest the cells by centrifugation at 5000*g* for 5 min at 4°C.
5. Wash the cells once, resuspend in 5 mL of in vitro BRET assay buffer (at 4°C) containing 10 mg/mL lysozyme, and incubate on ice for 30 min.
6. Aliquot 0.5 mL of the cells into microcentrifuge tubes, and put into a –70°C ethanol bath (or liquid nitrogen), to quickly freeze the cells.
7. Place the frozen tubes on ice, and add an equal volume of fresh in vitro BRET assay buffer (at 4°C) on top of the frozen extracts: Allow to thaw. Most cells will be broken by this freeze–thaw cycle.
8. Spin the mixtures for 5 min in a microcentrifuge at 8,000*g* at 4°C, and save the supernatants.
9. For monitoring protein–protein interaction, mix equal amounts of the LUC donor protein with the YFP acceptor protein, to a 1–2-mL final volume. For testing the positive BRET control, take 1–2 mL RLUC•EYFP fusion protein supernatant, without any other additions.
10. Incubate the mixed or unmixed supernatants at room temperature for a specific time, e.g., 0, 30, and 120 min (*see* **Note 17**).
11. To measure, transfer the supernatants to the 4-mL cuvet placed inside the SPEX Fluorolog spectrofluorometer, with the xenon arc lamp turned off.
12. Immediately add fresh coelenterazine to the extracts, to a final concentration of 1 μ*M*, and mix well by gentle pipeting up and down with a pipetman.

13. Detect the luminescence emission spectrum between wavelengths 440 and 580 nm, in 2-nm steps with 1-s integrations (*see* **Note 17**).
14. Analyze the BRET signals as described in **Subheading 3.5.**

4. Notes

1. For measuring total luminescence, without regard to color, the FB12 Luminometer from Zylux Corporation is sensitive and convenient. The advantage of the SPEX Fluorolog spectrofluorometer is that it can measure the full spectrum of light emission, but it is not very sensitive to low levels of luminescence. If the luminescence signal is too weak, the SPEX Fluorolog will not produce a clear spectrum. For measuring BRET ratios (light emission at 530 nm, divided by light emission at 480 nm), we use a modified, custom-built luminometer, which allows one to alternately interpose 480 and 530 nm interference filters between the sample and photon-counting photomultiplier tube, without disturbing the sample. The same operation could be performed manually with the FB12 Luminometer, but it would necessitate a slight disturbance of the sample, with every filter change. Two high-throughput, multi-sample apparatuses can be used for BRET assays: the Victor2 (Wallac) and the Fusion (Packard). Both instruments have the dual luminescence mode necessary to perform BRET assays, and are capable of reading 96- or 384-well plates.
2. The substrate coelenterazine is sensitive to light and oxygen. We first dissolve coelenterazine in ethanol, to prepare a stock solution (e.g., 250 μ*M*), then small volumes are aliquoted (e.g., 40 μL) to microcentrifuge tubes. The samples are dried in the tubes with a Speed-Vac (Savant). To remove oxygen, each tube is gently (to avoid the sample being blown out of the tube) gassed with either nitrogen or argon gas, then the tubes are sealed, and stored in a –80°C freezer in a black box. Just before use, the coelenterazine sample is redissolved in a small volume (e.g., 40 μL or less) of EtOH, then in BRET assay buffer to a final concentration of 100 μ*M*. The coelenterazine solution tube should be wrapped with aluminum foil, to protect it from light until it is used.
3. For BRET constructs, one candidate protein is genetically fused to the donor RLUC or GLUC, and the other protein of interest is fused to the acceptor fluorophore YFP. To make a construct for a fusion protein, the stop codon in the coding region for the N-terminal partner should be removed, and the coding region of the C-terminal partner should be in frame with that of the N-terminal one. Considering the protein folding and orientation, sometimes a linker might be useful between the two proteins. In addition, it is worthwhile to test various combinations of protein fusions with LUC or YFP, such as N- vs, N-, N- vs, C-, C- vs, N-, as well as C- vs C-terminal fusions, to discover which combination allows the donor and acceptor to form into the optimal orientation/distance. GLUC (19.9 kDa) is significantly smaller than RLUC (36 kDa) in size, and is therefore less likely to cause a steric hindrance problem.
4. The LUC and YFP fusion partners are best expressed at roughly equal levels for the BRET assay in living cells. Therefore, the co-transferred plasmids not only

need to carry distinct selection markers, but also should be compatible (especially in terms of their origins of replication) in the transformed cells. For expression of LUC/EYFP fusion partners with the Kai proteins, we used a modified pEGFP-N1 vector for kanamycin selection and a modified pRSET vector for ampicillin selection (modifications are described in **ref. *1***). Promoter activity is also an important factor that may affect the BRET. In the case of our studies with *E. coli*, a weakly artifactual BRET signal, e.g., a small second peak at 530 nm in the luminescence spectrum, will sometimes occur even in the negative-control combinations if the fusion proteins are expressed at too high a level. Hence, it is better not to use too strong a promoter to drive expression of BRET fusion constructs. In our case, we found that using the T7 promoter, but not inducing with IPTG, gave enough expression to get good BRET ratios in the experimental combinations, without causing an artifactual peak in the negative-control combinations.

5. The goal of using two plasmids in *E. coli* is to maintain an equal copy number for each plasmid, so as to maintain roughly equivalent expression levels, but the ratio of these two plasmids may fluctuate. If this causes a serious co-expression problem, there are two alternative strategies that may help to maintain roughly equal expression levels in *E. coli*. The first strategy is to put the two BRET fusion proteins into a construct in which they are expressed from the same operon. For construction of this expression operon, minimizing the length of the untranslated sequence between the two fusion genes should help to reduce polar effects. In other words, the untranslated sequence should be designed to include a ribosome-binding sequence (Shine Dalgarno sequence) between the two fusion protein sequences, and otherwise have the N-terminal end of the second fusion protein sequence as close as possible to the C-terminal end of the first one. If using the same Shine-Dalgarno sequence, the expression levels of the two fusion proteins should be comparable. The second strategy is to put the two BRET transcription units on the same plasmid. For example, the construct could be designed with divergent promoters in the middle of the plasmid, so that the LUC fusion protein is transcribed in one direction, while the YFP fusion protein is transcribed in the other direction.
6. To confirm that a BRET signal can be obtained under the available experimental conditions, it is best to use a positive BRET control. We have confirmed that fusion of RLUC directly to EYFP (an enhanced YFP) through a linker of 11 amino acids yields a good BRET control, i.e., the luminescence profile of the *E. coli* cells expressing the RLUC•EYFP fusion construct yielded a bimodal spectrum, with one peak centered at 480 nm (as for RLUC), and a new peak centered at 530 nm (as for EYFP fluorescence) ***(1)***. To make this construct, the RLUC coding region, without its stop codon TAA but with a T7 promoter, was amplified using *Pwo* DNA polymerase (Boehringer Mannheim), using pRL-null (Promega) as the template. The amplified fragment with an *Nde*I and an *Apa*I linker was then inserted into the *Nde*I/*Apa*I site of the vector pEGFP-N1 (Clontech), containing the EGFP, to give the plasmid, pT7/RLUC•EGFP. The EGFP coding region in the pT7/RLUC•EGFP was replaced with the *Bam*HI/*Not*I

fragment containing the EYFP coding sequence from the vector pEYFP (Clontech), to produce the plasmid, pT7/RLUC•EYFP, in which the EYFP open reading frame is in frame with that of RLUC, and there are 11 codons between RLUC and EYFP in this gene fusion *(1)*.

7. In BRET assays, two types of negative controls are important for evaluation of the BRET signals. One control is to co-express LUC and YFP alone within a cell, and the other one is to co-express a BRET donor fusion and an acceptor fusion containing protein(s) that do not interact with each other. These controls allow the experimenter to gage the maximal expression level from the T7 promoter, which does not result in an artifactual second peak at 530 nm. In our assays, we found that KaiB and KaiA did not interact, so KaiA fusion constructs were used as the latter type of negative control.
8. To test BRET as a protein–protein assay, we first tried the N-terminal fusions of KaiB to RLUC and to EYFP. The luminescence spectra of *E. coli* expressing these fusions showed a second peak in the cells expressing both RLUC•KaiB and EYFP•KaiB (**Fig. 1**). We then created various constructs for expression of other KaiB BRET fusions, including KaiB•RLUC (i.e., C-terminal fusion), EYFP•KaiB, and GLUC•KaiB, and further tested all other possible co-expressions of KaiB fusions with RLUC, GLUC, or EYFP, such as GLUC•KaiB and EYFP•KaiB, KaiB•RLUC and EYFP•KaiB, RLUC•KaiB and KaiB•EYFP, and KaiB•RLUC and KaiB•EYFP. All of these combinations of the KaiB fusion proteins showed BRET signals. Construction of pT7/RLUC•KaiB, pT7/EYFP•KaiB, pT7/EYFP•KaiA, pT7/RLUC, and pT7/EYFP was described previously *(1)*.
9. All BRET constructs are designed with either kanamycin or ampicillin resistance, to allow co-transformations. The T7 promoter needs T7 RNA polymerase for expression. *E. coli* BL21 (DE3) is used as an expression host strain, because it contains a chromosomal copy for T7 RNA polymerase under the control of *lacUV5*, whose expression can be induced by IPTG. In our laboratory, when this strain was grown in LB medium, the expression levels of the BRET fusion proteins were sufficient to achieve good BRET signals without IPTG expression (probably the LB medium contains enough lactose to induce T7 expression to sufficient levels). Adding IPTG to induce overexpression of the gene fusions sometimes caused an artifactual second peak in the negative controls (*see* **Note 4**).
10. This is a fast and easy way to confirm whether the RLUC or GLUC is expressed in transformed strains. Using the FB12 Luminometer, we usually consider a reading of 5×10^5 to 1×10^7 counts/s to be a strong signal. Alternatively, the luminescence activity and/or spectrum can also be determined by using the SPEX Fluorolog spectrofluorometer. With the SPEX Fluorolog, 5000 (or more) counts/s/2 nm is a strong signal.
11. Because LB medium has some component(s) that is fluorescent, the *E. coli* cultures are washed and resuspended in in vivo BRET assay buffer (M9 medium salts) prior to luminescence and fluorescence assay.
12. YFP fluorescence is usually strong enough that it can be seen by eye, upon ultraviolet (UV) illumination with a transilluminator. YFP-expressing cell cultures in

M9 medium salts look yellow-green under UV illumination. UV radiation is dangerous for the eyes, so the observer should wear UV-safety goggles, and minimize the exposure time.

13. It is advisable to use carbenicillin for liquid cultures in place of ampicillin to prevent overgrowth of cultures by cells that have lost the relevant plasmids. For good aeration of shaken cultures, add no more than 20% of the total flask volume with medium.
14. The luminescence reaction catalyzed by RLUC or GLUC requires oxygen and coelenterazine. Gentle air-bubbling into the cuvet, with small polyethylene tubing, can supply sufficient oxygen for the RLUC or GLUC reaction, and helps to maintain a relatively stable luminescence signal. In the case of small volumes of cells in a microwell plate, however, gas exchange is sufficient so that air bubbling is not essential.
15. It may be possible to image BRET signals using a microscope with interchangeable 480- and 530-nm filters, but we have not yet tested this. Also, it should be possible to develop an in vivo BRET system for screening libraries for protein–protein interaction by designing a LUC-fusion expression "bait" vector with one selection marker and a YFP-fusion "prey" expression library containing vector with another selection marker; plates of transformed *E. coli* could be screened for colonies whose 530 nm/480 nm ratio indicates BRET interaction *(2)*.
16. The magnitude of BRET can be expressed in a number of ways, but it is important to always correct the BRET signals of the experimental samples by comparison with the negative-control samples. If two filters are used (e.g., one interference filter centered on 480 nm and another filter centered on 530 nm), the ratio of emission from the 530 nm filter can be compared with that from the 480 nm filter for the experimental and control samples, as described previously *(1)*:

$$\text{BRET ratio} = [Em_{530}] / [Em_{480}]$$

To correct the experimental samples for the control emission spectrum, the magnitude of BRET ($BRET_{mag}$) can be calculated as the emission ratio of the experimental sample ($[Em_{530}]_E$ / $[Em_{480}]_E$) minus the emission ratio of the control sample ($[Em_{530}]_C$ / $[Em_{480}]_C$), or:

$$BRET_{mag} = ([Em_{530}]_E / [Em_{480}]_E) - ([Em_{530}]_C / [Em_{480}]_C)$$

Angers et al. *(3)* define the magnitude of BRET by a similar calculation.
Another way to measure the BRET ratio is to use only one filter (e.g., the 440–500 nm filter), and compare the light emitted in that range with the total spectrum of light emission (using no filter). This second method of calculation has not been as extensively used, but it is probably accurate, and requires the use of only one filter.

17. In in vitro experiments with extracts from *E. coli*, the overall luminescence of the samples decreased with time of incubation, especially at higher temperatures (e.g., 30–35°C). This effect may result from protein degradation within the extracts. Therefore, we recommend incubating at room temperature. Fortunately,

in the luciferase-containing cell extracts, the luminescent activity is so strong that a much shorter integration (e.g., 1 s or less) could be used. Hence, air-bubbling was not required for the in vitro assay; in fact, air-bubbling of the extracts can cause undesirable soap-like bubbles to form.

Acknowledgments

We appreciate the comments on the manuscript by Dr. Erik Joly.

References

1. Xu, Y., Piston, D. W., and Johnson, C. H. (1999) A bioluminescence resonance energy transfer (BRET) system: application to interacting circadian clock proteins.*Proc. Natl. Acad. Sci. USA* **96,** 151–156.
2. Xu, Y., Piston, D. W., and Johnson, C. H. (1999) Resonance energy transfer as an emerging strategy for monitoring protein-protein interactions in vivo: BRET *vs.* FRET. *The Spectrum* **12,** 9–14.
3. Angers, S., Salahpour, A., Joly, E., Hilairet, S., Dennis, M., and Bouvier, M. (2000) Detection of β_2-adrenergic receptor dimerization in living cells using bioluminescence resonance energy transfer (BRET). *Proc. Natl. Acad. Sci. USA* **97,** 3684–3689.
4. Wu, P. and Brand, L. (1994) Resonance energy transfer: methods and applications. *Analyt. Biochem.* **218,** 1–13.
5. Clegg, R. M. (1995) Fluorescence resonance energy transfer. *Curr. Opin. Biotechnol.* **6,** 103–110.
6. Clegg, R. M. (1996) Fluorescence resonance energy transfer, in *Fluorescence Imaging Spectroscopy and Microscopy* (Wang, X. F. and Herman, B., eds.), Wiley, New York, pp. 179–252.
7. Heim, R., Prasher, D. C., and Tsien, R. Y. (1994) Wavelength mutations and post-translational autoxidation of green fluorescent protein. *Proc. Natl. Acad. Sci. USA* **91,** 12,501–12,504.
8. Heim, R. and Tsien, R. Y. (1996) Engineering green fluorescent protein for improved brightness, longer wavelengths and fluorescence resonance energy transfer. *Curr. Biol.* **6,** 178–182.
9. Miyawaki, A., Llopis, J., Heim, R., et al. (1997) Fluorescent indicators for Ca^{2+} based on green fluorescent proteins and calmodulin. *Nature* **388,** 882–887.
10. Mahajan, N. P., Linder, K., Berry, G., Gordon, G. W., Heim, R., and Herman, B. (1998) Bcl-2 and Bax interactions in mitochondria probed with green fluorescent protein and fluorescence resonance energy transfer. *Nature Biotechnol.* **16,** 547–552.
11. Periasamy, A. and Day, R. N. (1998) FRET imaging of Pit-1 protein interactions in living cells. *J. Biomed. Opt.* **3,** 154–160.
12. Xu, X., Gerard, A. L., Huang, B. C., Anderson, D. C., Payan, D. G., and Luo, Y. (1998) Detection of programmed cell death using fluorescence energy transfer *Nucl. Acids Res.* **26,** 2034–2035.

13. Gadella, T. W., Jr., van der Krogt, G. N. M., and Bissseling, T. (1999) GFP-based FRET microscopy in living plant cells. *Trends in Plant Sci.* **4,** 287–291.
14. Ishiura, M., Kutsuna, S., Aoki, S., et al. (1998) Expression of a gene cluster *kaiABC* as a circadian feedback process in cyanobacteria. *Science* **281,** 1519–1523.
15. Sambrook, J., Fritsch, E. F., and Maniatis, T. (1989) Preparation and transformation of competent *E. coli*, in *Molecular Cloning*. Cold Spring Harbor Laboratory Press, Cold Spring Harbor, New York, pp. 1.74–1.84.

10

Whole-Body Fluorescence Imaging with Green Fluorescence Protein

Robert M. Hoffman

1. Introduction

1.1. Whole-Body Imaging of Green Fluorescent Protein-Expressing Tumors

Current methods for external imaging of internally growing tumors include X-rays, magnetic resonance imaging (MRI), and ultrasonography. Although these methods are well suited for the noninvasive imaging of large-scale structures in the human body, they have limitations in the investigation of internal, actively growing tumors. In particular, monitoring growth and metastatic dissemination by these methods is impractical, since they either use potentially harmful irradiation, or require harsh contrast agents, and therefore cannot be repeated on a frequent or real-time basis.

Previous attempts to endow tumors with specific, detectable spatial markers have mostly met with mediocre success. These included labeling with monoclonal antibodies and other high-affinity vector molecules targeted against tumor-associated markers ***(1–6)***. However, results were limited as a result of achieving only a low tumor–background contrast, and because of the toxicity of the procedures.

Intravital videomicroscopy is another approach to optical imaging of tumor cells, which allows direct observation of cancer cells ***(7)***. Even in this limited arena, intravital videomicroscopy does not lend itself to following tumor growth, progression, and internal metastasis in a live intact animal.

A major conceptual advance in optical imaging was to make the tumor the source of light. This renders the incident light scattering much less relevant. One early attempt inserted the luciferase gene into tumors, so that they emit

From: *Methods in Molecular Biology, vol. 183: Green Fluorescent Protein: Applications and Protocols*
Edited by: B. W. Hicks © Humana Press Inc., Totowa, NJ

light *(8)*. However, luciferase enzymes transferred to mammalian cells require the exogenous delivery of their luciferin substrate, an essentially impractical requirement in an intact animal. Also, it is not known whether luciferase genes can function stably over significant time periods in tumors and in the metastases derived from them.

A more practical approach to tumor luminance is to make the target tissue selectively fluorescent. In one attempt, tumor-bearing animals were infused with protease-activated, near-infrared fluorescent probes *(9)*. Tumors with appropriate proteases could activate the probes and be imaged externally. However, the system proved to have severe restrictions. The selectivity was limited, because many normal tissues have significant protease activity. In fact, the normal activity in liver is so high as to preclude imaging in this most important of metastatic sites. The short lifetime of the fluorescence probes would appear to rule out growth and efficacy studies *(9)*. The requirements of appropriate tumor-specific protease activity, and of effective tumor delivery of the probes, also limit this approach *(9)*.

A new approach of whole-body external imaging makes use of green fluorescent protein (GFP)-expressing tumors in intact animals. Stable GFP expression in cancer cells is an extremely effective marker *(10–18)*. The fluorescence illuminates tumor progression and allows visualization of tumor growth and metastases by whole-body imaging *(18)*. A major advantage of GFP-expressing tumor cells is that imaging requires no preparative procedures and are, therefore, uniquely suited for visualizing in live tissue *(10–18)*. Using stable GFP-expressing tumor cells *(10–18)*, external, noninvasive, whole-body, real-time, fluorescence optical imaging, is possible of internally growing tumors and metastases in transplanted animals *(18)*.

1.2. Whole-Body Imaging of GFP Gene Expression

Studies of gene expression in whole living animals involve spatial, as well as temporal and scalar, dimensions. The regional distribution of gene activity is of fundamental importance, as is the timing of response to physiological signals. Making such measurements in animals has been difficult. Every data point required sacrificing and dissecting the experimental animal and measuring the distribution of a reporter gene. Following a time-course in a single subject was, of course, impossible.

New techniques can visualize transgene expression noninvasively in intact animals, and promise a veritable revolution in genetic and physiological studies *(19–24)*. The methods are a significant extension of the century-old, noninvasive imaging of the internal tissues of intact animals. From Röntgen's X-rays to modern computed X-ray tomography and MRI, the static distribution of tissue mass has been visualized with ever-increasing resolution. Recent developments, such as MRI, have made possible visualizing dynamic processes. Now, noninvasive imaging has been extended to imaging the spatial distribution of transgene expression in living animals. Marker gene products have been visualized by MRI ***(20,21)***, by emitted γ rays in micropositron emission tomography ***(22)***, and single-photon emission computed tomography ***(23)***, or by luciferin fluorescence ***(24)***. Unfortunately, the procedures require complex and expensive apparatus, as well as the administration either of contrast agents or of substrates that are radioactive or fluorescent. Also, the signals are generally weak and so require long processing times, which limits detailed time-course measurements.

We have found that GFP can be used to visualize gene expression in small mammals ***(25)***. Such imaging requires only that the gene under study or its promoter be coupled to GFP. The measurements are sufficiently rapid as to allow video recording for real-time measurements ***(25)***.

2. Materials

2.1. Whole-Body Imaging Apparatus (see Note 1)

1. A Leica fluorescence stereo microscope, model LZ12, equipped with a 50 W mercury lamp, was used for high-magnification imaging. Selective excitation of GFP was produced through a D425/60 band-pass filter and 470 DCXR dichroic mirror. Emitted fluorescence was collected through a long-pass filter GG475 (Chroma Technology, Brattleboro, VT) on a Hamamatsu C5810 three-chip, thermoelectrically cooled, color charge-coupled device camera (Hamamatsu Photonics, Bridgewater, NJ).
2. Images were processed for contrast and brightness, and analyzed with the use of Image Pro Plus 3.1 software (Media Cybernetics, Silver Spring, MD).
3. Images of 1024 × 724 pixels were captured directly on an IBM PC, or continuously through video output on a high-resolution Sony VCR model SLV-R1000 (Sony, Tokyo, Japan).
4. Imaging at lower magnification, which visualized the entire animal, was carried out in a light box illuminated by blue light fiber optics (Lightools Research, Encinitas, CA) and imaged using the charge-coupled device camera described in **Subheading 1.**

2.2. Tumor Models and Gene Expression

1. 6-wk-old B57CL/6 or BALB/c *nu/nu* nude mice.
2. GFP-expressing tumor cells: Lewis lung carcinoma cells stably expressing GFP, murine melanoma B16F0-GFP.
3. Trypsin.
4. Ice-cold, serum-free modified Eagle's medium (MEM).
5. 1-mL latex-free syringe (Becton Dickinson, Franklin Lakes, NJ), 27- and 39-gauge needles.

6. GFP-expressing colon tumor fragments (1 mm^3).
7. 6-0, 7-0, and 8-0 surgical sutures.
8. Isoflurane anesthesia. Ketsel anesthesia.
9. 7X magnification microscope (Leica MZ6, Nussloch, Germany).
10. The adenovirus vector, AdCMV5GFPAE1/AE3 (vAd-GFP) (Quantum, Montreal, Canada), expressing enhanced green fluorescent protein and the ampicillin resistance gene.
11. Bone wax.
12. Sterile cotton

3. Methods

3.1. Tumor Models

3.1.1. Cell Injection

GFP-expressing cancer cells were made and isolated by growth in levels of Geneticin (G418) up to 800 µg/mL, as previously described ***(15–18)***. The selected cancer cells have a strikingly bright GFP fluorescence that remains stable in the absence of selective agent after numerous passages. There is no difference in the doubling times of parental cells and GFP-expressing cells, as determined by comparison of proliferation in monolayer culture.

Metastasis in the brain, bone, liver, pancreas, lung and lymph nodes were externally imaged by GFP expression in intact mice ***(18,31)***. External fluorescent images were acquired throughout the axial skeleton, including the skull, scapula, femur, tibia, and pelvis. A series of external fluorescence images of a tumor in the tibia were obtained from d-14 to d-25, after tail vein injection.

1. Harvest GFP-expressing tumor cells by trypsinization, using 0.25% trypsin for 3 min at 37°C.
2. Wash cells 3× with cold serum-free MEM.
3. Resuspend the cells in ~0.2 mL MEM.
4. Within 30 min of harvesting, inject 6-wk-old B57CL/6 or BALB/c *nu/nu* mice with 10^6 GFP-expressing tumor cells into the lateral tail vein, in a total volume of 0.2 mL, using a 1-mL 27G2 latex-free syringe (Becton Dickinson). External images from brain are shown in **Fig. 1**.
5. For liver expression (**Fig. 2**), cells can be injected directly into the portal vein.

3.1.2. Surgical Orthotopic Implantation

1. Perform all procedures of the operation under a 7× magnification microscope (Leica MZ6, Nussloch, Germany).
2. GFP-expressing tumor fragments (1 mm^3) are isolated by mincing tumor tissue that was growing subcutaneously in nude mice.
3. After proper exposure of the target organ, implant three tumor fragments per mouse.

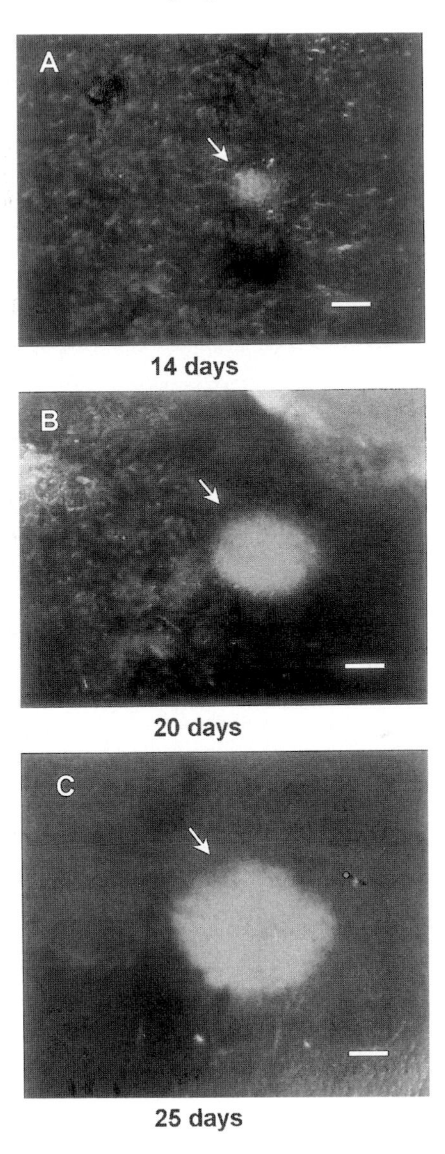

Fig. 1. External images of murine melanoma (B16F0-GFP) metastasis in brain *(**18**)*. Murine melanoma metastasis in the mouse brain were imaged by fluorescence microscopy using GFP expression, after injection of 10^6 B16F0-GFP cells in the tail vein. A clear image of a metastatic lesion in the brain can be visualized through the scalp and skull. **(A)** External image obtained of the tumor in the brain of nude mouse, d-14 after GFP tumor-cell injection. Bar = 1280 m. **(B)** Same as A, d-19 after injection. Bar = 1280 m. **(C)** Same as A and B, d-25 after injection. Bar = 1280 m. (For optimal, color representation please see accompanying CD-ROM.)

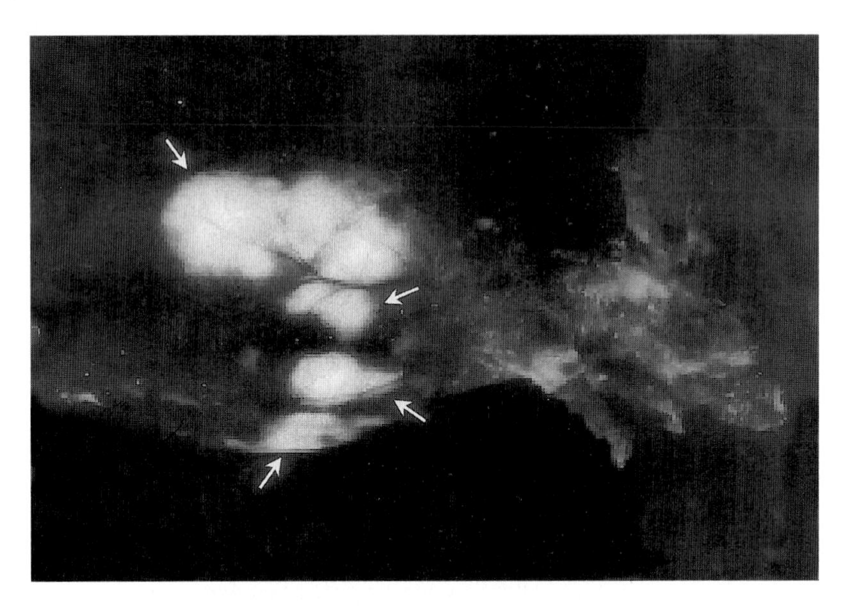

Fig. 2. External images of B16F0-GFP colonizing the liver *(18)*. Metastatic lesions of B16F0-GFP in the liver growing at a depth of 0.8 mm after portal vein injection. External image through the abdominal wall of the intact nude mouse. (For optimal, color representation please see accompanying CD-ROM.)

4. Using an 8-0 surgical suture, penetrate the tumor fragment, and suture the fragments onto the target organ.
5. Animals are kept in a barrier facility under high-efficiency particulate air filtration *(26)*.

3.1.3. Angiogenesis Model (see **Note 2**)

1. Inject Lewis lung carcinoma cells, stably expressing GFP subcutaneously, into a site of the footpad of 6-wk-old nude mice.
2. The relative transparency of the footpad reduces the scatter of green fluorescent light emitted from the tumor, and the relatively few resident blood vessels in the footpad makes it an excellent tumor transplantation site for tumor angiogenesis imaging.
3. The strong tumor cell GFP fluorescence contrasts well with the vessels, which are nonfluorescent, enabling their efficient imaging. The initiation of angiogenesis could be imaged externally when the tumor reaches approx 2 mm^2 *(31)*.

3.2. DNA Expression Models: Delivery of vAd-GFP to Various Organs for Whole-body Imaging (see Notes 3 and 4)

GFP expression in intact mice in the brain, liver, pancreas, prostate, and bone was externally imaged.

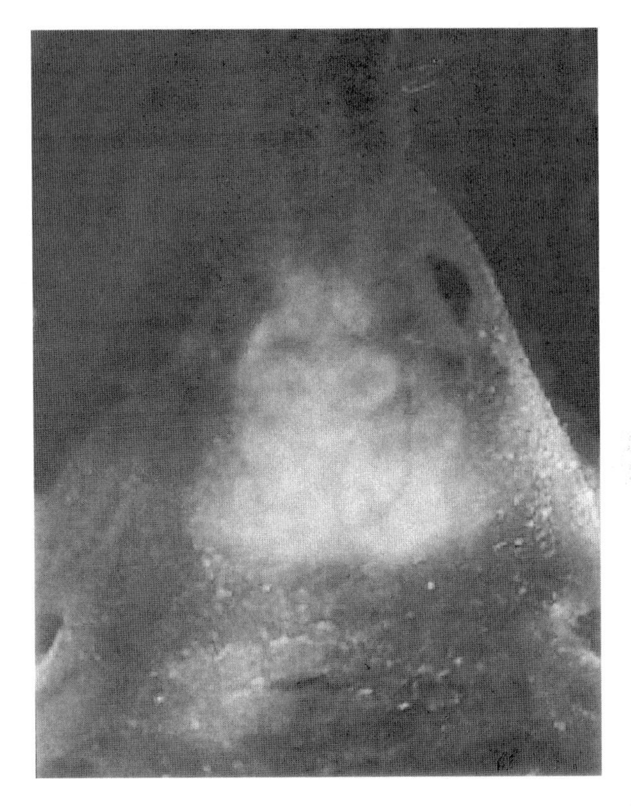

Fig. 3. External whole-body image of vAd-GFP gene expression in the brain *(25)*. An external image of vAd-GFP gene expression in the brain acquired from a nude mouse in the lightbox 24 h after gene delivery. Clear image of transgene expression in the brain can be visualized through the scalp and skull. (For optimal, color representation please see accompanying CD-ROM.)

3.2.1. Brain

1. The animals are kept under isofluorane anesthesia during surgery.
2. Following an upper midline scalp incision, expose the parietal bone of the skull.
3. Inject 20 μL recombinant adenovirus in phosphate-buffered saline (PBS), with 10% glycerol containing 8×10^{10} pfu/mL vAd-GFP/mouse, into the skull, using a 27G1/2 needle on a 1-mL latex-free syringe (Becton Dickinson).
4. Plug the puncture hole in the skull with bone wax.
5. Close the incision in the scalp with a 7-0 surgical suture in one layer.
6. Images can be collected beginning ~7 h after gene delivery (**Fig. 3**).

3.2.2. Liver

1. Keep the animals under Ketsel anesthesia during surgery.
2. The portal vein is exposed following an upper midline abdominal incision.

3. Inject 100 μL PBS with 10% glycerol containing 8×10^{10} pfu/mL vAd-GFP/mouse into the portal vein, using a 1-mL 39G1 latex-free syringe (Becton Dickinson).
4. For hemostasis, press the puncture hole in the portal vein with sterile cotton for ~10 s.
5. Close the incision in the abdominal wall with a 7-0 surgical suture in one layer.
6. All procedures of the operation described above were performed with a 7× magnification stereo microscope (Leica MZ12).

3.2.3. Pancreas

1. Keep the animals under Ketsel anesthesia during surgery.
2. All procedures of the operation described below are performed with a 7× magnification stereomicroscope.
3. Expose the pancreas following an upper midline abdominal incision.
4. Inject 100 μL PBS with 10% glycerol, containing 8×10^{10} pfu/mL vAd-GFP/mouse, into the pancreas, using a 39G1 needle on a 1-mL latex-free syringe (Becton Dickinson).
5. Press the puncture hole for ~10 s with sterile cotton for hemostasis.
6. Close the incision with a 7-0 surgical suture in one layer.

3.2.4. Prostate

1. Keep the animals under isoflurane anesthesia during surgery.
2. Expose the bladder and prostate, after making a lower midline abdominal incision.
3. Inject 30 μL PBS with 10% glycerol, containing 8×10^{10} pfu/mL vAd-GFP/mouse, into the prostate, using a 1-mL 39G1 latex-free syringe (Becton Dickinson).
4. Press the puncture hole in the prostate for ~10 s with sterile cotton for hemostasis.
5. Close the incision in the abdominal wall with a 6-0 surgical suture in one layer.
6. All procedures of the operation described above are performed with a 7× magnification stereomicroscope.

3.2.5. Bone Marrow

1. Anesthetize the animals by inhalation of isofluorane.
2. Open the skin on the hind leg with a 1-cm incision to expose the tibia.
3. A 27-gage needle with a 1-mL latex-free syringe (Becton Dickinson) is then inserted into the bone marrow cavity. Inject a total volume of 20 μL PBS with 10% glycerol (8×10^{10} pfu/mL) vAd-GFP/mouse into the bone marrow cavity.
4. Plug the puncture hole in the bone with bone wax and close the incision with a 6-0 surgical suture.

3.3. Measuring the Intensity of GFP Expression: Imaging Sensitivity and Resolution (see Note 5)

Estimating the intensity of GFP fluorescence is complicated by variations in the exciting illumination with time, and across the imaging area. These factors are corrected for by using the intrinsic red fluorescence of mouse skin as a base line to correct the increase over intrinsic green fluorescence caused by GFP *(25)*. This can be done because there is little red luminance in the GFP radiance. Consequently, the green fluorescence was calculated relative to red, based on red and green channel composition in the skin images *(25)*.

1. Produce a ratio (γ) of green to red channel emissions for each pixel in the images of skin without and with GFP.
2. Values of γ for mouse skin throughout the image in the absence of GFP should be fairly constant, varying between 0.7 and 1.0. The contribution of GFP fluorescence from within the animal increases the green component, compared to red, and is reflected in higher γ values.
3. Approximate the total amount of GFP fluorescence by multiplying the number of pixels in which the γ value was higher than 1× by the γ value of that pixel. Such a product roughly corresponds to the integral GFP fluorescence [I'_{GFP}] above the maximum value of γ for skin without GFP. The number of pixels in mouse skin images with γ value >1.0 without GFP was less than 0.02%.
4. GFP-expressing primary and metastatic lesions were considered to be externally measurable if the average fluorescence of the GFP-expressing tumor was at least 20% above the average fluorescence of the surrounding skin *(18)*. The level of background dorsal and abdominal skin fluorescence of nude mice was in a range of 6–9% of the exposed tumor fluorescence. The intensity of GFP fluorescence of a tumor (1 mm diameter) growing at a depth of ~0.8 mm was approx ~25% that of the exposed tumor. The minimum tumor size that could be imaged was a function of depth. The range of minimal size of GFP-expressing tumors that have been externally imaged thus far was from ~59 μm in diameter, at a depth of 0.5 mm, to ~1.86 mm in diameter, at a depth of 2.2 mm, in various tissues.
5. GFP transgene expression in various organs and tissues was considered to be externally measurable if the average fluorescence of the GFP-expressing organs was at least 20% above the average fluorescence of the surrounding skin *(25)*. The fluorescence intensity at maximal level of expression GFP in the liver exceeded more than 100× backdorsal and abdominal skin fluorescence. The intensity of GFP fluorescence of vAd-GFP expression in the mouse liver at a depth of 0.8 mm under the skin was approx ~25% of that of the exposed organ *(25)*.

3.4. Comparison of External Whole-Body Direct Images of GFP-Expressing Brain Metastasis

3.4.1. Comparison of Tumor Metastasis

A comparison was made between an external and direct image of a brain and metastasis of murine melanoma B16F0-GFP *(18)*.

1. Using a C57BL/6 mouse, inject 10^6 GFP-expressing B16F0-GFP tumor cells in 0.2 mL serum-free MEM into the tail vein.
2. Allow 25 d for tumor development.
3. Obtain an external fluorescent image (5.5 mm diameter and 0.8 mm depth) of B16F0-GFP cells through the scalp and skull of the mouse, on d 14, 19, and 25.
4. On d 25, after obtaining an external fluorescence image, dissect the animal, and remove the scalp and skull.
5. Record fluorescent images of the exposed brain for comparison.
6. The externally acquired images closely matched the images acquired from the open brain after the scalp and skull were removed. A series of external fluorescence images of the B16F0-GFP brain tumor in a single animal was obtained from d-14 to -25 after tail vein injection of B16F0-GFP in a nude mouse.
7. As determined by external imaging, the size of the metastatic lesion grew progressively with time (**Fig. 1**). The sizes of the tumors from external images at d 14, 19, and 25 were ~1.2, 2.25, and 3.5 mm, respectively.

3.4.2. Comparison of Viral Gene Expression in Liver

External images of vAd-GFP fluorescence from labeled mouse organs in living, intact animals were compared to the fluorescence of the organs viewed directly after sacrifice and dissection *(25)*. The fluorescence mapped the region of gene expression in the brain, liver, pancreas, prostate, and bone. The images made external to the animal appear similar to those of the exposed organs, reproducing much of the detailed structure of the direct image.

1. The simplest and most rapid method of obtaining whole-body fluorescent images of vAd-GFP gene expression is to place a freely moving mouse in a fluorescence light box (**Fig. 4**).
2. This system suggests the feasibility of high-throughput screening of agents that affect specific gene expression.

3.5. Real-Time Quantitative Whole-Body Imaging of vAd-GFP Gene Expression

Another important advantage of the GFP fluorescence assay for gene expression is its rapid data acquisition. Under the conditions used, images could be obtained at video rates, i.e., with exposure times in the order of one-thirtieth of a second *(25)*. The fluorescence from vAd-GFP gene expression in the brain of a single animal was visible within 6 h after local delivery of the vAd-GFP gene in a nude mouse, and by 24 h was very bright (**Fig. 3**). Liver fluorescence first became detectable at ~7 h after the injection of vAd-GFP into the tail vein *(25)*.

Fig. 4. External whole-body image of vAd-GFP gene expression in the liver ***(25)***. An external image of vAd-GFP gene expression accquired from a nude mouse in the light box 72 h after gene delivery. Lateral, whole-body image of transgene expression in the liver can be clearly visualized through the abdominal wall. (For optimal, color representation please see accompanying CD-ROM.)

4. Notes

1. The GFP-based fluorescent optical tumor imaging system presents many powerful features ***(18)***: Only the tumors and metastases contain the heritable GFP gene and are therefore selectively imaged with high intrinsic contrast to other tissues. GFP expression in the tumor cells is stable over long, indefinite time periods, which is the key feature allowing the quantitative imaging of tumor growth and metastasis formation, as well as their inhibition by agents of all types. The very bright GFP fluorescence enables internal tumors and metastases to be externally observed in critical organs such as colon, liver, bone, brain, pancreas, lymph nodes, and presumably breast, prostate, and so on. Blue-light illumination is the only requirement. No contrast agents, other chemical compounds, or additional treatments need to be administered to the animals.
2. Simultaneous, real-time, visual imaging of angiogenesis and tumor growth in intact animals is also enabled by establishment of human and rodent tumors that stably express high levels of GFP ***(31)***. Vessels are highly visible by their natural contrast to the GFP expression in the tumor cells. The GFP whole-body imaging technology enables the quantitative imaging of the onset and development of tumor angiogenesis, which can be applied to high-throughput, in vivo screening of antiangiogenic agents.
3. vAd-GFP delivered to various organs was induced rapidly and was stable over long time periods, allowing real-time quantitative imaging of transgene expression ***(25)***. These results indicate that gene induction and other kinetic studies can be visualized by whole-body imaging. The high intensity of GFP fluorescence

makes transgene expression externally observable from internal organs, including brain, liver, pancreas, prostate and bone and presumably in many other organs, such as breast, lymph nodes, and so on. No contrast agents, radioactive sources, or enzyme substrates need be administered to the animals; only blue-light illumination is necessary. The images can be acquired in real time, because of the strong GFP signal. The technology reported here can be applied to any gene or promoter fused or operatively linked to GFP in any organ.

4. We chose vAd-GFP as a vector, since it can transduce many normal tissues efficiently *(25)*. It was found that vAd-GFP gene is stably expressed in the brain and liver of nude mice at least for a number of months. Similar studies could be performed in transgenic animals in which GFP or other fluorescent proteins were fused or operatively linked to any gene or promoter.
5. Current sensitivity is limited, in part, by the nonoptimum spectrum of the GFP fluorescence (520 nm). At this relatively short wavelength, the emitted radiation is strongly scattered by surrounding tissue. However, powerful new techniques of using ultrafast lasers *(27)*, dual-photon imaging *(28)*, and ballistic photon imaging *(29,30)* may offer large gains in sensitivity, increased depth of detection, and spatial resolution.

References

1. Tearney, G. J., Brezinski, M. E., Bouma, B. E., et al. (1997) In vivo endoscopic optical biopsy with optical coherence tomography. *Science* **276,** 2037–2039.
2. Baum, P. R. and Brummendorf, T. H. (1998) Radioimmunolocalization of primary and metastatic breast cancer. *Q. J. Nucl. Med.* **42,** 33–42.
3. Teates, C. D. and Parekh, J. S. (1993) New radiopharmaceuticals and new applications in medicine. *Curr. Probl. Diagn. Radiol.* **22,** 229–226.
4. Dessureault, S. (1997) Pre-operative assessment of axillary lymph node status in patients with breast adenocarcinoma using intravenous ^{99}technetium mAb-170H. 82. *Breast Cancer Res. Treat.* **45,** 29–37.
5. Pasqualini, R., Koivunen, E., and Ruoslahti, R. (1997) Alpha v integrins as receptors for tumor targeting by circulating ligands. *Nat. Biotechnol.* **15,** 542–546.
6. Neri, D., Carnemolla, B., Nissim, A., et al. (1997) Targeting by affinity-matured recombinant antibody fragments on an angiogenesis associated fibronectin isoform. *Nat. Biotechnol.* **15,** 1271–1275.
7. Chambers, A. F., MacDonald, I. C., Schmidt, E. E., et al. (1995) Steps in tumor metastasis: new concepts from intravital videomicroscopy. *Cancer Metastasis Rev.* **14,** 279–301.
8. Sweeney, T. J., Mailander, V., Tucker, A. A., et al. (1999) Imaging brain structure and function, infection and gene expression in the body using light. *Proc. Natl. Acad. Sci. USA* **96,** 12,044–12,049.
9. Weissleder, R., Tung, C. H., Mahmood, U., Bogdanov, Jr., A. (1999) In vivo imaging of tumors with protease-activated near-infrared fluorescent probes. *Nat. Biotechnol.* **17,** 375–378.

10. Chishima, T., Miyagi, Y., Wang, X., et al. (1997) Cancer invasion and micrometastasis visualized in live tissue by green fluorescent protein expression. *Cancer Res.* **57,** 2042–2047.
11. Chishima, T., Miyagi, Y., Wang, X., Tan, Y., Shimada, H., Moossa, A. R., and Hoffman, R. M. . (1997) Visualization of the metastatic process by green fluorescent protein expression. *Anticancer Res.* **17,** 2377–2384.
12. Chishima, T., Miyagi, Y., Wang, X., et al. (1997) Metastatic patterns of lung cancer visualized live and in process by green fluorescence protein expression. *Clin. Exp. Metastasis* **15,** 547–552.
13. Chishima, T., Miyagi, Y., Li, L., et al. (1997) Use of histoculture and green fluorescent protein to visualize tumor cell host interaction. *In Vitro Cell Dev. Biol.-Anim.* **33,** 745–747.
14. Chishima, T., Yang, M., Miyagi, Y., et al. (1997) Governing step of metastasis visualized in vitro. *Proc. Natl. Acad. Sci. USA* **94,** 11,573–11,576.
15. Yang, M., Hasegawa, S., Jiang, P., et al. M. (1998) Widespread skeletal metastatic potential of human lung cancer revealed by green fluorescent protein expression. *Cancer Res.* **58,** 4217–4221.
16. Yang, M., Jiang, P., Sun, F.-X., et al. (1999) A fluorescent orthotopic bone metastasis model of human prostate cancer. *Cancer Res.* **59,** 781–786.
17. Yang, M., Jiang, P., An, Z., et al. (1999) Genetically fluorescent melanoma bone and organ metastasis models. *Clin. Cancer Res.* **5,** 3549–3559.
18. Yang, M., Baranov, E., Jiang, P., et al. (2000) Whole-body optical imaging of green fluorescent protein-expressing tumors and metastases. *Proc. Natl. Acad. Sci. USA* **97,** 1206–1211.
19. Herschman, H. R., MacLaren, D. C., Iyer, M., et al. (2000) Seeing is believing: noninvasive, quantitative and repetitive imaging of reporter gene expression in living animals, using positron emission tomography. *J. Neurosci. Res.* **59,** 699–705.
20. Louie, A. Y., Huber, M. M., Ahrens, E. T., et al. (2000) In vivo visualization of gene expression using magnetic resonance imaging. *Nat. Biotechnol.* **18,** 321–325.
21. Weissleder, R., Moore, A., Mahmood, U., et al. (2000) *In vivo* magnetic resonance imaging of transgene expression. *Nat. Med.* **6,** 351–354.
22. Gambhir, S. S., Barrio, J. R., Phelps, M. E., et al. (1999) Imaging adenoviral-directed reporter gene expression in living animals with positron emission tomography. *Proc. Natl. Acad. Sci. USA* **96,** 2333–2338.
23. Tjuvajev, J. G., Finn, R., Watanabe, K., et al. (1996) Noninvasive imaging of herpes virus thymidine kinase gene transfer and expressions potential method for monitoring clinical gene therapy. *Cancer Res.* **56,** 4087–4095.
24. Contag, P. R., Olomu, I. N., Stevenson, D. K., and Contag, C. H. (1998) Bioluminescent indicators in living mammals. *Nat. Med.* **4,** 245–247.
25. Yang, M., Baranov, E., Moossa, A. R., Penman, S., and Hoffman, R. M. (2000) Visualizing gene expression by whole-body fluorescence imaging. *Proc. Natl. Acad. Sci. USA* **97,** 12,278–12,282.

26. Fu, X., Besterman, J. M., Monosov, A., and Hoffman, R. M. (1991) Models of human metastatic colon cancer in nude mice orthotopically constructed by using histologically-intact patient specimens. *Proc. Natl. Acad. Sci. USA* **88,** 9345–9349.
27. Alfano, R. R., Demos, S. G., and Gayen, S. K. (1997) Advances in optical imaging of biomedical media. *Ann. NY Acad. Sci.* **820,** 248–270.
28. Masters, B. R., So, P. T., and Gratton, E. (1998) Multiphoton excitation microscopy of in vivo human skin. Functional and morphological optical biopsy based on three-dimensional imaging, lifetime measurements and fluorescence spectroscopy. *Ann. NY Acad. Sci.* **838,** 58–67.
29. Wu, J., Perelman, L., Dasari, R., and Feld, M. (1997) Fluorescence tomographic imaging in turbid media using early-arriving photons and Laplace transforms. *Proc. Natl. Acad. Sci. USA* **94,** 8783–8788.
30. Alfano, R. R., Demos, S. G., Galland, P., et al. (1998) Time-resolved and nonlinear optical imaging for medical applications. *Ann. NY Acad. Sci.* **838,** 14–28.
31. Yang, M., Baranov, E., Li, X.-M., et al. (2001) Whole-body and intra-vital optical imaging of angiogenesis in orthotopically implanted tumors. *Proc. Natl. Acad. Sci. USA* **98,** 2616–2621.
32. Bouvet, M., Yang, M., Nardin, S., et al. (2000) Chronologically-specific metastatic targeting models. *Clin. Experim. Metastat.* **18,** 213–218.

III

Green Fluorescent Protein to Monitor Protein Distribution and Trafficking

11

Drug-Induced Translocation of Nucleolar Proteins Fused to Green Fluorescent Protein

Benigno C. Valdez and Laszlo Perlaky

1. Introduction

Protein translocation in a cell is a dynamic process. Proteins move from one part of the cell to another in a nonrandom fashion, which involves transport receptors and adapters *(1,2)*. Such protein movement is regulated during cell growth, development, and apoptosis *(1,3)*. Protein translocation may also occur in response to drugs, viral infection, or abnormal cellular metabolisms.

Subcellular redistribution of proteins may be used as an assay for screening and developing cytotoxic drugs. For example, topotecan, a drug used to treat anaplastic astrocytoma, causes translocation of DNA topoisomerase I from the nucleolus to the nucleoplasm *(4)*. Cytotoxic drugs cause redistribution of other nucleolar proteins including nucleophosmin/B23 *(5)*, poly-adenosinediphosphate-ribose polymerase *(6)*, and RNA helicase II/Gu *(7)*.

There is a need for an efficient, economical, and real-time assay to screen for cytotoxic drugs that cause protein translocation. One approach is to tag a protein that is known to redistribute and observe its localization in the presence of a drug. We recently applied this methodology to tag RNA helicase II/Gu with green fluorescent protein and monitor its translocation from the nucleolus to the nucleoplasm in the presence of cytotoxic drugs *(8)*. This procedure is described below, and a similar method may be applied to other nucleolar proteins, such as nucleophosmin/B23 *(9)* and nucleolin/C23 *(10)*.

From: *Methods in Molecular Biology, vol. 183: Green Fluorescent Protein: Applications and Protocols*
Edited by: B. W. Hicks © Humana Press Inc., Totowa, NJ

2. Materials

2.1. Preparation of GFP-Expression Constructs

1. Expression vectors and cDNA:
   a. pEGFP-C1 vector (Clontech, Palo Alto, CA).
   b. cDNA encoding for nucleolar protein RNA helicase II/Gu, nucleolin/C23 (both available from B. Valdez), or nucleophosmin/B23 (available from P. K. Chan, Department of Pharmacology, Baylor College of Medicine, Houston, TX).
2. Polymerase chain reaction (PCR) primers:
   a. To amplify RNA helicase II/Gu cDNA:
5'-ATTCGCGGCCATGGGATCCGCGGTTGAGAAGACCGGT-3' (GU-5' has *Bam*HI site).
5'-AACATCATTCTCGAGTTCTATATAAATCTTCT-3' (GU-3' has *Xho*I site).
   b. To amplify nucleophosmin/B23 cDNA:
5'-CCGATGGAAGGATCCATGGACATGGACATGAGCCC-3' (B23-5' has *Bam*HI site).
5'-TAACAAATTGTGTCGACTATTTTCTTAAAGAGACT-3' (B23-3' has *Sal*I site).
   c. To amplify nucleolin/C23 cDNA:
5'-GCCGCCATCAGATCTAAGCTCGCGAAGGCAGGTAAA-3' (C23-5' has *Bgl*II site).
5'-AGGGAAAGCAGGTCGACAGAAGCTATTCAAACTTC-3' (C23-3' has *Sal*I site).
3. GeneAmp PCR kit (Perkin Elmer, Foster City, CA); GeneAmp DNA Thermal Cycler 480 (Perkin Elmer).
4. Restriction enzymes and digest buffers: *Bam*HI, *Bgl*II, *Xho*I, *Sal*I (Invitrogen™ Life Technologies, Carlsbad, CA, or other companies).
5. T4 DNA ligase with 5X ligation buffer (Invitrogen™ Life Technologies).
6. Bacterial cell line: We routinely use subcloning efficiency competent DH5α *Escherichia coli* cells (Invitrogen™ Life Technologies) for cloning purposes.
7. Selection media for bacteria: Premixed Luria-Bertani (LB) agar or LB broth base (Invitrogen™ Life Technologies) sterilized, and kanamycin added to a final concentration of 50 μg/mL.
8. DNA isolation: QIAprep Spin Miniprep kit from Qiagen, Valencia, CA (*see* **Note 1**).
9. 1% Agarose gel.
10. 10X TBE (1 L): 108 g Tris, 55 g boric acid, 9.3 g EDTA.
11. Agarose gel loading buffer: 38% sucrose, 0.1% bromophenol blue, 67 m*M* EDTA.
12. QIAquick gel extraction kit (Qiagen).

2.2. Transfection of Human Cell Line

1. The cell line we used was the human osteogenic sarcoma cell line (U-2 OS, HTB 96, American Type Culture Collection [ATCC], Rockville, MD) (*see* **Note 2**).
2. Complete cell culture media: McCoy's 5A modified medium (Invitrogen™ Life Technologies) complete with 10% fetal bovine serum (Sigma) and 1% penicillin-streptomycin solution (Invitrogen™ Life Technologies) (*see* **Note 3**). The culture medium for selection of stable clones contains the complete cell culture media, along with 500 μg/mL active geneticin (antibiotic G418, Invitrogen™ Life Technologies) (*see* **Note 4**).

3. Kit for transfection: Lipofectin® Reagent from Invitrogen™ Life Technologies (*see* **Note 3**); Lipofectin and DNA are diluted into phosphate buffered saline (PBS) (1 L): 0.2 g KCl, 8.0 g NaCl, 0.2 g KH_2PO_4, 1.15 g Na_2HPO_4, pH 7.4.
4. 0.25% Trypsin-EDTA in Hank's balanced salt solution (HBSS), without Ca^{2+} and Mg^{2+} (Invitrogen™ Life Technologies).
5. Opti-MEM-I Reduced Serum Medium (Invitrogen™ Life Technologies).

2.3. Drugs

1. Stock solution (1000×) of 0.2 mg/mL actinomycin D (Sigma) dissolved in absolute ethyl alcohol and stored at 4°C in a dark container.
2. Stock solution (200×) of 10 m*M* toyocamycin dissolved in dimethyl sulfoxide, and stored at 4°C in a dark container. Toyocamycin is available from the Drug Synthesis Branch of the National Cancer Institute, Bethesda, MD.

2.4. Microscopic Analysis

1. Microscope: Nikon Diaphot TMD-EF2 inverted, reflected-light phase contrast, fluorescence microscope (*see* **Note 5**).
2. Filters: Nikon B-2E that matches the microscope, and supplied by the manufacturer of the microscope (*see* **Note 6**).
3. Camera for digital fluorescence imaging: high-sensitivity charge-coupled device (CCD) camera (COHU model no. 4915, San Diego, CA) with a frame storage (Colorado Video model no. 440, Boulder, CO) and a LG-3 frame grabber (Scion, Frederick, MD) (*see* **Note 7**).
4. Software for quantitative analysis of translocation: NIH Image Ver. 1.55 (NIH Image was written by Wayne Rasband at the US National Institutes of Health and is available on a floppy disk from National Technical Information Service, 5285 Port Royal Rd., Springfield, VA 22161; http://rsb.info.nih.gov/nih-image; *see* **Note 8**).

3. Methods

3.1. Preparation of GFP-Expression Constructs

1. Amplify RNA helicase II/Gu cDNA, using the GeneAmp kit, by mixing 20 µL 10X PCR buffer, 20 µL 25 m*M* $MgCl_2$, 4 µL of each 10 m*M* deoxyribonucleoside triphosphate, 200 pmol Gu-5' primer, 200 pmol of Gu-3' primer, and 10 ng pGEX-Gu ***(11)*** template in a GeneAmp thin-walled reaction tube. Complete the volume to 198 µL with water. Mix the contents of the tube. Briefly centrifuge the tube and add 2 µL 5 U/µL AmpliTaq DNA polymerase. Mix the contents and briefly centrifuge before adding two drops of mineral oil (*see* **Note 9**). After heating for 5 min at 95°C, perform the PCR amplification for 35 cycles in a GeneAmp DNA Thermal Cycler as follows: 1 min at 95°C, 1.5 min at 50°C, and 2.5 min at 72°C. Extend the polymerization during the last cycle to 10 min at 72°C, then soak the reaction mixture at 4°C.
2. Analyze 10 µL of the PCR product on a 1 % agarose gel to check the amplification of a 2400-bp DNA. Purify the remainder of PCR product using a QIAQuick

PCR purification kit, and digest with *Bam*HI and *Xho*I in reaction 2 buffer at 37°C for at least 2 h. Similarly, digest 5 µg pEGFP-C1 vector DNA with *Bgl*II and *Sal*I in reaction 2 buffer at 37°C for at least 2 h. *Bam*HI and *Bgl*II produce compatible cohesive ends; *Xho*I and *Sal*I ends are also compatible.

3. Add loading buffer to the digestion mixtures and load them onto a 1% agarose gel for electrophoresis.
4. Excise the proper DNA bands and purify with a QIAQuick gel extraction kit. Elute the DNA from the column with water instead of TE or Tris-HCl buffer. Determine the concentration of the DNA (*see* **Note 10**).
5. Incubate 50 ng linearized pEGFP-C1 DNA with 50 ng of the purified digested RNA helicase II/Gu cDNA (1:2 molar ratio) in 10 µL ligation buffer in the presence of 1 U T4 DNA ligase overnight at 16°C. Transform 100 µL DH5α *E. coli* cells with 3 µL ligation mixture on ice for 30 min. Heat at 42°C for 1 min, and incubate on ice for 2 min. Add 1 mL LB and shake at 37°C for 1 h. Briefly centrifuge to pellet the cells and decant the supernatant leaving approx 100 µL LB. Resuspend the cells and spread them on LB-agar plates containing 50 µg/mL kanamycin. Incubate the plates overnight at 37°C.
6. Pick a couple of colonies and grow them at 37°C in LB medium containing 50 µg/mL kanamycin overnight. Purify the constructed plasmid, using a QIAprep Spin Miniprep kit, and digest the purified DNA with *Bgl*II and *Bam*HI. Positive clones should contain a 1.4-kb *Bgl*II-*Bam*HI DNA fragment (*see* **Note 11**).

3.2. Stable Transfection of Human Cell Line

1. Grow a monolayer of human osteogenic sarcoma cells (U-2 OS 96-HTB) in complete McCoy's 5A modified medium, with 10% fetal bovine serum and antibiotics, in a 75-cm^2 Corning T75 vent-cap flask at 37°C, in a humidified water-jacketed incubator containing 5% CO_2 (*see* **Note 12**). The doubling time of U-2 OS cells is 36 h, when grown as monolayer. Cells should be passed at a 1:5 dilution weekly.
2. 2 d prior to transfection, seed 1×10^6 U-2 OS cells in a 10 cm tissue culture dish. Prepare three plates, one each for the negative control, the pEGFP-C1 vector, and the pEGFP-Gu construct.
3. Grow the cells until they are 40–50% confluent, which usually takes 24–48 h. The exact time will vary with cell type.
4. 2 h prior to transfection, change the medium with a fresh complete medium, and continue incubating the cells.
5. Transfect the cells, using Lipofectin Reagent according to the protocol for stable transfection of adherent cells supplied by Invitrogen™ Life Technologies (*see* **Note 3**).
6. Use polystyrene 6-well dish to prepare the DNA–Lipofectin Reagent complex.
7. For each transfection prepare A and B solutions in one well of a 6-well plate. Solution A: Dilute 10 µg DNA into 50 µL PBS in one well of a 6-well plate. Solution B: Dilute in another well, 50 µL Lipofectin Reagent in 50 µL PBS.

8. For each transfection, transfer solution B into solution A and mix well by pipeting up and down.
9. Incubate the mixture at slightly tilted position in a moist chamber at 37°C for 15 min, to form the DNA–Lipofectin Reagent (cationic lipid) complex.
10. While complexes are forming, take the cells out of the incubator, and aspirate the complete medium. Wash the cells twice with 10 mL Opti-MEM-I (*see* **Note 13**).
11. Remove the polystyrene 6-well dish, containing the DNA- Lipofectin Reagent complexes, from the moist chamber, and add 5 mL Opti-MEM-I for each complex. Mix gently, and overlay onto the previously washed cells. Make sure the transfection medium completely covers the cells.
12. Incubate the cells for 4 h at 37°C in a humidified CO_2 incubator.
13. Aspirate the transfection medium, and replace with 10 mL complete culture medium. Incubate the cells at 37°C in a humidified CO_2 incubator for 48 h, before selecting the transfected clones.

3.3. Geneticin Selection and Cloning of the Colonies

1. Check the cells for cell growth and for GFP fusion protein expression, under a Diaphot TMD inverted fluorescence microscope (Nikon, Melville, NY) with an HBO100 mercury short arc lamp and a filter B-2E with excitation wavelength of 450–490 nm and a barrier filter of 520–560 nm. The maximum emission wavelength for GFP is 507 nm.
2. Prepare the selection culture medium, which is a complete cell culture medium including 500 μg/mL active geneticin (*see* **Note 4**). Replace the complete medium with 10 mL selection medium. Include a Lipofectin Reagent alone sham transfected plate, as a negative control for the antibiotic selection process.
3. Incubate the cells at 37°C in a humidified CO_2 incubator for approx 2 wk, but feed the cells with 10 mL fresh selection medium every 3 d.
4. Look for the appearance of microcolonies (8–16 cells, 3–4 cell divisions) and colonies (more than 32 cells, more than 5 cell divisions).
5. Examine the colonies for fluorescence of the GFP fusion protein and mark them. Continue growing the cells at 37°C in a humidified CO_2 incubator, until the positive fluorescent colonies reach a minimum of 128 cells.
6. Remove the selection medium, leaving just enough so the cells will not dry.
7. Dip the thicker edge of a 5-mm cloning cylinder (Scienceware, Pequannock, NJ) into sterile silicone grease (*see* **Note 14**).
8. Apply the cloning cylinder around the selected clone by pressing it gently against the Petri dish using sterile forceps. Select at least 12 colonies from each plate.
9. Place a couple of drops of serum-free medium into the cloning cylinder to wash the colony, remove the medium, and add a drop or two of 0.25% Trypsin-EDTA in HBSS, without Ca^{2+} and Mg^{2+}, to each cloning cylinder.
10. Incubate the plate at 37°C for 5 min.
11. Add a few drops of complete medium into the cloning cylinders, suspend the cells using a pipet, and transfer the cells into an 24-well cell culture dish. Trans-

fer each colony into a separate well. Add 1 mL selection medium per well and grow the cells at 37°C in a humidified CO_2 incubator.

12. Change the selection medium every 3 d, and examine the cells for fluorescence of the GFP fusion protein. Mark the clones that show a homogeneous fluorescence (i.e., nucleolar fluorescence for GFP-RNA helicase II/Gu, and cytoplasmic fluorescence for GFP). Remove the medium and add 0.5 mL 0.25% trypsin into the marked colonies and keep at room temperature for 30 s. Carefully remove the trypsin solution without disturbing the cells. Incubate the trypsin-treated colonies at 37°C for 3–4 min, or until the cells no longer adhere onto the dish. Resuspend the cells in 3 mL selection medium and grow them in a 6-well dish. Expand the cells later, using a similar procedure into 10-cm dishes. At this time, 0.5–1 × 10^6 cells may be frozen in 2.5% dimethyl sulfoxide + 40% fetal bovine serum containing complete medium, and stored in liquid nitrogen. The remaining cells need to be serially plated and checked for expression of GFP fusion protein.

3.4. Drug Treatment of Transfected Cells

1. Using a 6-well dish, seed 2–4 × 10^5 cells/well of logarithmically growing stable clones.
2. Continue growing the cells until they are 50% confluent, which may take 1–2 d.
3. Change to a fresh medium 2 h prior to drug treatment.
4. Prepare a complete medium containing 0.2 μg/mL actinomycin D or 50 μ*M* toyocamycin.
5. Exchange the growth medium for a drug-containing medium, and incubate the cells for 2 h at 37°C. Be sure to include cells not treated with drugs. At this time, the cells can be observed for translocation of GFP-RNA helicase II/Gu fusion protein from the nucleolus to the nucleoplasm as described below.
6. Cells may be allowed to recover by rinsing out the drug with a fresh complete medium and incubating them at 37°C in 5% CO_2 for 24–48 h. Examine the cells and observe the relocalization of the GFP-RNA helicase II/Gu fusion protein from the nucleoplasm to the nucleoli.

3.5. Microscopic Analysis

3.5.1. Qualitative Microscopic Analysis

Figure 1 shows examples of phase contrast images (**A** and **C**), nucleolar fluorescence (**B**), and nucleoplasmic fluorescence induced by toyocamycin (**D**).

1. We normally use a Diaphot TMD inverted fluorescence microscope (Nikon), with an HBO100 mercury short arc lamp and a filter B-2E with excitation wavelength of 450–490 nm and a barrier filter of 520–560 nm for observation of the cells for fluorescence distribution (*see* **Notes 5–8**). The maximum emission wavelength for GFP is 507 nm.
2. Living cells, in tissue culture plates that are either untreated or drug-treated, are viewed with long-working distance 40× or 64× phase contrast FL lenses. After

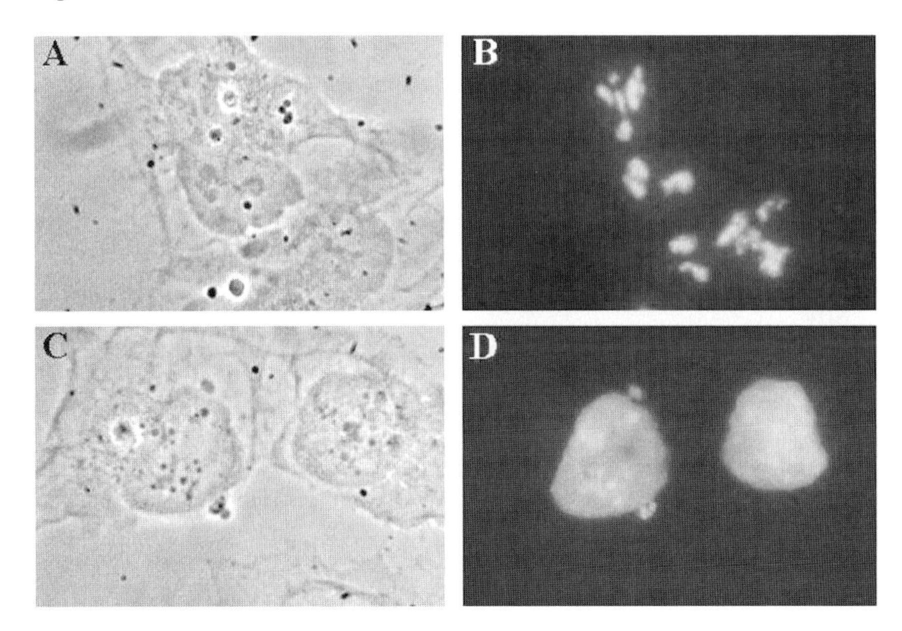

Fig. 1. Subcellular localization of GFP-RNA helicase II/Gu. Untreated U-2 OS cells expressed GFP-RNA helicase II/Gu fusion protein which localized to the nucleoli (**A**: phase; **B**: fluorescence). Treatment of the cells with 50 μ*M* toyocamycin for 2 h resulted in translocation of the fusion protein to the nucleoplasm (**C**: phase; **D**: fluorescence). (For optimal, color representation please see accompanying CD-ROM.)

obtaining sharp phase-contrast images, open the fluorescent shutter and switch off the visible light to monitor GFP fusion protein distribution.

3. Examine the control plates for expression of GFP-RNA helicase II/Gu fusion protein in the nucleoli and the drug-treated plates for translocation of the GFP-fusion protein to the nucleoplasm.
4. Record the observations and capture 10–20 digital images of cells for objective quantitative analysis.

3.5.2. Quantitative Microscopic Analysis

Figure 2 shows examples of three-dimensional plots for translocation of GFP-RNA helicase II/Gu from the nucleolus to the nucleoplasm.

1. The captured and stored images for the drug-induced translocation or relocalization of GFP-RNA helicase II/Gu fusion protein, may be quantitatively analyzed according to the previous work of Dr. P. K. Chan ***(12)***, using the NIH Image software. The full gray scale is 256 U; complete darkness at 0, to the brightest at 255.
2. Measure the fluorescence intensity across a cell image, and generate a three-dimensional plot on the computer screen, similar to **Fig. 2**.

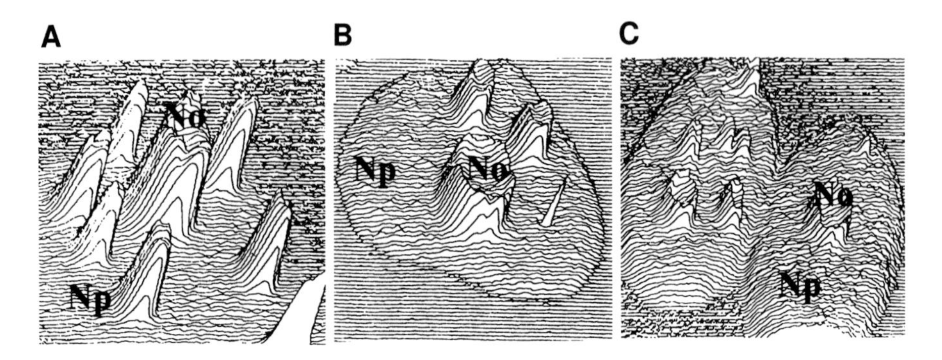

Fig. 2. Three-dimensional plot of the fluorescence intensity showing the drug-induced translocation of GFP-RNA helicase II/Gu in stably transfected U-2 OS cells. **(A)** Before addition of drug, GFP-RNA helicase II/Gu fusion protein was in the nucleoli as shown by the small peaks. **(B)** Cells were treated with 0.2 µg/mL actinomycin D for 30 min, which resulted in a partial translocation; the nucleolar fluorescence decreased and the nucleoplasmic fluorescence increased. **(C)** After 2 h, the nucleoplasmic fluorescence became more intense and the nucleolar fluorescence decreased. A residual fluorescence was visible at the periphery of the nucleoli. No, nucleolus; Np, nucleoplasm. (Also on CD-ROM.)

3. Measure the average fluorescence intensity in the nucleoli and in the nucleoplasm for each individual cell.
4. Calculate the ratio of nucleolar over nucleoplasmic fluorescence intensity. This ratio is called the "localization index" (LI) (*see* **Note 15**). The localization index for untreated control is between 9 and 11, which indicates that the GFP-RNA helicase II/Gu fusion protein localizes to the nucleoli **(Fig. 2A)**. The value of the localization index decreases as nucleoplasmic fluorescence increases relative to the nucleolar fluorescence **(Fig. 2B)**. A localization index 1–5 shows a strong nucleoplasmic translocation **(Fig. 2C)**.

4. Notes

1. We have used miniprep kits from other companies to prepare plasmid DNA for transfection of mammalian cells but the result was either cell death or low transfection efficiency.
2. We used HeLa S3 (ATCC CCL 2.2, human epitheloid cervix carcinoma) cells but got variable expression of the recombinant protein. Some selected, stable HeLa clones maintained resistance to geneticin after several passages, but they later lost expression of the recombinant protein.
3. We used Lipofectin Reagent in previous experiments. Since then, Lipofectamine Plus (Invitrogen™ Life Technologies), Transfectace™ (Invitrogen™ Life Technologies) and SuperFect Transfection Reagent (Qiagen) have been used, all work well for transient and stable transfection of adherent cells. However, the increased

transfection efficiency sometimes results in increased cytotoxicity of the liposomes. It is recommended testing the cytotoxicity of liposome using colony-forming assay or another methods before large scale transfections. The optimized nontoxic and effective concentration of Lipofectin Reagent for U-2 OS cells is 10 µg/mL (10 µL 1 mg/mL stock solution added to 1 mL medium) in Opti-MEM-I (Invitrogen™ Life Technologies) for 4 h at 37°C in a humidified CO_2 incubator.

4. Geneticin (antibiotic G418) cytotoxicity differs among cell lines, e.g., the active geneticin concentration for HeLa S3 cells is 900 µg/mL. The optimum active geneticin concentration for 99% cell kill of untransfected U-2 OS cells is 500 µg/mL. Always plate a liposome-only transfected control to test the cytotoxicity of geneticin.
5. We previously used a Nikon Diaphot TMD-EF2 reflected-light fluorescence microscope. Since then we have used a Nikon Diaphot TE-300 inverted-phase contrast fluorescence research microscope as well as other brands (Zeiss, Leica, Olympus) with similar specifications. All work well for visualizing the green fluorescent fusion protein expression and localization.
6. The Nikon Diaphot TMD-EF2 is equipped for fluorescence microscopy with a Nikon epifluorescence filter block B-2E, with excitation wavelength of 450–490 nm, a barrier filter of 520–560 nm, and a dichroic mirror of 510 nm.
7. Fluorescence images are best captured by a CCD camera (COHU model no. 4915), if quantitative analysis is to be performed. The authors' camera is equipped with frame storage (Colorado Video model 440A), and images are ported to a Macintosh computer using an LG-3 frame grabber (Scion). We also used a Nikon Diaphot TE-300 microscope equipped with Coolsnap (RS Photometrics, Tucson, AZ) high-quality CCD, with 12-bit scientific digitizer, and low noise electronics, to produce 36 bit, digital color images with > 1 megapixel resolution.
8. We have used NIH Image version 1.55. Image requires a Macintosh computer with at least 16 MB memory, 24-bit color. Ours is supported with a large monitor, flatbed scanner, film recorder, PostScript laser printer, photo typesetters, and color dye sublimation color printer. We have also used Coolsnap camera equipped with PCI card for Power Macintosh or PCI card for Windows-based Pentium III computers. Photoshop (Adobe) can also be used to manipulate the images, and analysis can also be performed using a SigmaScan Pro (Jandel) or comparable software.
9. Keep all reagents on ice to prevent premature nonspecific amplification.
10. If the concentration of the DNA is less than 5 ng/µL, concentrate it using a Speed Vac concentrator. This explains why we prefer to elute DNA from the column with water.
11. RNA helicase II/Gu cDNA has a *Bgl*II site at nucleotide 1038, and pEGFP-C1 vector has a *Bam*HI site at nucleotide 1390. The constructs must be sequenced to ensure that the insert is in frame with the GFP and no mutation is introduced by the PCR. EGFP-N and EGFP-C sequencing primers, from Clontech, may be used to sequence the N-terminal and C-terminal gene fusions, respectively. We usually request Seqwright (Houston, TX) to sequence our constructs.
12. All cell culture media and trypsin solution should be equilibrated at 37°C water bath at least 30 min prior to use.

13. The Opti-MEM-I is a serum-free medium. Do not use any serum-containing medium during transfection. The serum reduces or blocks the transfection efficiency. Different serum-free media or even PBS alone can also be used to wash the cells.
14. Cloning cylinders and silicone grease may be put in glass beakers and autoclaved at 15 psi for 20 min.
15. In order to reduce variability and to establish the same basis for fluorescent measurements, take pictures using the same conditions for each experiment and correct the measured data accordingly, such as time of exposure to the excitation light source, the gain setting in the camera, the numbers of frames collected, and internal positive and negative controls.

Acknowledgment

This work was supported by grant DK52341 from the National Institute of Diabetes and Digestive and Kidney Diseases to Benigno C.Valdez.

References

1. Moroianu, J. (1999) Nuclear import and export: transport factors, mechanisms and regulation. *Crit. Rev. Eukaryot. Gene Expression* **9,** 89–106.
2. Bauer, M. F., Hofmann, S., Neupert, W., and Brunner, M. (2000) Protein translocation into mitochondria: the role of TIM complexes. *Trends Cell Biol.* **10,** 25–31.
3. Porter, A. G. (1999) Protein translocation in apoptosis. *Trends Cell Biol.* **10,** 394–401.
4. Danks, M. K., Garret, K. E., Marion, R. C., and Whipple, D. O. (1996) Subcellular redistribution of DNA topoisomerase I in anaplastic astrocytoma cells treated with topotecan. *Cancer Res.* **56,** 1664–1673.
5. Chan, P.-K. (1992) Characterization and cellular localization of nucleophosmin/B23 in Hela cells treated with cytotoxic agents (studies of B23-translocation mechanism). *Exp. Cell Res.* **203,** 174–181.
6. Desnoyers, S., Kaufmann, S. H., and Poirier, G. G. (1996). Alteration of the nucleolar localization of poly(ADP-ribose) polymerase upon treatment with transcription inhibitors. *Exp. Cell Res.* **227,** 146–153.
7. Perlaky, L., Valdez, B. C., and Busch, H. (1997). Effects of cytotoxic drugs on translocation of nucleolar RNA helicase RH-II/Gu. *Exp. Cell Res.* **235,** 413–420.
8. Valdez, B. C., Perlaky, L., Cai, Z.-J., Henning, D., and Busch, H. (1998) Green fluorescent protein tag for studies of drug-induced translocation of nucleolar protein RH-II/Gu. *Biotechniques* **24,** 1032–1036.
9. Chan, W.-Y., Liu, Q.-R., Borjigin, J., Busch, H., Rennert, O. W., Tease, L. A., and Chan, P.-K. (1989) Characterization of the cDNA encoding human nucleophosmin and studies of its role in normal and abnormal growth. *Biochemistry* **28,** 1033–1039.
10. Srivastava, M., Fleming, P. J., Pollard, H. B., and Burns, A.L. (1989) Cloning and sequencing of the human nucleolin cDNA. *FEBS Lett.* **250,** 99–105.

11. Valdez, B. C., Henning, D., Busch, R. K., Woods, K., Flores-Rozas, H., Hurwitz, J., Perlaky, L., and Busch, H. (1996) A nucleolar RNA helicase recognized by autoimmune antibodies from a patient with watermelon stomach disease. *Nucl. Acids Res.* **24,** 1220–1224.
12. Chan, P.K., Qi, Y., AmLey, J., and Koller, C.A. (1996) Quantitation of the nucleophosmin/B23-translocation using imaging analysis. *Cancer Lett.* **100,** 191–197.

12

Light-Induced Nuclear Targeting of PhytochromeB–sGreen Fluorescent Protein in Plants

Akira Nagatani and Tomonao Matsushita

1. Introduction

Phytochrome is a ubiquitous plant photoreceptor, that regulates various aspects of plant development *(1)*. Phytochrome is a soluble chromoprotein of ~120 kDa, to which a linear tetrapyrrole chromophore is covalently attached. The holoprotein undergoes photoreversible conversion between two spectrally distinct forms, the far-red light-absorbing form (Pfr) and the red light-absorbing form (Pr). Of these two forms, Pfr is biologically active. Red light induces phototransformation of inactive Pr to active Pfr. Conversely, far-red light converts Pfr back to Pr. In this way, phytochrome acts as a molecular switch.

Green Fluorescent Protein (GFP) is widely used as a cytological fusion tag to study intracellular distribution of proteins of interest. Fusion GFPs have been targeted successfully to practically every major organelle of plant cells *(2)*. More recently, random GFP::cDNA fusions have been expressed, to identify various subcellular structures in plants *(3)*. The authors constructed a cDNA library in which cDNAs were inserted at the 3' end of the GFP coding sequence. *Arabidopsis* was transformed with this library in a large scale. The resulting transgenic plants exhibited various subcellular localization patterns of GFP at high frequency.

Because immunochemical analyses indicate that the major molecular species of phytochrome reside in the cytoplasm in darkness, it had long been thought that the phytochrome action takes place in the cytoplasm *(4,5)*. However, fusion proteins consisting of β-glucuronidase and C-terminal fragments of phytochrome have been found to reside in the nucleus *(6)*. To confirm this,

From: *Methods in Molecular Biology, vol. 183: Green Fluorescent Protein: Applications and Protocols*
Edited by: B. W. Hicks © Humana Press Inc., Totowa, NJ

fusion proteins consisting of phytochromes and GFP were expressed in transgenic plants *(7,8)*.

A major molecular species of phytochrome, phyB, was fused to sGFP(S65T) and expressed in the *phyB*-deficient mutant of *Arabidopsis* ***(7)***. Physiological analysis of the resulting transgenic plants indicated that the introduced protein is biologically active. Immunoblot analysis confirmed that a fusion protein of the predicted size was accumulated. As expected, the phyB–sGFP fusion protein was detected in the nucleus in the light. The fluorescence exhibited a speckled distribution within the nucleus. In contrast, the fusion protein was evenly distributed throughout the cell in darkness. Hence, it is suggested that phyB translocates to specific sites within the nucleus upon photoreceptor activation. A similar approach has been taken to elucidate the intracellular distribution of phyB and another molecular species, phytochrome A (phyA), in transgenic tobacco ***(8,9)***. The results indicate that not only phyB but also phyA accumulate in the nucleus under certain light conditions. Hence, the nucleus is highlighted as the site of phytochrome action.

In the above experiments, transgenic plants were prepared to examine intracellular distribution of phy–sGFP. However, preparation of transgenic plants is time-consuming and laborious. One alternative approach is transient expression of the fusion protein. A particle-delivery system, in which plant cells are bombarded with micro-particles coated with DNA, is widely used for this purpose. The expression of introduced protein can be observed as early as 2–24 h after the bombardment. Using this method, the authors have successfully expressed phyB-sGFP in onion and *Arabidopsis* epidermal cells (unpublished). We herein describe detailed methods for the expression and observation of the phyB–sGFP fusion protein in plant cells.

2. Materials

1. Details of construction of the *phyB–sGFP* fusion gene is described elsewhere ***(7)***. The *phyB–sGFP* chimeric cassette was inserted between the constitutive cauliflower mosaic virus 35S promoter ***(10)*** and the Nos terminator of an *Agrobacterium* transformation vector. This vector was used both for the stable transformation and transient expression. An oligoamino acid sequence (GGGGIDKLDP) was inserted between the *phyB* and *sGFP* sequences to avoid possible interference between the two proteins. However, this appears to be unnecessary because other groups have reported that phyA and phyB directly fused to the GFP sequence are biologically active ***(8,9)***.
2. *Escherichia coli* DH5α containing the phyB–sGFP plasmid DNA.
3. Onion bulbs for transient expression were purchased from a local market. Bulbs freshly harvested gave higher expression. The bulbs, once purchased, can be stored in a dry, cool, and dark place for at least a few weeks.
4. 1.5% agar plates containing Murashige-Skoog salt mixture without sucrose.

5. Particle delivery system Biolistic PDS-1000/He (Bio-Rad, Hercules, CA).
6. Tungsten particles of ~1.1 µm diameter (Tungsten M17; Bio-Rad).
7. Gold particles of ~1.0 µm diameter (1.0 Micron Gold, Bio-Rad).
8. QIAfilter Plasmid Maxi or Midi Kit (Qiagen, Hilden, Germany).
9. Rupture disks 1350 (Bio-Rad).
10. *Arabidopsis thaliana*. For transient expression of phyB-sGFP in hypocotyl cells, seeds of *A. thaliana* were sown on 1.5% agar plates containing Murashige-Skoog salt mixture without sucrose. Seeds were sown along a circle of ~3 cm diameter at the center of a 9 cm plate (**Fig. 1**). The transformation efficiency is highest at this position. After the cold treatment of 3–4 d at 4°C, plates were placed vertically under light from white fluorescent tubes in a growth chamber at 23°C (**Fig. 1**). This arrangement allows the seedlings to grow along the surface of the plate. Seeds were grown for 2 d then subjected to treatment with a particle delivery system (*see* **Note 1**).
11. Fixative solution: 3.7% formaldehyde in 0.1 *M* PIPES, 5 m*M* EGTA, 5 m*M* $MgCl_2$, and 1 *M* sorbitol.
12. Olympus BX60 microscope (Olympus, Tokyo) equipped with X20, X40, and X100 objectives, differential interference contrast (DIC) optics and a 100-W mercury arc light source. Ultraviolet (U-MWU) or fluorescein isothiocyanate (U-MNIBA) filter sets were used (Olympus).
13. Inverted laser-scanning confocal microscope (Model LSM410 invert, Carl Zeiss Jena, Jena,Switzerland) equipped with X40 and X63 objectives was used for confocal microscopy. The laser scan images were obtained with a combination of 488 nm laser excitation and 515-nm long pass emission filter (LP515, Carl Zeiss Jena).
14. 5% Low-melting-point agarose (Seaplaque, FML BioProducts, Rockland, ME).
15. Vibrating blade microtome (Microslicer DTK-1000, Dousaka EM, Kyoto, Japan).

3. Methods

Preparation of transgenic *Arabidopsis* expressing phyB–sGFP fusion proteins by *Agrobacterium*-mediated method is described elsewhere (*7*, *see* **Note 2**). Transient transfection of onion epidermal cells and *Arabidopsis* with a particle delivery system are described below.

3.1. Transient Expression in Onion Cells (see Note 3)

1. Purify the DNA plasmid encoding for phyB-sGFP from *Escherichia coli* cells using Qiafilter Plasmid Maxi or Midi Kit (Qiagen), according to manufacturer's instructions (*see* **Note 4**).
2. Coat tungsten particles of ~1.1 µm diameter with plasmid DNA, by reacting 5 µg purified phyB–sGFP plasmid DNA with 1.5 mg tungsten particles according to the manufacturer's instruction.
3. Load the tungsten particles onto a macrocarrier, and place it in the Biolistic equipment for a bombardment (*see* **Note 5**).

4. Peel an onion scale from a bulb and place it on a 9-cm agar plate with the inside surface upwards.
5. Place the plate into the chamber of PDS-1000/HE. Evacuate the chamber to 27 in of mercury.
6. Bombard the scales at 1350 psi using Rupture Disks 1350.
7. After the bombardment, put lids on the plates and seal them with tape.
8. Incubate the plates in a plant-growth chamber (23°C) for 24 h (*see* **Note 6**). Keep some plates in total darkness and others under white light.

3.2. Transient Expression in Arabidopsis Hypocotyl Cells

1. Epidermal cells of *Arabidopsis* hypocotyls are bombarded with a particle delivery system as described for onion cells, unless otherwise stated.
2. We use gold particles of ~1.0 μm diameter instead of tungsten particles.
3. Place the seedlings on agar plates into the chamber of PDS-1000/HE and bombarded at 1550 psi instead of 1350 psi (**Fig. 1**).
4. Incubate the seedlings, either in darkness or under white light for 24 h at 23°C. It is important here that the plates are placed in the vertical position. Otherwise, hypocotyls bend upwards and are detached from the agar surface, and this complicates microscopic observation.

3.3. Observation of GFP Fusion Proteins in Onion Epidermal Cells (see Note 7)

1. Detach the epidermis from the onion scales using forceps. Mount the peel, which consists of one layer of epidermal cells, on a glass slide with water.
2. In case of photoreceptors such as phytochrome, the intracellular distribution pattern may be altered by exposure to excitation light for fluorescence observation. Hence, it is recommended to fix the tissue in darkness before observation.
3. For this purpose, soak the peels in 3.7% formaldehyde fixative solution for 1–2 h in the dark at room temperature.
4. After fixing, wash the peels twice with water.
5. Observe the specimens by DIC and fluorescence microscopy. We used an Olympus BX60 microscope equipped with DIC optics and a 100 W mercury arc light source. Fluorescence can be observed using ultraviolet or fluorescein isothiocyanate filter sets.
6. Typically, transiently transformed onion epidermal cells grown in the presence of white light, have a speckled nuclear distribution of the fusion protein; those grown in darkness have the majority of the fluorescence signal in the cytoplasm (**Fig. 2A**).
7. Alternatively, an inverted laser-scanning confocal microscope can be used for confocal microscopy. Obtain the laser-scan images with a combination of 488 nm laser-excitation and 515 nm long-pass emission filter.

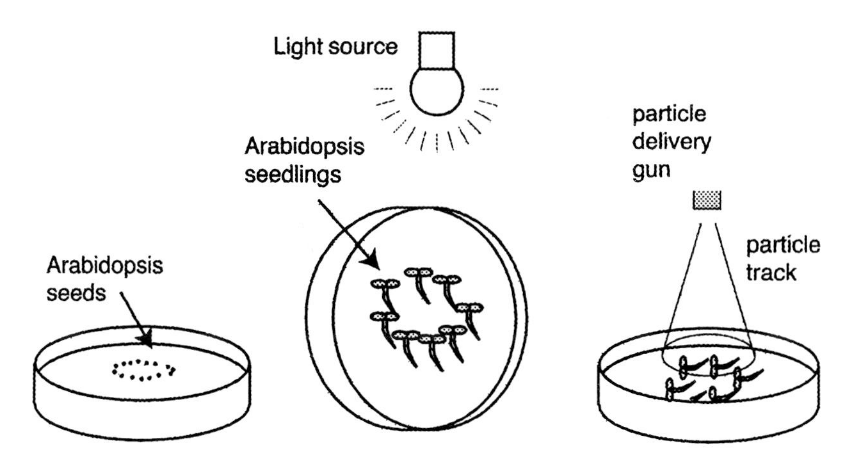

Fig. 1. Growth of *Arabidopsis* seedlings for transient expression. About 100 seeds/plate were sown on 9-cm agar plates (*left*). Plates were positioned vertically during the growth period (*center*). Plates were placed in the chamber of a particle delivery system and subjected to bombardment (*right*).

3.4. Observation of GFP Fusion Proteins Expressed in Arabidopsis (see Notes 7–9)

1. Place the hypocotyls on glass slides with water and press gently for the observation.
2. As is the case with onion cells, *Arabidopsis* hypocotyls can be fixed before observation. The hypocotyls were soaked in the fixing solution described previously, and are vacuum-infiltrated. The hypocotyls were then incubated overnight at room temperature in the fixing solution, and washed twice with water.
3. Microscopic observation was performed as described previously (**Fig. 2B**).
4. As seen in onion and in transgenic cells, the phyB–sGFP fusion protein is essentially all translocated to the nucleus when grown in the presence of white light, but remains in the cytoplasm when grown in darkness.

4. Notes

1. Various *Arabidopsis* mutants are known. Seeds of those mutants can be obtained from the Arabidopsis Biological Resource Center (http://www.biosci.ohio-state.edu/~plantbio/plantbio.html) or Nottingham Arabidopsis Stock Centre (http://nasc.life.nott.ac.uk/home.html).
2. Epidermal and cortex cells can be observed easily in transgenic *Arabidopsis* seedlings; trichome and stomata cells are also easily observed. Mesophyll cells are relatively difficult, because of background fluorescence from chloroplasts, inner tissues, such as vascular bundles are most difficult to observe. For those tissues, sections should be prepared.

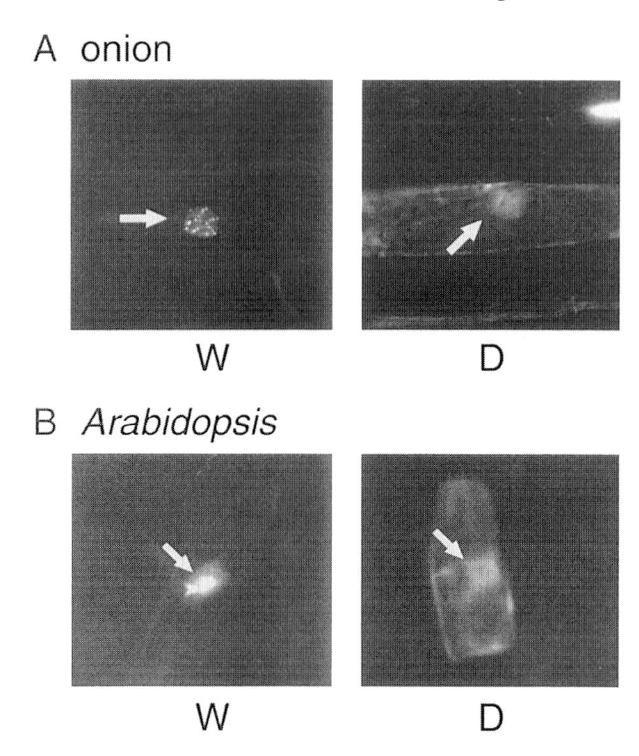

Fig. 2. Intracellular distribution of phyB–sGFP protein transiently expressed in onion **(A)** and *Arabidopsis* **(B)** cells. The introduced proteins were expressed either under white light (W) or in darkness (D). Epidermal cells were observed using an Olympus BX60 microscope as described in the text. Arrows indicate positions of the nuclei. As shown, Distribution patterns of phyB–sGFP are different under different light conditions.

3. Although the transient expression is advantageous in many respects, attention should be paid to its limitations. The expression levels of introduced proteins are very difficult to control. Levels deviated substantially in individual cells in one experiment. It is also difficult to confirm that the fusion protein is not fragmented in the cell. Hence, it is recommended to confirm important results by stable transformation.
4. Several GFP variants are available for plants ***(11,12)***. Among them, mGFP4 ***(13)*** and sGFP(S65T) ***(14)*** have been widely used. A cryptic *Arabidopsis* intron sequence is removed in both mGFP4 and sGFP(S65T). Codon usage is adapted for plants in sGFP.
5. Both gold and tungsten particles worked well for the expression of phyB–GFP in onion cells. However, we found that gold particles give much better results than tungsten ones for *Arabidopsis* seedlings. Hence, the conditions of bombardment

(types and sizes of the particles, bombardment pressure and so on) should be optimized for different plant materials.

6. It is also important to choose appropriate time and temperature for the expression of introduced proteins. Some GFP fusion proteins accumulate in the cell within a few hours; others take much longer (e.g., 24 h for phyB–sGFP). Some fusion GFPs exhibit aggregation (probably caused by overaccumulation) in the cytoplasm, when expressed transiently in plant cells. In some cases, decreasing the incubation temperature solves this problem.
7. Evaluating the nuclear localization signal (NLS) activity in relatively small proteins is difficult, for at least two reasons. First, molecules smaller than the size-exclusion limit of ~40–60 kDa can enter the nucleus by passive diffusion through the nuclear pore complex (NPC) ***(15)***. A good example is GFP itself. GFP expressed in plant cells is detected both in the cytoplasm and nucleus. Hence, small GFP fusion proteins may enter the nucleus passively, even if they do not contain NLS. Second, small GFP fusion proteins, whose molecular mass is less than 50 kDa may diffuse from the nucleus through NPC, even if they have NLS ***(16)***. Hence, we recommend fusing a larger protein (e.g., B-glucuronidase is suitable for this purpose), in addition to GFP, when NLS activity is examined for small proteins (or small fragments of proteins).

   When GFP-fused proteins, larger than the size exclusion limit of NPC, show the signal in both cytoplasm and nucleoplasm, we speculate that the protein contains both NLS and nuclear export signal. Such proteins shuttle actively between the two compartments through NPC. However, there are some exceptional cases, in which GFP-fused proteins that appear to be larger than the exclusion limit diffuse passively through NPC ***(17,18)***.
8. Since *Arabidopsis* seedlings are small (~5 mm height), a whole seedling can be mounted on a glass slide. However, detach organs of interest from the seedlings before observation. The specimens (whole seedlings or detached organs) were placed on glass slides with water and pressed gently. The seedlings may be fixed before the observation as described previously.
9. Some tissues cannot be observed easily in a whole organ specimen. In such cases, sections should be prepared. Fresh seedlings (or organs) were embedded in 5% low-melting-point agarose (Seaplaque, FML BioProducts) made in water. Specimens of ~50 μm thickness were sectioned with a vibrating blade microtome (Microslicer DTK-1000, Dousaka). The specimens were mounted on glass slides with water and subjected to the observation.

Acknowledgments

This study was supported in part by a grant from the Program for Promotion of Basic Research Activities for Innovative Biosciences, a grant from Special Coordination Funds for Promoting Science and Technology from the Science and Technology Agency, and a Grants-in-Aid for Scientific Research (B) (no. 08454253) from the Ministry of Education, Science, Sports and Culture of Japan.

References

1. Furuya, M. (1993) Phytochromes: their molecular species, gene families and functions. *Annu. Rev. Plant Physiol. Plant Mol. Biol.* **44,** 517–645.
2. Köhler, R. H. (1998) GFP for in vivo imaging of subcellular structures in plant cells. *Trends Plant Sci.* **3,** 317–320.
3. Cutler, S. R., Ehrhardt, D. W., Griffitts, J. S., and Somerville, C. R. (2000) Random GFP::cDNA fusions enable visualization of subcellular structures in cells of *Arabidopsis* at a high frequency. *Proc. Natl. Acad. Sci. USA* **97,** 3718–3723.
4. Pratt, L. H. (1994) Distribution and localization of phytochrome within the plant, in *Photomorphogenesis in Plants* (Kendrick, R. E. and Kronenberg, G. H. M. ed.), Kluwer, Dordrecht, The Netherlands, pp. 163–185.
5. Nagatani, A. (1997) Saptial distribution of phytochromes. *J. Plant Res.* **110,** 123–130.
6. Sakamoto, K. and Nagatani, A. (1996) Nuclear localization activity of phytochrome B. *Plant J.* **10,** 859–868.
7. Yamaguchi, R., Nakamura, M., Mochizuki, N., Kay, S. A., and Nagatani, A. (1999) Light-dependent translocation of a phytochrome B-GFP fusion protein to the nucleus in transgenic *Arabidopsis. J. Cell Biol.* **145,** 437–435.
8. Kircher, S., Kozma-Bognar, L., Kim, L., et al. (1999) Light quality-dependent nuclear import of the plant photoreceptors phytochrome A and B. *Plant Cell* **11,** 1445–1456.
9. Gil, P., Kircher, S., Adam, E., et al. (2000) Photocontrol of subcellular partitioning of phytochrome-B:GFP fusion protein in tobacco seedlings. *Plant J.* **22,** 135–145.
10. Benfy, P. N. and Chua, N.-H. (1989) Regulated genes in transgenic plants. *Science* **244,** 174–181.
11. Kawakami, S. and Watanabe, Y. (1997) Use of green fluorescent protein as a molecular tag of protein movement in vivo. *Plant Biotech.* **14,** 127–130.
12. Haseloff, J. (1999) GFP variants for multispectral imaging of living cells. *Meth. Cell Biol.* **58,** 139–151.
13. Haseloff, J., Siemering, K. R., Prasher, D. C., and Hodge, S. (1997) Removal of a cryptic intron and subcellular localization of green fluorescent protein are required to mark transgenic Arabidopsis plant brightly. *Proc. Natl. Acad. Sci. USA* **94,** 2122–2127.
14. Chiu, W.-L., Niwa, Y., Zeng, W., Hirano, T., Kobayashi, H., and Sheen, J. (1996) Engineered GFP as a vital reporter in plants. *Curr. Biol.* **6,** 325–330.
15. Nigg, E. A. (1997) Nucleocytoplasmic transport: signals, mechanisms and regulation. *Nature* **386,** 779–787.
16. Grebenok, R. J., Pierson, E., Lambert, G. M., et al. (1997) Green-fluorescent protein fusions for efficient characterization of nuclear targeting. *Plant J.* **11,** 573–586.
17. Haasen, D., Kohler, C., Neuhaus, G., and Merkle, T. (1999) Nuclear export of proteins in plants: AtXPO1 is the export receptor for leucine-rich nuclear export signals in *Arabidopsis thaliana. Plant J.* **20,** 695–705.
18. Kudo, N., Wolff, B., Sekimoto, T., et al. (1998) Leptomycin B inhibition of signal-mediated nuclear export by direct binding to CRM1. *Exp. Cell Res.* **242,** 540–547.

13

Mechanisms of Protein Trafficking

Two Different Signal Sequences Fused to Green Fluorescent Protein to Study Mitochondrial Import

Henry Weiner

1. Introduction

Soon after cDNAs were obtainable, it became apparent that proteins found in the mitochondrial matrix space were coded by nuclear genes and synthesized as precursor proteins. That is, they have an extension of amino acids at their N-terminal end. This extension is removed after import by the action of a specific protease. Thus, when a mitochondrial protein was isolated, the extension, called a leader sequence, was missing from most, but not all, matrix space proteins. During the past two decades, much work was done to learn how the leader helped bring the protein through the membranes and into the matrix space. Initial studies were focused on trying to learn what features were common among the leaders since all those investigated were found to have different primary sequences. A good general review covering most aspects of mitochondrial protein import can be found in **ref *1***.

Inspection of the sequences led investigators to postulate that the leaders could form amphilphatic helices. Leaders all have a net positive charge, because to the presence of many arginine residues. Initially, circular dichroism was used to show that the peptides that corresponded to the leader could form a helix in a buffer containing a co-solvent, such as trichloroethanol. Eventually, nuclear magnetic resonance techniques were employed to identify the portions of the leader peptide that were actually in the helix *(2–4)*.

Investigators tried to determine if the entire leader was necessary for it to function. Some of the studies involved in vivo work. Using the tools of molec-

From: *Methods in Molecular Biology, vol. 183: Green Fluorescent Protein: Applications and Protocols*
Edited by: B. W. Hicks © Humana Press Inc., Totowa, NJ

ular biology, leaders were fused to a carrier protein, such as β-galactosidase or dihydrofolate reductase. Typically, yeast was employed for these were easy to transform and grow. Investigators truncated or altered the sequence and structure of the leaders. In general, it was found that the crucial segment was the most N-terminal end of the leader. This was the portion that typically possessed a helix and had at least one excess positive charge. Most import experiments, however, were performed using an in vitro import system. For these, the mRNA coding for the precursor protein was translated in a rabbit reticulocyte lysate in the presence of 35[S]-methionine. Proteins were visualized after import by first separating them by sodium dodecylsulfate-polyacrylamide gel electrophoresis (SDS-PAGE), then visualizing them using autoradiography. For the import assay, mitochondria, isolated from either rat liver or yeast, were incubated with the newly synthesized protein. After a period of time, typically 30 min, the mitochondria were separated from the mixture by centrifugation. Protease K was added to the mitochondria to digest any of the proteins that were bound to the surface. Finally, the mitochondria were lysed with SDS and the protein mixture subjected to electrophoresis *(5)*. Results from a typical assay are shown in **Fig. 1**. For the in vitro experiments, both fusion proteins and naturally occurring precursor proteins were employed.

Investigators found that it was necessary to have some components from the reticulocyte lysate cell free synthesis system present in an in vitro import system. Many of the components were characterized and found to be proteins that helped keep the precursor in an unfolded state *(6)*. Consistent with this were reports that urea would increase the rate of import of precursor proteins *(7)*. Thus, to be imported, a protein must have a leader and be unfolded. Because the in vitro import was done after the proteins were synthesized, it became obvious that import could be a posttranslational event, i.e., the precursor protein is completely synthesized prior to its import. This is in contrast to import into the endoplasmic reticulum (ER), where in some cases a docking system exists so that the protein, while being synthesized, is bound to the organelle and is being imported as synthesis continues.

The discovery of the green fluorescent protein (GFP) has allowed investigators to study many in vivo cellular events. Only a few studies have been published concerning mitochondrial import. One of the earliest publications showing that GFP can be used to study mitochondrial import was from Mori's laboratory *(8)*. They used a fusion protein to actually visualize mitochondria from patients with different diseases. They went on to perform microinjections of cDNA and measure incorporation of protein in mitochondria as a function of time *(9)*, and finally showed that they could use chimeric proteins to study the role of mitochondrial receptor proteins in COS cells *(10)*.

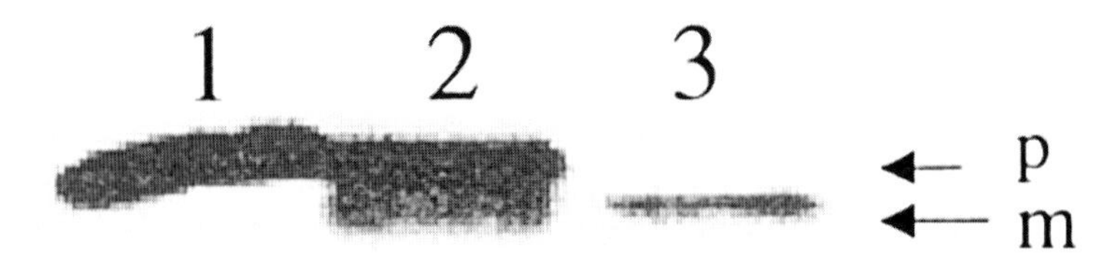

Fig. 1. Example of protein import data. Lane 1 shows the precursor that was synthesized by the in vitro system in the presence of 35[S]-methionine. After synthesis, the sample is subjected to SDS-PAGE and the radioactivity visualized by audoradiography. Lane 2 shows the labeled protein in intact mitochondria after import. Some of the precursor protein is bound to the surface; some protein is imported and processed to the size of the mature protein. Lane 3 shows the mitochondria after treatment with Protease K, a nonspecific endoprotease that destroys all proteins not protected by the mitochondrial membranes. Thus, all that remains is the imported mature protein. (p, precursor; m, mature.)

My laboratory used GFP to address a mechanistic question *(11)*. In vitro mitochondrial protein import occurs as a posttranslational event, i.e., the precursor protein is made then is subsequently imported into mitochondria. When doing in vitro import we, like others, have observed that large portions of the newly synthesized precursors actually become import-incompetent, most likely because of them becoming folded. We argued that it is possible that import could be a co-translational in vivo event since precursor proteins are not found in the cytosol. Cotranslation implies that the protein is never free in the cytosol because it might be imported while being made or is at least brought to the mitochondrial membrane during its synthesis. This means that the N-terminal segment might be bringing the protein into the mitochondria before the C-terminal end is made. Data supporting the notion that import could be a cotranslational event includes finding that the leader is susceptible to proteolysis in a cytosol extract, and that many mitochondrial proteins appear to have a long half-life. Further, it has been reported that there are between, perhaps, 100 and 1000 receptor complexes on the mitochondrial outer membrane *(12)*. These facts imply that, because of the low rate of synthesis and low copy number of receptor complexes, the probability of a precursor finding the mitochondrial translocation apparatus prior to improper folding or proteolysis would be low. Other investigators, as reviewed in the paper from this laboratory *(11)*, have argued that import could be a cotranslational event, but definitive data was lacking.

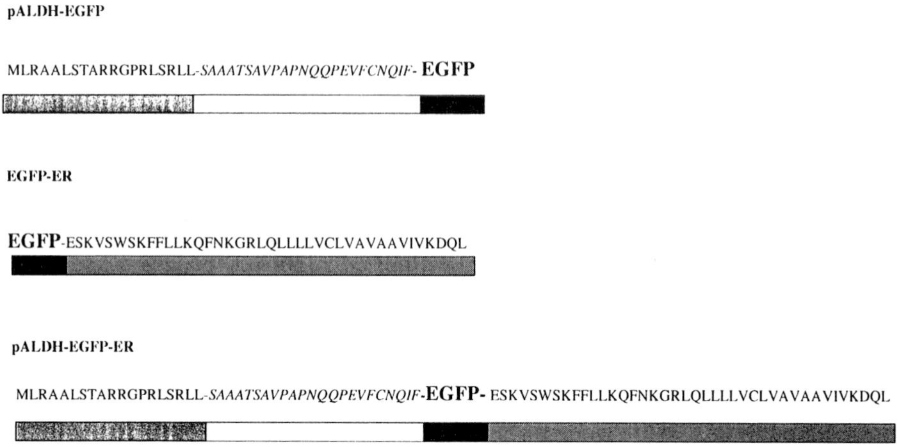

pALDH-EGFP

MLRAALSTARRGPRLSRLL-*SAAATSAVPAPNQQPEVFCNQIF*- **EGFP**

EGFP-ER

EGFP-ESKVSWSKFFLLKQFNKGRLQLLLVCLVAVAAVIVKDQL

pALDH-EGFP-ER

MLRAALSTARRGPRLSRLL-*SAAATSAVPAPNQQPEVFCNQIF*-**EGFP**- ESKVSWSKFFLLKQFNKGRLQLLLVCLVAVAAVIVKDQL

Fig. 2. The three major constructs used in the study. "pALDH" refers to the N-terminal mitochondrial leader from rat liver aldehyde dehydrogenase. The first 19 residues are the leader; the ones in italics are from the mature portion of the protein. Constructs are also used in which the leader is fused directly to one of the enhanced green fluorescent proteins, EGFP *(11)*. ER refers to the C-terminal 35 amino acid signal from a rat microsomal aldehyde dehydrogenase that targets that protein to the outer membrane of the endoplasmic reticulum. pALDH-EGFP is found only in the mitochondria, after the HeLa cells are transfected. The EGFP-ER is found only in the mircosomes; pALDH-EGFP-ER is found only to be associated with the mitochondria.

GFP allowed convenient investigation of the nontraditional concept of cotranslation. To do so, we attached two different signals to GFP. One would direct the protein to the mitochondria, the other would lead it to the ER. These constructs are illustrated in **Fig. 2**. Photographs of the fluorescent data are presented as figures in the supplemental CD-ROM that accompanies the book or can be seen in the original publication *(11)*. The mitochondrial signal used was found at the N-terminus *(3)*, while the ER signal was located at the C-terminus *(13)*. The results were as follows. If HeLa cells were transformed with just GFP, protein would be found in the entire cell. If only the mitochondrial leader were attached, protein would be found in the mitochondria and lastly, if only the ER signal was present protein was found in that organelle. The crucial experiment was with the double-labeled GFP. If both signals were available, the protein should have been distributed between the two organelles, if a postranslational event were occurring. If co-translation were occurring, then the C-terminal ER-targeting signal would not be available and all the protein would be found in mitochondria. This is what was found. The most logical interpretation of finding all the protein in the mitochondria is that the precursor protein never was completely free in the cytosol, so the C-terminal ER targeting peptide was not available. A co-translational import model could logically be used to explain the data.

The approach we employed would not be applicable to study all cellular protein trafficking. What is necessary is that the signals must be associated with different parts of the carrier protein. One could envision addressing a question related to membrane anchoring of a protein. For example, if one had a membrane-spanning domain for a protein destined for one organelle located at the N-terminal end of the natural protein, one would then need a membrane-spanning domain from a different protein be located at the C-terminal end of that protein. Both these domains could be fused to GFP to determine if the chimeric protein was now found associated with one or two organelles. Finding it located in just the one organelle governed by the domain fused to the N-terminus would show that the protein was associating with the organelle as it was being synthesized. Analogous to the import experiments, the only way to find the protein associated with two different organelles would be to have the protein synthesized completely prior to it being associated with the organelle targeted by the C-terminal signal.

In addition to investigating co- and posttranslational import questions, one can use a dually labeled GFP for other types of studies. These would be approachable only if the import occurred in a posttranslational manner. For example, one could investigate an ion or pH effect on import. Similarly, a drug effect on two different transport systems could be studied simultaneously. The role of selected chaperones or heat shock proteins could be investigated by monitoring for a different distribution of trafficking if one of the components was missing.

2. Materials

Recombinant plasmids were created using standard molecular biology techniques (***11***; *see* Chapters 1–5 of this volume). Furthermore, the specific cloning strategy must change, depending upon the targeting sequences employed and their sources. Cell transformation efficiency will differ for each cell type used.

1. Recombinant GFP plasmids without targeting sequence (TS): pEGFP-N1 (Clontech), with N-terminal mitochondrial TS, with C-terminal ER TS, and with both mitochondrial and ER TS (**Fig. 2**).
2. HeLa cells transfected and expressing the above constructs (*see* **Note 1**).
3. HeLa cell growth medium: Dulbecco's modified Eagle's medium (DMEM) (Life Technologies) containing 10% calf serum, 100 µg/mL streptomycin, 100 U/mL penicillin, and 0.25 µg/mL amphotericin B.

4. SuperFect Transfection Reagent (Qiagen).
5. In vitro transcription-coupled translation kits (Promega).
6. Protease K.
7. Phenylmethanylsulfonyl fluoride.
8. Phosphate-buffered saline (PBS) (1 L): 0.2 g KCl, 8.0 g NaCl, 0.2 g KH_2PO_4, 1.15 g Na_2HPO_4 pH 7.4.
9. 4% Paraformaldehyde in PBS.
10. Isolated yeast or rat liver mitochondria.
11. SDS-PAGE apparatus, gels, and Coomassie blue for gel staining.
12. Autoradiographic film.
13. Six-well plates, 60-mm culture dish, and cover slips for HeLa culture and microscopy.
14. Olympus BX60 Fluorescent Microscope or equivalent.
15. Modified medium for observation of fluorescence containing 10 mL glycerol, 10 mL PBS, and 0.1 g propyl gallate. The pH is adjusted to 8.5 with 1 *N* NaOH.

3. Methods

3.1. In Vitro Import of Proteins into Mitochondria

1. Prior to transforming the HeLa or other cell, an in vitro import assay should be performed. Use a commercial kit, such as the Promega TNT kit to make the protein, labeled with an amino acid such as 35[S] methionine.
2. Use the TNT kit essentially as described by the manufacturer. After a 60–90 min time period to allow for the translation of the protein, subject an aliquot of the solution to SDS-PAGE to verify that a protein of the expected molecular weight was synthesized in vitro.
3. To check for import, add 6 μL of either yeast or liver mitochondria (7 mg protein/mL) to 3 μL of the expression system, in a total volume of 100 μL, and allow import to proceed for 30 min at 30°C. Divide the incubation solution into two portions. Additional descriptions can be found in **refs. *11*, *14*,** and ***15***.
4. After import, add 4 μL protease K (2 mg/mL) to one portion to destroy proteins not imported or that are just bound to the outer membrane of the mitochondria. It should take 15 min on ice to digest the proteins. Terminate the reaction by adding 2 μL phenylmethylsulfonyl fluoride (stock solution is 200 m*M*). This is a fast reaction, requiring less than 1 min. Pellet both the protease K-treated and untreated portions, by subjecting the samples to centrifugation. Since no other organelles are present the velocity is not important. Wash the mitochondria by gently stirring them with buffer to resuspend them; again isolate the organelles by subjecting the suspension to centrifugation. Next lyse the pellet in SDS and subject an aliquot to SDS-PAGE.
5. Follow SDS-PAGE separation by autoradiography. The protease K-treated sample contains protein that is imported into mitochondria. The nontreated sample will contain any protein that is just bound the outer membrane as well as those imported.

6. There is much variation between batches of mitochondria isolated from different animals. Qualitatively, similar results are obtained with mitochondria isolated from rats or yeast. The advantage of using mitochondria from yeast is that the organelles can be stored ***(16)***; those from rats must be prepared fresh each time they are needed.

3.2. In Vivo Visualization of Targeted GFP by Fluorescence Microscopy

The author's laboratory has had no previous experience in using microscopy to study a cellular event. Thus, special expertise was not required to do this study. The department's Olympus BX60 Fluorescent Microscope was used, and the images were recorded on 35 mm film, which is acceptable, but the more modern digital camera is more desirable.

1. One day prior to transient transfection, seed 2–8 × 10^5 Hela cells on cover slips in 6-well plates, or simply in a cell culture dish, and grow in 3 mL growth medium (DMEM plus 10% calf serum and antibiotics), incubated at 37°C under an atmosphere of 5% CO2 and 95% air. The cell number seeded should produce 40–80% confluent on the day of transfection (*see* **Note 1**).
2. Dilute 5 μg of DNA (minimum concentration 0.1 μg/mL) dissolved in 10 m*M* Tris-chloride buffer containing 1 m*M* EDTA with cell growth medium containing no serum, protein, or antibiotics. The final volume should be 150 μL.
3. Transfect the cells with 2.5 μg plasmid at 37°C, by adding 30 μL SuperFect Transfection Reagent from Qiagen, essentially as described by the manufacturer.
4. Incubate the samples for 5–10 min at room temperature.
5. Add 1 mL growth medium containing serum and antibiotics. Mix by pipeting a few times, and transfer the total volume to the cells.
6. Remove the medium by aspiration, and wash the cells with 4 mL PBS.
7. Next, replace the transfection medium with fresh growth medium.
8. Between 48 and 72 h after transfection, wash the cells twice with PBS.
9. For fixation, incubate the cells on the cover slips with 1 mL 4% paraformaldehyde in PBS, 37°C for 15–30 min, then wash the cells twice using PBS (*see* **Note 2**).
10. Place 20 μL of modified medium on a microscope slide and place the cover slide on it, cell-side down.
11. Observe the GFP fluorescence from the cells on the cover slips with an Olympus BX60 Fluorescence Microscope or equivalent.

4. Notes

1. Use of the two-signal approach has been applied to investigate a limited number of systems. HeLa cells were the only cell line used, but there is no reason that yeast or other mammalian cell lines could not be used. Mori has used COS cells for a different study ***(10)***. For each signal sequence employed, it would be useful to determine whether or not the signals were stable to the cytosolic environment

of the cell line to be studied. After the precursor is translated in the in vitro system mentioned previously, some cytosol lysate from the cell line should be added to it. Over the next 30 min, aliquots should be removed and analyzed by SDS-PAGE, to determine whether or not the signals are proteolyzed by actions of nonspecific proteases that were present in the cytosol. If the signal fused to the N-terminal end of GFP were destroyed, then the only way the protein could come to the organelle in vivo would be by a co-translational event. If the signal fused to the C-terminal end of GFP were unstable, then it would not be possible for the carrier protein to become associated with that organelle. One could not use a negative finding to prove co-translation, unless it was shown that the signal at the C-terminal end was stable. If the signal at the N-terminal was very stable to the cytosolic environment, then it could be possible that import occurred in a posttranslational manner. To prove that a co-translational trafficking event took place, one would have to show that the signal located at the N-terminal was less efficient than was the other. Thus, finding GFP only in the organelle that used the N-terminal signal could only occur if import occurred before the C-terminal signal was available.

2. Observing whether 100% of the cellular events occurred in an identical manner is simple, but it is not easy to tell by visual observation, if, for example, 80% of the protein were in one organelle and 20% in another. Isolation of subcellular organelles is often advisable just to verify that the presence of fluorescence was not artifactual. Isolating the various subcellular organelles and Western blotting, or similar techniques can be used to identify the GFP carrier protein in the isolated fractions.

Supplemental Material

Four color photographs showing the actual subcellular localization in HeLa cells are presented on the CD-ROM.

EGFP is a photograph of a HeLa cell transformed with just EGFP. SP1-EGFP corresponds to the transfection with a mitochondria leader fused to N-terminal of EGFP. EGFP-ER corresponds to the transfection with an endoplasmic reticulum signal fused to the C-terminal of EGFP. SP-EGFP-ER refers to the construct with both signals on GFP. The subcellular organelles were verified by isolation and analyzing for markers proteins. The data reveal that EGFP is found throughout the cell. When the mitochondrial leader is attached, all the fluorescence is found associated with the mitochondria. When the ER signal is attached, the fluorescence is found to be associated with the endoplasmic reticulum. When both signals were fused to EGFP, the fluorescence was found only to be associated with the mitochondria, consistent with import being a co-translational event. *See also* Ni, L., Heard, T. S., and Weiner, H. (1999) *In vivo* mitochondrial import: a comparison of leader sequence charge and structural relationships with the in vitro model resulting in evidence for co-translational import. *J. Biol. Chem.* **274,** 12,685–12,691.

Acknowledgment

The author thanks two of his former graduate students, Li Ni and Thomas Heard who did more than the work. They introduced him to thinking about GFP and using cell biology as a tool. The research was support by grants from the National Institutes of Health.

References

1. Neupret, W. (1997) Protein import into mitochondria. *Annu. Rev. Biochem.* **66,** 863–917.
2. Endo, T., Shimada, I., Roise, D., and Inagaki, F. (1989) N-terminal half of a mitochondrial presequence peptide takes a helical conformation when bound to dodecylphosphocholine micelles: a proton nuclear magnetic resonance study. *J. Biochem.* **106,** 396–400.
3. Karslake, C., Piotto, M. E., Pak, Y. K., Weiner, H., and Gorenstein, D. G. (1990) 2D-NMR and structural model for a mitochondrial signal peptide bound to a micelle. *Biochemistry* **29,** 9872–9878.
4. Hammen, P. K., Gorenstein, D. G., and Weiner, H. (1994) Structure of the signal sequences for two mitochondrial matrix proteins that are not proteolytically processed upon import. *Biochemistry* **33,** 8610–8617.
5. Waltner, M. and Weiner, H. (1995) Conversion of a nonprocessed mitochondrial precursor protein into one that is processed by the mitochondrial processing peptidase. *J. Biol. Chem.* **270,** 16,311–26,317.
6. Mori, M. and Terada, K. (1998) Mitochondrial protein import in animals. *Biochim. Biophys. Acta* **1403,** 12–27.
7. Eilers, M., Hwang, S., and Schatz, G. (1988) Unfolding and refolding of a purified precursor protein during import into isolated mitochondria. *EMBO J.* **7,** 1139–1145.
8. Kanazawama, M., Yano, M., Namchai, C., Yamamoto, S., Ohtake, A, Takayanagi, M., Mori, M., and Niimi, H. (1997) Visualization of mitochondria with green fluorescent protein in cultured fibroblasts from patients with mitochondrial diseases. *Biochem. Biophys. Res. Commun.* **239,** 580–584.
9. Yano, M., Kanazawa, M., Terada, K., Namchai, C., Yamaizumi, M., Hanson, B., Hoogenraad, N., and Mori, M. (1997) Visualization of mitochondrial protein import in cultured mammalian cells with green fluorescent protein and effects of overexpression of the human import receptor Tom20. *J. Biol. Chem.* **272,** 8459–8465.
10. Yano, M., Kanazawa, M., Terada, K., Takeya, M., Hoogenraad, N., and Mori, M. (1998) Functional analysis of human mitochondrial receptor Tom20 for protein import into mitochondria. *J. Biol. Chem.* **273,** 26,844–26,851.
11. Ni, L., Heard, T. S., and Weiner, H. (1999) In vivo mitochondrial import: a comparison of leader sequence charge and structural relationships with the in vitro model resulting in evidence for co-translational import. *J. Biol. Chem.* **274,** 12,685–12,691.

12. Vestweber, D. and Schatz, G. (1988) A chimeric mitochondrial precursor with internal disulfide bridges blocks import of authentic precursors into mitochondria and allows quantitation of import sites. *J. Cell Biol.* **107,** 2037–2043.
13. Masaki, R., Yamamoto, A., and Tashiro, Y. (1994) Microsomal aldehyde dehydrogenase is localized to the endoplasmic reticulum via its carboxyl-terminal 35 amino acids. *J. Cell Biol.* **126,** 1407–1420.
14. Wang. T. T. Y., Farrés, J., and Weiner, H. (1989) Liver mitochondrial aldehyde dehydrogenase: *in vitro* expression, *in vitro* import and effects of alcohol in import. *Arch. Biochem. Biophys.* **272,** 440–449.
15. Pak, Y.-K. and Weiner, H. (1990) Import of chemically synthesized peptides into rat liver mitochondria. *J. Biol. Chem.* **265,** 14,298–14,307.
16. Glick, B. and Pon, L. A. (1995) Isolation of highly purified mitochondria from *Saccharomyces cerevisiae*. *Meth. Enzymol.* **260,** 213–223.

14

Analysis of Nucleocytoplasmic Transport Using Green Fluorescent Protein

Roland H. Stauber

1. Introduction

A hallmark of eukaryotic cells is their spatial and functional separation into the nucleus and the cytoplasm by the nuclear envelope. Although this separation introduces a potent and sophisticated level of regulation, it also requires highly effective and selective transport machinery. All known transport between the nucleus and the cytoplasm occurs through the nuclear pore complex *(1–3)*. Theoretically, proteins <40 kDa can enter and leave the nucleus by passive diffusion. However, even most of the smaller proteins and nucleic acids appear to be transported by signal-mediated pathways, probably because signal-mediated transport is more efficient and more amenable to specific regulation than diffusion. Nuclear import is mediated by short stretches of basic amino acids, called "nuclear localization signals" (NLS), which interact with the importin α/β-axis, or directly with alternative import receptors **(Fig. 1A)**. Following transfer through the NPC binding of Ran-guanosine triphosphate (GTP) to the import complex, causes cargo release **(Fig. 1B**; *4–9***)**. Nuclear export of proteins, on the other hand, is mediated by nuclear export signals (NES), mostly characterized by clusters of leucines, as identified in a variety of cellular and viral proteins *(4,5,9)*. Recent studies indicate that individual import and export signals differ in their activity, and that this may represent a level of regulation for the biological activity of shuttle proteins (*10*, Stauber, unpublished observations). NESs interact with the export receptor, CRM1, in the presence of Ran-GTP and the trimeric complex of cargo, CRM1, and Ran-GTP is subsequently translocated to the cytoplasm **(Fig. 1C)**. Ran-GTP hydrolysis

From: *Methods in Molecular Biology, vol. 183: Green Fluorescent Protein: Applications and Protocols*
Edited by: B. W. Hicks © Humana Press Inc., Totowa, NJ

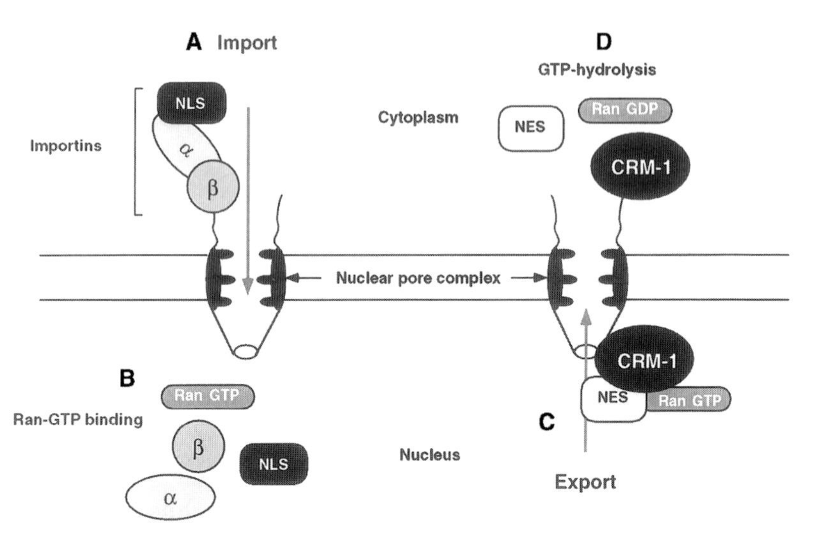

Fig. 1. Nucleocytoplasmic transport. **(A)** Import receptors bind to NLS containing substrates and mediate their transport into the nucleus. **(B)** Binding of Ran-GTP to the import complex causes cargo release. **(C)** NES-containing macromolecules interact with the export receptor, CRM1, promoted by Ran-GTP. **(D)** Following transfer through the nuclear pore, the export complex is disassembled via GTP hydrolysis.

triggers the dissociation of the complex that is induced by the cytoplasmic Ran-GTPase-activating protein and its accessory Ran-binding protein 1 (**Fig. 1D**; *6,11,12*). The Ran-GTPase cycle creates, therefore, a Ran-GTP gradient that determines transport directionality by regulating the stability of the transport receptor–cargo complexes *(13)*.

Since the controlled transport of molecules between the cytoplasm and the nucleus is critical for cellular homeostasis, there is great emphasis on the identification of shuttle proteins and the characterization of specific transport signals and pathways. To date, several experimental in vitro and in vivo systems have been established to study the requirements for nucleocytoplasmic trafficking. Autofluorescent proteins (AFPs) are ideal tools to investigate these dynamic processes because AFPs, in contrast to other bioluminescent molecules, operate independently of cofactors and can be detected in living cells. This chapter focuses on several AFP applications to investigate nucleocytoplasmic transport and to characterize transport signals.

2. Materials

2.1. Microscopy, Cell Transfection, and Drug Treatment

1. Green fluorescent protein (GFP) signals were observed using an inverted fluorescence microscope, e.g., Axiovert®135 Inverted Fluorescence Microscope (Carl

Zeiss, Thornwood, NY) equipped with a fluorescein isothiocyanate-fluorescence filter set (Zeiss O9, excitation, 450–490 nm; beam splitter, 510 nm; emission filter, >520 nm; Carl Zeiss). Blue fluorescent protein (BFP) signals were obtained using a broad band DAPI-filterset (Zeiss 02, excitation maximum, 365 nm; beam splitter, 460 nm; emission filter, >470 nm).
2. The use of a charge-coupled device camera (e.g., MicroMax, Princeton Instruments, Stanford, CA) to record images of the same cell at different time-points. Image analysis and presentation was performed using the IPLab 3.1.1c software package (Scanalytics, Vienna, VI).
3. For direct microscopic observation, cells were seeded into 35-mm glass bottom dishes (no. P35G-C-R, MatTek, Ashland, MA).
4. Plasmids encoding AFP-fusions. For fusions to the N-terminus of GFP/BFP, the coding regions of the specific genes were inserted into the unique *Nhe*I-site in the plasmids pCMV-GFPsg25 or pCMV-BFPsg50 ***(14)***. To generate fusions to the C-terminus of GFP, plasmid pF25GFP-Hyg was used ***(14)***. However, commercially available AFP vectors (e.g., pQBI 25/50, Quantum Appligene Biotech., PQ) can be used as well (*see* **Note 1**).
5. HeLa, Vero or NIH3T3 cells were cultured in Dulbecco's modified Eagle's medium (DMEM) supplemented with 10% fetal bovine serum.
6. Transfections were performed using the ProFection® Calcium Phosphate Transfection Kit (Promega, Madison, WI).
7. Prepare a stock solution of 1 µ*M* leptomycin B (LMB) (Sigma) in dimethyl sulfoxide (DMSO), and store aliquots at –20°C.
8. Prepare a stock solution of 1 mg/mL actinomycin D (ActD), or 10 mg/mL 5,6-dichlororibofuranosylbenzimidazole (DRB) (Sigma) in 70% EtOH, and store aliquots at –20°C.

2.2. Polyethylene Glycol-Induced Cell Fusion Assays

1. Polyethylene glycol-1500 (PEG) (Roche, Mannheim, Germany).
2. Prepare a stock solution of 1 mg/mL cycloheximide (Sigma) in 70% EtOH and store aliquots at –20°C.
3. Hoechst 33258 (Molecular Probes) (stock solution: 1 mg/mL in DMSO) for nuclear stain.
4. Phosphate-buffered saline (PBS), pH 7.3: 137 m*M* NaCl, 2.68 m*M* KCl, 7.3 m*M* Na_2HPO_4, 1.47 m*M* KH_2PO_4, 0.91 m*M* $CaCl_2$, 0.49 m*M* $MgCl_2$.

2.3. Preparation and Microinjection of Recombinant GFP Fusion Proteins

A detailed description of the individual apparatus and techniques necessary for computer-assisted microinjection would exceed this chapter (for specific details, *see* **ref. *26*** and references therein). The described experiments were performed using a CompiC INJECT computer-assisted injection system (Cellbiology, Hamburg, Germany) and a horizontal pipet puller (Sutter, Novato, USA).

1. *Escherichia coli* (strain BL21) containing the appropriate (GST)-AFP expression plasmid.
2. Luria-Bertani medium (LB)-ampicillin: LB medium containing 50 µg/mL ampicillin.
3. Isopropyl β-D-thiogalactooside (IPTG), freshly prepared in H_2O.
4. 96-well plates (Costar® Opaque Black Solid 96 well plate, Corning, Corning, NY).
5. Victor® 1420 Multilabel Fluorescence Plate Reader (Wallace Oy, Turku, Finland).
6. Injection medium: Cell culture medium without serum, buffered with 25 m*M* HEPES (1 *M* stock solution, sterile-filtered, stored at room temperature.
7. Prepare a 100X protease inhibitor stock solution: 200 m*M* phenylmethylsulfonyl fluoride, 5 mg/mL pepstatin, 10 mg/mL chymostatin (Sigma). Dissolve in DMSO, and store aliquots at –70°C.
8. Hen egg lysozyme (Sigma): 10 mg/mL stock solution in water.
9. Dnase I grade II (Roche): 1 mg/mL stock solution in 20 m*M* Tris-HCl, pH 7.5, 1 m*M* $MgCl_2$; store aliquots at –20°C.
10. 1 *M* $MnCl_2$.
11. 1 *M* $MgCl_2$.
12. Branson Sonifier 450 with a microtip probe (Schwäbisch Gmünd, Germany).
13. 20% Triton X-100 (Sigma).
14. 5 *M* NaCl.
15. Glutathione-Separose® 4B (Pharmacia Biotech., Freiburg, Germany).
16. Elution buffer (freshly prepared): 50 m*M* Tris-base, 150 m*M* NaCl, 15 m*M* reduced glutathione (Sigma).
17. Slide-A-Lyzer® 10K Dialysis Cassette (Pierce, Rockford, IL).
18. Sodium dodecylsulfate-polyacrylamide gel electrophoreses (SDS-PAGE) gels and Coomassie Blue stain.
19. Poly Prep Chromatography Column (Bio-Rad, Munich, Germany).

3. Methods

3.1. Visualizing Nucleocytoplasmic Trafficking by Drug Treatment

The first step in studying the requirements of nuclear export for a given protein is to demonstrate that the protein is indeed capable of nucleocytoplasmic shuttling. A simple approach is the use of chemical compounds that cause a change in the steady-state localization of the protein under investigation. Before and after drug treatment, the protein is visualized by indirect immunofluorescence or, in living cells, by using fusions to autofluorescent proteins (e.g., GFP, BFP, red fluorescent [RFP]) ***(14–16)***. Many fusion proteins with GFP and BFP have been studied and in general, AFP fusions assume the localization and functional properties of the fusion partner ***(14)***. A variety of eukaryotic expression vectors have been developed and are also commercially available (Clontech, Palo Alto, CA; Quantum Appligene). As an example, **Fig. 2** illustrates the expression vector, pER1 (Stauber, unpublished observa-

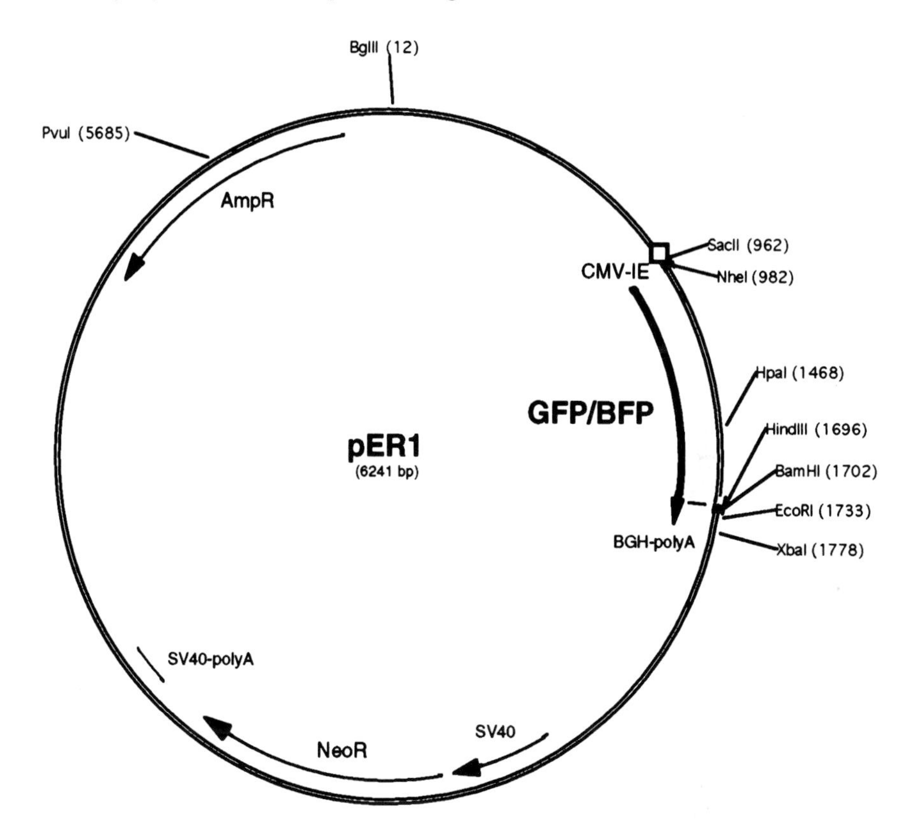

Fig. 2. Schematic representation of a plasmid for the expression of AFP fusion proteins in mammalian cells. Fusions to the N- or C-terminus of GFP/BFP can be constructed by inserting the coding regions of the specific genes into appropriate unique restriction sites. In the resulting plasmids, the expression of the hybrid gene is under the control of the cytomegalovirus immediate-early promoter and the HIV-1 *tat* gene translation initiation sequence and contains the bovine growth hormone polyadenylation signal and the aminoglycosyl phosphotransferase gene (*Neo*) as a selection marker.

tions), which permits expression of the gene of interest as a N- or C-terminal fusion to GFP. Cloning is performed by standard recombinant DNA techniques ***(17)***, and expression of the hybrid protein is usually confirmed by microscopic observation in living cells following transient transfection.

For shuttle proteins displaying a predominantly nuclear localization, transcription inhibitors have been used to demonstrate export. As reported, ActD or DRB appear to block nuclear import of hnRNPA1 ***(18)*** or the HIV-1 Rev protein (***19,20***; **Fig. 3A,B**). Alternatively, for shuttle proteins displaying a predominantly cytoplasmic localization, the use of export inhibitors (e.g., LMB)

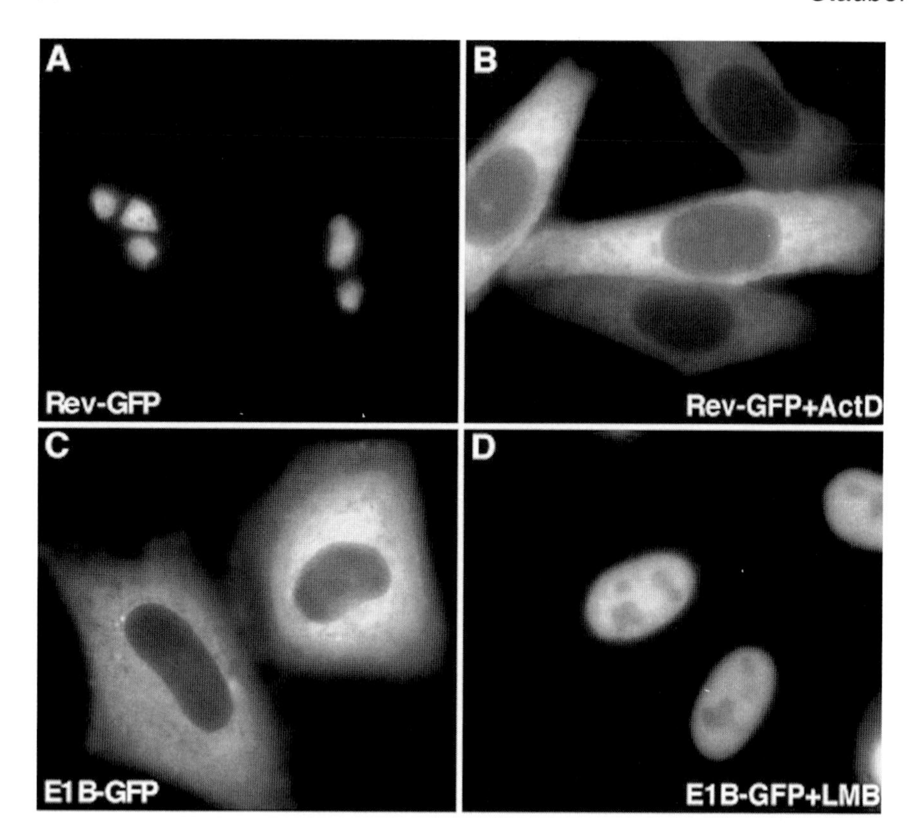

Fig. 3. Drug treatment to detect nucleocytoplasmic trafficking. Cells transfected with the indicated plasmids were analyzed by fluorescence microscopy 16 h later. In living HeLa cells, the nucleolar HIV-1 Rev-GFP protein **(A)** accumulated in the cytoplasm following treatment with ActD **(B)**. The adenovirus type 5 E1B 55K-GFP hybrid localized predominantly to the cytoplasm **(C)**, and accumulated in the nucleus following LMB treatment **(D)**, which blocks nuclear export. (For optimal, color representation please see accompanying CD-ROM.)

will result in nuclear accumulation (**Fig. 3 C,D**; *21–24*). LMB binds and irreversibly inactivates the CRM1 export receptor, thereby blocking the nuclear export of proteins containing a leucine-rich NES *(25)*.

3.1.1. Expression of GFP-Tagged Proteins in Eukaryotic Cells

To verify expression of the GFP-tagged protein, transient expression is straightforward. In addition to electroporation, several transfection methods, including calcium phosphate, diethylaminoethyls-dextran, and liposome-based transfections, have been developed and are commercially available from several companies along with their recommended protocols (Clontech, Palo Alto, CA; Gibco-BRL, Bethesada, MD; Promega). However, it is advisable to optimize and compare different transfection protocols for the individual cell lines and applications.

1. Seed HeLa cells in the appropriate cell culture dishes in DMEM containing 10% fetal bovine serum.
2. 24 h later, wash cells with fresh medium 2 h prior to transfection with calcium phosphate-DNA precipitates.
3. Transfections were performed using the ProFection Calcium Phosphate Transfection kit, according to the manufacturer's instructions. In standard transfection reactions, 5 µg plasmid DNA, encoding AFP fusions, was used. AFP signals were usually observed 16 h posttransfection.

3.1.2. Treatment with LMB or Transcription Inhibitors

1. Transfect cells with plasmids expressing the AFP-hybrid proteins (*see* **Subheading 3.1.1**, **step 1**.).
2. Treat cells 16 h after transfection with LMB (10 n*M* final concentration) for up to 6 h and directly monitor the effect on the localization of the tagged protein by fluorescence microscopy, after various time-points (*see* **Note 2**).
3. Alternatively, incubate transiently transfected cells for 1–4 h with medium containing either 2–10 µg/mL ActD or 50–200 µg/mL DR, and observe the effect on the localization of the tagged protein, by fluorescence microscopy (*see* **Note 2**).

3.2. Cell Fusion Assay to Investigate Nucleocytoplasmic Shuttling

For nuclear proteins not responding to drug treatment the heterokaryon fusion assay represents an efficient in vivo approach ***(27,28)***. In a standard fusion assay, cells expressing the protein of interest (e.g., stable cell lines or by transient expression) are mixed with nonexpressing cells. After PEG treatment, the plasma membranes of the individual cells start to fuse, resulting in a multinuclear syncytium. If the protein of interest is constantly exported from and imported into the donor nucleus, it will be imported also into the acceptor nuclei over time (**Fig. 4A**,**B**). The use of AFP-tagged proteins allows monitoring of shuttling in living cells and comparison of transport kinetics between different proteins. However, fusion assays represent a mixture of nuclear export (from the donor nucleus) and nuclear import (into the donor and acceptor nucleus). To facilitate the discrimination of donor and acceptor nuclei, mouse cells are often used as acceptor cells, because the nuclei of human and mouse cells can easily be distinguished by staining with Hoechst dye, which produces a more punctate staining of mouse nuclei compared to HeLa nuclei ***(29,30)***. Because prolonged or extensive treatment with PEG can also cause the disruption of

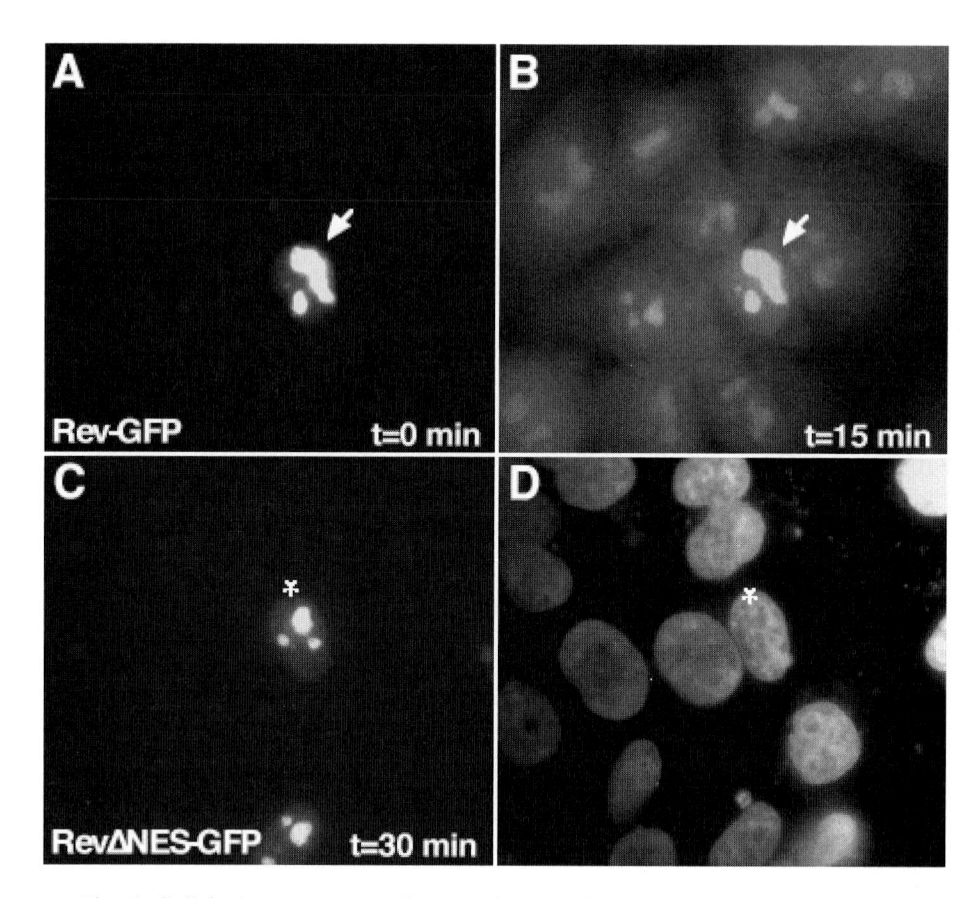

Fig. 4. Cell fusion assay to analyze nucleocytoplasmic transport. Cells transfected with the indicated plasmids were fused 16 h later with PEG and analyzed by fluorescence microscopy. Rev-GFP was efficiently exported and accumulated in the nuclei of the acceptor cells, but RevΔNES-GFP harboring an inactive NES remained in the nucleus of the donor cells. (**A** and **B**) Rev-GFP prior to or 15 min postfusion, respectively. **(C)** RevΔNES-GFP 30 min postfusion. **(D)** Staining of surrounding nuclei in **(C)** with Hoechst dye, to visualize the presence of acceptor nuclei. Arrows indicate donor nuclei. Asterisks mark the RevΔNES-GFP-expressing cell in **(C)** and **(D)**. (For optimal, color representation please see accompanying CD-ROM.)

the nuclear envelope, the experimental conditions have to be controlled by including a nonshuttling protein as a negative control **(Fig. 4C,D)**. In addition, new protein synthesis might flaw the results. Therefore, appropriate inhibitors should be present during the experiment.

1. Transfect cells with the specific expression plasmid, as done in **Subheading 3.1.1.**
2. To block new protein synthesis, cycloheximide (50 µg/mL) should be added 30 min before fusion and should be present during the experiment. Additionally, all solutions should be prewarmed to 37°C prior to use.

3. 16 h posttransfection seed cells with a five-fold excess of untransfected cells (e.g., HeLa or mouse NIH3T3 cells).
4. The following day, wash the cells with 2 mL PBS, and add 1 mL PEG for 2 min.
5. Wash the cells 3× with 3 mL PBS, add 2 mL of the appropriate cell culture medium and continue further incubation at 37°C.
6. Immediately after fusion, cells should be observed under phase contrast and fluorescent illumination and scanned for fusion events involving one donor cell and surrounding acceptor cells. Ideally, the same fusion event should be recorded over time (**Fig. 4**; *see* **Note 3**).
7. Because the degree of fusion may vary, depending on the PEG preparation and the cell lines used, it is advisable to optimize the experimental parameters in pilot experiments.
8. For Hoechst staining, the cells are incubated with Hoechst 33258 (final concentration 1 µg/mL in PBS) for 15 min and the dye removed by several washes with PBS prior to microscopic observation.

3.3. Microinjection of Recombinant Proteins to Characterize Transport Signals

The previously described assays indicate whether the proteins of interest are capable of nucleocytoplasmic transport. However, the identification of transport signals is essential, in order to characterize and understand specific transport pathways and the regulation thereof. Capillary microinjection of recombinant proteins has been proven an efficient approach to directly investigate nuclear export independent of import or vice versa. A limitation is often the efficient expression and purification of large or toxic proteins in sufficient amounts to study transport of the full-length proteins. Thus, this technique has been mostly used to identify and characterize in detail nuclear export or import signals ***(31,32)***. NLS/NESs are either expressed as a fusion with a heterologous protein (e.g., GST) or NLS/NES-peptides are conjugated to bulky in vitro fluorescently labeled carrier proteins (e.g., bovine serum albumin), to avoid passive intracellular diffusion. Subsequently, the substrates are injected into the nuclei of somatic cells to study export or into the cytoplasm to investigate nuclear import, respectively. Microinjection also allows one to introduce specific inhibitors (e.g., antibodies, drugs, fluorescent lectins, and so on), together with the transport substrates in order to analyze their effects on individual pathways ***(33–35)***. The localization of fusion proteins is monitored by immunostaining or by direct fluorescence. An elegant and stringent approach represents the use of transport signals linked to a chimeric protein composed of GST fused to GFP or BFP (*see* **Fig. 5**), respectively, which allows recording of real-time kinetics of transport ***(35,36)***. Transport of the stable and highly fluorescent substrates can be observed directly by fluorescence microscopy in living cells following microinjection. In addition, the size of GST-AFP

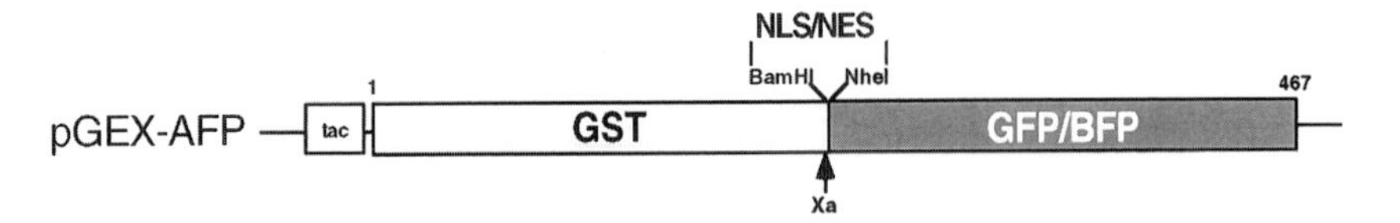

Fig. 5. Schematic representation of the plasmid used for the bacterial expression of GST-AFP fusion proteins. Expression of the GST-GFP/BFP hybrid proteins is under the control of the IPTG-inducible *tac*-promotor/*lac*-operator system. The indicated unique *Bam*HI- and *Nhe*I-restriction sites were used for cloning of the NLS/NES sequences. Xa indicates the sequence recognized by the factor Xa-protease, in order to cleave off the GST part.

(~54 kDa, as a monomer) prevents intracellular passive diffusion. Strictly speaking, transient expression of a GFP-NLS fusion will only indicate if the tested signal can mediate nuclear retention, since GFP is able to enter the nucleus by itself. Thus, microinjection of the GST-NLS/NES-AFP fusions is a highly stringent approach to investigate whether the signal under investigation is capable of mediating nuclear entry or export, respectively. In addition, microinjection allows one to compare and quantify the activities of different import and export signals, which may be important for the biological function of the protein of interest.

The combination of GST–AFP-tagging also allows efficiently control of protein expression and monitoring of protein purification, amenable especially for toxic or weakly expressed proteins. In addition, a variety of parameters essential for the expression of the recombinant protein in *E. coli* can be tested rapidly by quantifying the GFP- signal in bacterial suspensions using a multiwell fluorescence plate reader ***(37)***. The attempt to include the newly discovered RFP from *Anthozoa* ***(16)*** in this system failed, because since a GST–RFP hybrid displayed extensive aggregation and could not be used in microinjection (Stauber, unpublished observations).

The recombinant GST-AFP proteins containing import/export signals can also be used to study the interaction with in vitro translated import/export receptors in pull-down experiments (Stauber, unpublished observations).

As outlined in **Fig. 6** the following section focuses on the production and microinjection of recombinant chimeric GST-AFP proteins containing specific transport signals (*see* **Note 9**).

3.3.1. Induction of Recombinant Protein Expression

1. Inoculate a single colony of *E. coli* (strain BL21), containing the expression plasmid, into 50 mL LB/ampicillin (50 µg/mL ampicillin), and grow 12–15 h at 37°C in a shaking incubator.

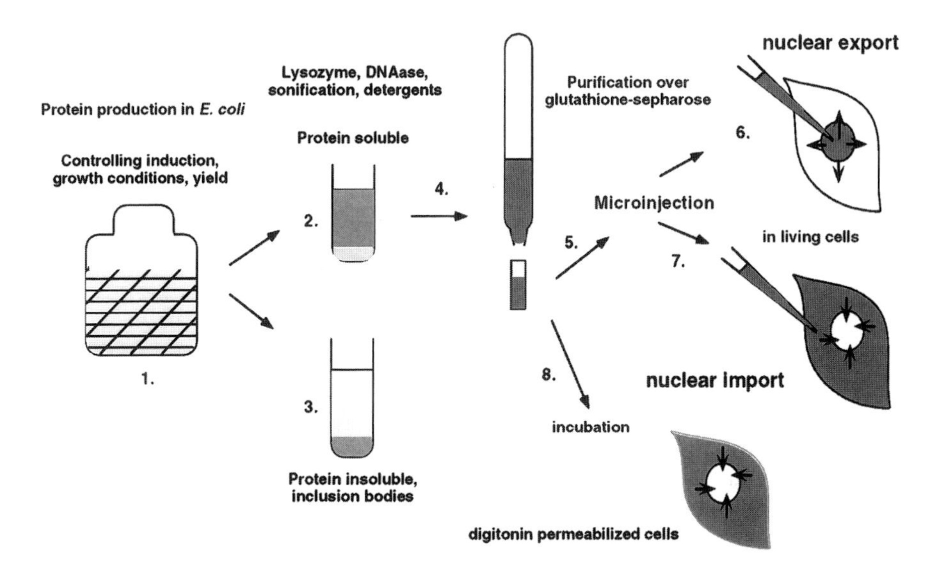

Fig. 6. Purification scheme for recombinant GST-AFP fusion proteins from bacteria. (For optimal, color representation please see accompanying CD-ROM.)

2. Add 450 mL LB/ampicillin, and grow until an OD600 of 0.5. Add IPTG, (freshly prepared in water) to a final concentration of 0.2 m*M*, and continue incubation for additional 3–4 h.
3. Harvest the bacterial cells by centrifugation at 5000*g* for 20 min at 4°C and discard the supernatant. The pellet can be stored at –70°C.

3.3.2. Optimization of Expression Conditions (see **Note 3**)

By optimizing growth conditions, the yield of fusion protein may be greatly improved. Investigate the effects of delaying the addition of IPTG, various IPTG concentrations, and of altering the induction period and growth temperature.

1. Inoculate 30 mL LB with a single colony of *E. coli* BL21 containing the expression plasmid.
2. Grow overnight at 37°C in a shaking incubator.
3. Add 5 mL of the culture to 45 mL LB, and grow until a OD600 of 0.5 at 20, 27, or 37°C. Add IPTG (freshly prepared in water) to a final concentration of 0.05, 0.2, or 0.6 m*M*. Continue incubation for the desired time-points (e.g., 4, 8, 16 h).
4. Following induction, harvest aliquots of the cultures at the desired time-points, and adjust to a OD600 of 0.4 with LB. Transfer 100 µL of the suspensions to a 96-well plate, and measure the GFP signal in a fluorescence plate reader. Quantification of the GFP signal will immediately reveal the optimal expression conditions as illustrated in **Fig. 7**.

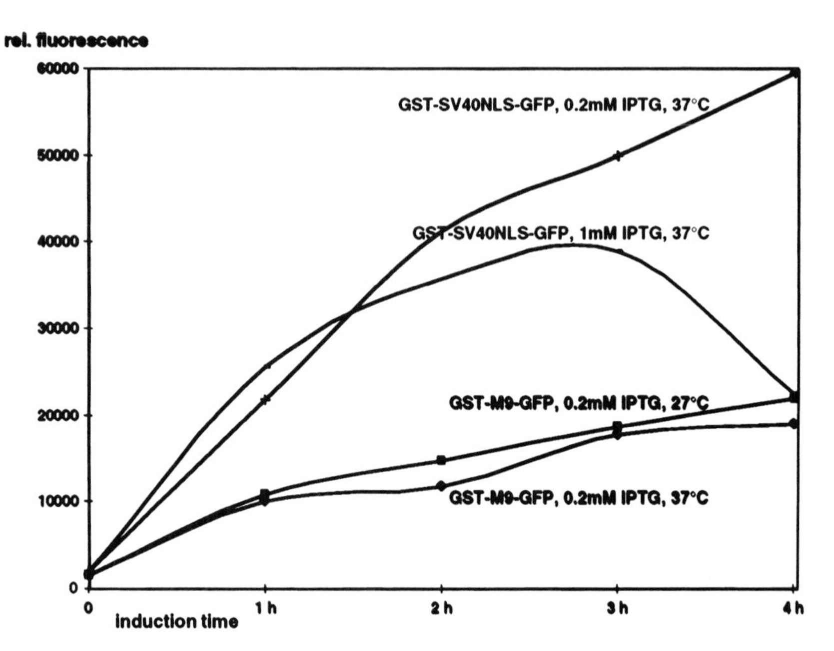

Fig. 7. Quantitation of recombinant AFP production under various growth conditions. The GFP signal in bacterial suspensions expressing a GST-SV40NLS-GFP or GST-M9-GFP *(37)*, respectively, was recorded in a Victor 1420 Multilabel Counter using the F485 nm excitation and the F510 nm emission filter with a lamp energy setting at 10,000 and a measure time of 0.2 s. Inducing the expression of a GST-SV40NLS-GFP hybrid protein, with 0.2 m*M* IPTG, resulted in a linear increase of the GFP signal over time; induction with 1 m*M* IPTG caused a drop of the GFP signal after 3 h, possibly because toxic side effects of IPTG. Induction at 27 or 37°C, respectively, resulted in similar expression yields of the GST-M9-GFP fusion protein.

3.3.3. *Lysis of* E. coli

1. Resuspend the bacterial cell pellet (500 mL culture) in 10 mL cold PBS containing 1X protease inhibitor mix. Add lysozyme to a concentration of 250 µg/mL, 26 µL $MnCl_2$ (1 *M* stock solution), and 260 µL $MgCl_2$ (1 *M* stock solution), DNase to a concentration of 100 µg/mL. Incubate for 20 min at 4°C.
2. Lyse cells using a probe sonicator with a 5-mm-diameter probe. For a Branson Sonifier 450, sonify for 3 × 10 s. The output frequency should be set to the microtip limit noted on the dial.
3. Add 1 mL 5 *M* NaCl + 0.5 mL 20% Triton X-100 and incubate for 15 min at 4°C.
4. Centrifuge the suspension at 10,000*g* for 20 min at 4°C to remove insoluble material and intact cells. Usually, the supernatant is green if the protein is soluble and well-expressed **(Fig. 6, no. 2)**.

3.3.4. Affinity Chromatography of GST-GFP Fusion Protein

1. Wash 250–750 µL glutathione-Sepharose 4B 3X with 10 mL PBS.
2. Remove PBS, add the bacterial lysate, and incubate for 2 h on a rotator shaker at 4°C. The capacity of glutathione-Sepharose 4B is ~6 mg protein/mL beads.
3. Spin down the sepharose at 1000*g* for 5 min at 4°C, remove supernatant, and add 10 mL PBS containing 1% Triton X-100. The sepharose pellet should be green, because of the bound GFP-hybrid protein. BFP will only be visible under UV illumination using, e.g., a UV-light box.
4. Spin down the sepharose at 1000*g* for 5 min at 4°C, remove supernatant, and perform two additional washing steps in 10 mL PBS–1%Triton X-100, followed by two additional washing steps with PBS.
5. Resuspend beads in 5 mL PBS, and transfer to a Poly Prep Chromatography Column (*see* **Note 5**).
6. Let the column run dry, add 1–2 mL of freshly prepared elution buffer and incubate for 10–20 min at room temperature with gentle agitation.
7. Collect the eluted protein fractions. Additional elution steps can be performed and fractions controlled by quantitation of the GFP signal.
8. Transfer the eluted protein to a Slide-A-Lyzer 10K Dialysis Cassette and dialyze extensively against PBS overnight at 4°C.
9. Freeze aliquots of the purified GST-GFP hybrid protein in liquid nitrogen and store at –70°C.
10. The integrity and concentration of the purified proteins should be analyzed by SDS-PAGE, using Coomassie blue stain.

3.3.5. Microinjection and Observation of the GST-AFP Fusion Protein

1. Seed cells into 35-mm glass bottom dishes and let them grow for 16 h (*see* **Notes 6** and **7**).
2. Spin down the recombinant protein (concentration ~2 mg/mL) for 30 min in a tabletop centrifuge at 13,000 rpm, 4°C for 30 min to remove any debris that might clog the injection needle.
3. Exchange the cell culture medium for injection medium, perform microinjection, and continue incubation in regular cell culture medium. If desired, co-inject immunoglobulinG (concentration ~1 mg/mL) as a microinjection control, which can be detected by appropriate rhodamine-conjugated secondary antibodies following fixation of the cells in 4% paraformaldehyde/PBS for 20 min.
4. Observe injected cells immediately after injection, at the desired time-points or after fixation **(Fig. 8)**. Fixed cells should be stored in PBS at 4°C and the GFP signal is usually stable for several days (*see* **Note 8**).

4. Notes

1. Working with BFP: BFP-hybrid proteins have been used in practical applications (e.g., transient transfection, recombinant protein production ***[21,35]***). However, one has to keep in mind that the BFP signal is generally weaker, compared to

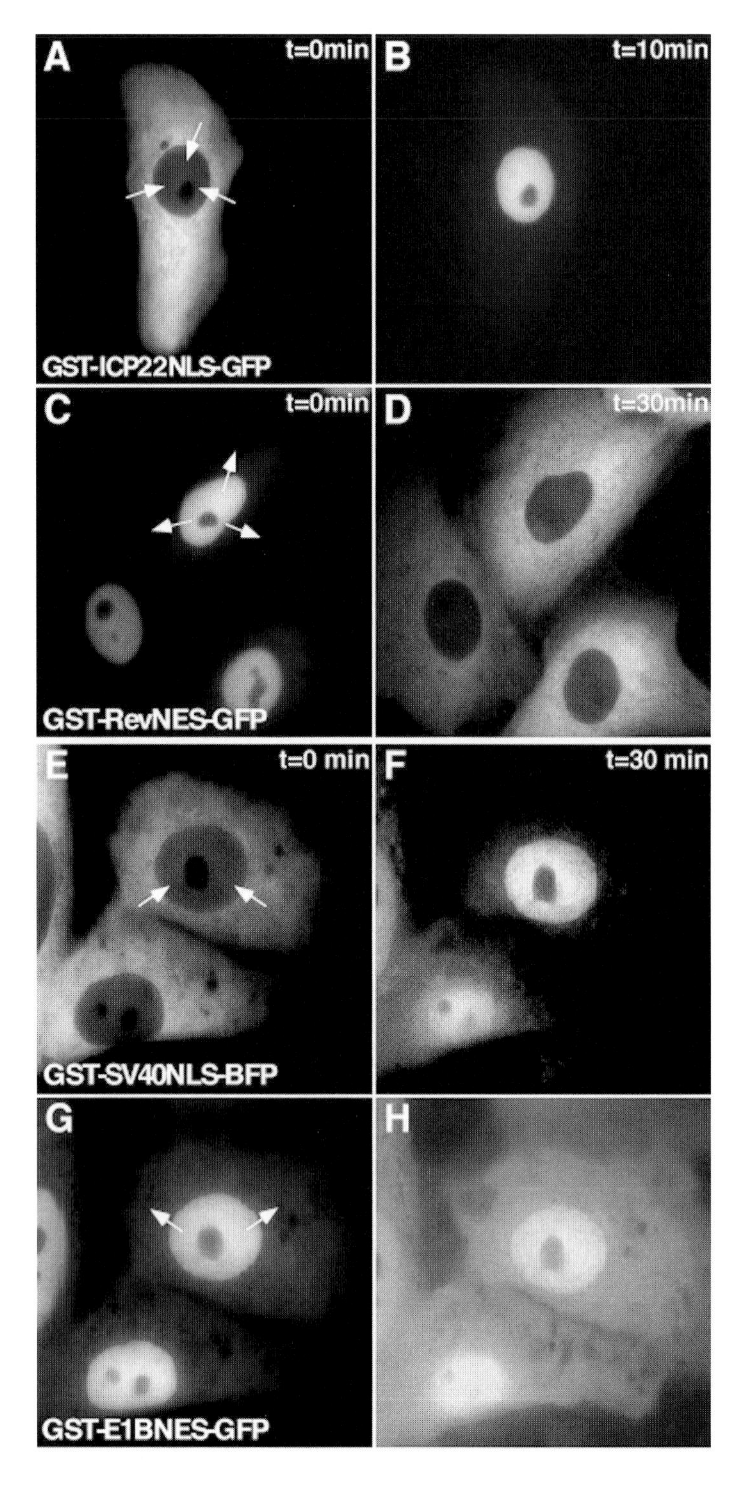
A
t=0min
GST-ICP22NLS-GFP
B
t=10min
C
t=0min
GST-RevNES-GFP
D
t=30min
E
t=0 min
GST-SV40NLS-BFP
F
t=30 min
G
GST-E1BNES-GFP
H

GFP, is more affected by bleaching and harder to detect because of the higher background fluorescence caused by the short-wavelength UV illumination necessary for BFP excitation.

2. For prolonged or high-dose treatment, the cytotoxic side effects of LMB should be taken into consideration. ActD or DRB are also cytotoxic drugs, especially when used in high doses. Since the activity of different LMB preparations appears to vary, the LMB preparations should be tested prior to use. Treatment of cells expressing the HTLV-1 Rex-GFP *(22)* or the E1B-55K-GFP protein *(23)* for 30 min should result in nuclear accumulation of the GFP-hybrid proteins (**Fig. 3C**,**D**).
3. The transport and accumulation of the AFP-tagged proteins into the acceptor nuclei may vary from minutes to hours, depending on the size of the tagged protein and its shuttling activity (**Fig. 40**; ***18,30***).
4. Proteins containing sequences rich in basic amino acids have a tendency to be insoluble (*see* **Fig. 6**, **no. 3**). Lowering the growth temperature, and the concentration of IPTG, and extending the induction time, can increase the percentage of soluble protein in the supernatant. Such fusion proteins also tend to aggregate and precipitate during dialysis. Therefore, it is advisable to elute and dialyze the less-concentrated recombinant proteins. In order to achieve the protein concentration necessary for detection, the protein solution can be concentrated using microconcentration spin columns (e.g., 30K spin columns, Pall Filtron Nanosep™ Micro-concentrations, Dreieich, Germany) immediately prior to microinjection. In addition, storage of highly concentrated protein solutions can result in aggregation and precipitation over time.
5. The GST part of the GST-AFP fusion protein can be removed by treatment with the factor Xa-protease (Roche). Cleavage can be performed "on the beads" in the buffer specified by the manufacturer. However, in contrast to GST-AFPs, GFP/BFP distributes equally between the nucleus and the cytoplasm following microinjection because of its size and compact structure.

Fig. 8. (*opposite page*) Real-time observation of nucleocytoplasmic transport in living cells. Purified import or export substrates were injected into the cytoplasm or nucleus of Vero cells, respectively, and transport was monitored directly by fluorescence microscopy. Microinjected GST-ICP22NLS-GFP *(36)* was imported into the nucleus in 10 min (**A** and **B**). Nuclear export of a GST-RevNES-GFP hybrid protein *(35)* was completed after 30 min (**C** and **D**). Simultaneous observation of nuclear import and export. Vero cells were microinjected into the cytoplasm with an import substrate (GST-SV40NLS-BFP) (**F**) and, subsequently, an export substrate (GST-E1BNES-GFP) (**G**) was injected into the nucleus of the same cells. Nuclear import (**F**) and export (**H**) occurred simultaneously and could be observed independently using the appropriate filters to detect BFP (**E** and **F**) or GFP (**G** and **H**), respectively. To prevent the onset of export in-between microinjections, cells were kept in ice-cold medium. (For optimal, color representation please see accompanying CD-ROM.)

6. The success of microinjection experiments not only depends on the quality of the protein preparation, but also on the conditions of the cells used for the experiments. Thus, care should be taken in cultivating cells (i.e., high-quality cell culture medium and good cell culture techniques).
7. Instead of using glass-bottomed dishes, cells can also be seeded on glass cover slips for microinjection, and can also be observed after fixation with 4% paraformaldehyde/PBS. However, the author prefers "live recording," since a loss of the AFP signal intensity was experienced following fixation.
8. If transport occurs to fast to be monitored at room temperature, the use of cooled microinjection stages or of cold injection medium will slow down transport processes.
9. In vivo export assays are important for the identification and characterization of export signals and pathways. To further dissect the molecular mechanism of transport, in vitro assays are required to reconstitute export/import from biochemically defined components. Mostly, digitonin-permeabilized cells serve as an in vitro assay to investigate the nucleocytoplasmic transport of fluorescent-labeled substrates. The described GST–AFP fusions are, however, not well-suited for digitonin-permeabilized cells, since the recombinant substrates tend to adhere to unknown structures in the cytoplasm. The reader is therefore advised to refer to the original publications for experimental details and to choose the most adequate assay system to address specific questions ***(38–41)***.

Acknowledgments

The author thanks Friedrich Krätzer and Prof. K. Rhino for critical reading of the manuscript, and Stefanie Müller for continuous encouragement. Research in the author's laboratory is supported by the Wilhelm-Sander Stiftung, the Johannes und Frieda Marohn Stiftung, and the Kalkhof-Rose Stiftung.

References

1. Pantè, N. and Aebi, U. (1996) Molecular dissection of the nuclear pore complex. *Crit. Rev. Biochem. Mol. Biol.* **31,** 153–199.
2. Fabre, E. and Hurt, E. (1997) Yeast genetics to dissect the nuclear pore complex and nucleocytoplasmic trafficking. *Annu. Rev. Genet.* **31,** 277–313.
3. Gant, T. M., Goldberg, M. W., and Allen, T. D. (1998) Nuclear envelope and nuclear pore assembly: analysis of assembly intermediates by electron microscopy. *Curr. Opin. Cell Biol.* **10,** 409–415.
4. Corbett, A. and Silver, P. A. (1997) Nucleocytoplasmic transport of macromolecules. *Microbiol. Mol. Biol. Rev.* **61,** 193–211.
5. Görlich, D. (1998) Transport into and out of the cell nucleus. *EMBO J.* **17,** 2721–2727.
6. Görlich, D. and Kutay, U. (1999) Transport between the cell nucleus and the cytoplasm. *Annu. Rev. Cell Dev. Biol.* **15,** 607–660.
7. Izzaurralde, E. and Adam, S. (1998) Transport of macromolecules between the nucleus and the cytoplasm. *RNA* **4,** 351–364.

8. Jäkel, S. and Görlich, D. (1998) Importin ß, transportin, RanBP5 and RanBP7 mediate nuclear import of ribosomal proteins in mammalian cells. *EMBO J.* **17,** 4491–4502.
9. Weis, K. (1998) Importins and exportins: how to get in and out of the nucleus. *TIBS* **23,** 185–189.
10. Henderson, B. R. and Eleftheriou, A. (2000) A comparison of the activity, sequence specificity, and CRM1–dependence of different nuclear export signals. *Exp. Cell Res.* **256,** 213–224.
11. Mattaj, I. W. and Englmeier, L. (1998) Nucleocytoplasmic transport: The soluble phase. *Ann. Rev. Biochem.* **67,** 265–306.
12. Wozniak, R. W., Rout, M. P., and Aitchison, J. D. (1998) Karyopherins and kissing cousins. *Trends Cell Biol.* **8,** 184–188.
13. Nachury, M. V. and Weis, K. (1999) The direction of transport through the nuclear pore can be inverted. *Proc. Natl. Acad. Sci. USA* **96,** 9622–9627.
14. Stauber, R. H., Horie, K., Carney, P., et al. (1998) Development and applications of enhanced green fluorescent protein mutants. *BioTechniques* **24,** 462–471.
15. Tsien, R. (1998) The green fluorescent protein. *Annu. Rev. Biochem.* **67,** 509–544.
16. Matz, M. V., Fradkov, A. F., Labas, Y. A., et al. (1999) Fluorescent proteins from nonbioluminiscent Antozoa species. *Nat. Biotechnol.* **17,** 969–973.
17. Sambrook, J., Fritsch, E. F., and Maniatis, T. (1989) *Molecular Cloning: A Laboratory Manual, 2nd ed.*, Cold Spring Harbor Laboratory, Cold Spring Harbor, New York.
18. Pinol-Roma, S. and Dreyfuss, G. (1992) Shuttling of pre-mRNA binding proteins between nucleus and cytoplasm. *Nature* **355,** 730–732.
19. Richard, N., Iacampo, S., and Cochrane, A. (1994) HIV-1 Rev is capable of shuttling between the nucleus and cytoplasm. *Virology* **204,** 123–131.
20. Meyer, B. E. and Malim, M. H. (1994) The HIV-1 Rev trans-activator shuttles between the nucleus and the cytoplasm. *Genes Dev.* **8,** 1538–1547.
21. Stauber, R. H., Afonina, E., Gulnik, S., et al. (1998) Analysis of intracellular trafficking and interactions of cytoplasmic HIV-1 Rev mutants in living cells. *Virology* **251,** 38–48.
22. Heger, P., Rosorius, O., Hauber, J., et al. (1999) Titration of cellular export factors, but not heteromultimerization, is the molecular mechanism of trans-dominant HTLV-1 Rex mutants. *Oncogene* **18,** 4080–4090.
23. Krätzer, F., Rosorius, O., Heger, P., et al. (2000) The adenovirus type 5 E1B-55K oncoprotein is a highly active shuttle protein and shuttling is independent of E4orf6, p53 and Mdm2. *Oncogene* **19,** 850–857.
24. Toyoshima, F., Moriguchi, T., Wada, A., et al. (1998) Nuclear export of cyclin B1 and its possible role in the DNA damage-induced G2 checkpoint. *EMBO J.* **17,** 2728–2735.
25. Fornerod, M., Ohno, M., Yoshida, M., et al. (1997) CRM1 is an export receptor for leucine-rich nuclear export signals. *Cell* **90,** 1051–1060.
26. Cid-Arregui, A. and García-Carrancá, A. (1998) Springer-Verlag, Heidelberg.

27. Izaurralde, E., Jarmolowski, A., Beisel, C., et al. (1997) A role for the M9 transport signal of hnRNP A1 in mRNA nuclear export. *J. Cell. Biol.* **137,** 27–35.
28. Afonina, E., Stauber, R., and Pavlakis, G. N. (1998) The human poly-A binding protein 1 shuttles between the nucleus and the cytoplasm. *J. Biol. Chem.* **273,** 13,015–13,021.
29. Moser, F. G., Dorman, B. P., and Ruddle, F. H. (1975) Mouse-human heterokaryon analysis with a 33258 Hoechst-Giemsa technique. *J. Cell. Biol.* **66,** 676–680.
30. Fan, X. C. and Steitz, J. A. (1998) HNS, a nuclear-cytoplasmic shuttling sequence in HuR. *Proc Natl Acad Sci USA* **95,** 15,293–15,298.
31. Bogerd, H. P., Benson, R. E., Truant, R., et al. (1999) Definition of a consensus transportin-specific nucleocytoplasmic transport signal. *J. Biol. Chem.* **274,** 9771–9777.
32. Wen, W., Meinkoth, J. L., Tsien, R. Y., et al. (1995) Identification of a signal for rapid export of proteins from the nucleus. *Cell* **82,** 463–473.
33. Ullman, K. S., Shah, S., Powers, M. A., et al. (1999) The nucleoporin nup153 plays a critical role in multiple types of nuclear export. *Mol. Cell Biol.* **10,** 649–664.
34. Elfgang, C., Rosorius, O., Hofer, L., et al. (1999) Evidence for specific nucleoplasmic transport pathways used by leucine-rich nuclear export signals. *Proc. Natl. Acad. Sci. USA* **96,** 6229–6234.
35. Stauber, R. H., Krätzer, F., Schneider, G., et al. (2000) Investigation of nucleocytoplasmic transport using UV-guided microinjection. *J. Cell. Biochem.* (in press).
36. Rosorius, O., Heger, P., Stelz, G., et al. (1999) Direct observation of nucleo-cytoplasmic transport by microinjection of GFP-tagged proteins in living cells. *BioTechniques* **27,** 350–355.
37. Rücker, E., Schneider, G., Steinhäuser, K., et al. (2000) Rapid evaluation and optimization of recombinant protein production using GFP tagging. *Protein Expression and Purification,* (in press).
38. Love, D. C., Sweitzer, T. D., and Hanover, J. A. (1998) Reconstitution of HIV-1 rev nuclear export: independent requirements for nuclear import and export. *Proc. Natl. Acad. Sci. USA* **95,** 10,608–10,613.
39. Kehlenbach, R. H., Dickmanns, A., and Gerace, L. (1998) Nucleocytoplasmic shuttling factors including Ran and CRM1 mediate nuclear export of NFAT In vitro. *J. Cell Biol.* **141,** 863–874.
40. Adam, E. J. H. and Adam, S. A. (1994) Identification of cytosolic factors required for nuclear localization sequence-mediated binding to the nuclear envelope. *J. Cell Biol.* **125,** 547–555.
41. Keminer, O., Siebrasse, J.-P., Zerf, K., and Peters, R. (1999) Optical recording of signal-mediated protein transport through single nuclear pore complexes. *Proc. Natl. Acad. Sci. USA* **96,** 11,842–11,847.

IV

Green Fluorescent Protein in Transgenic Organisms

15

Transgenic Bovine Embryo Selection Using Green Fluorescent Protein

Anthony W. S. Chan, Kowit-Yu Chong, and Gerald Schatten

1. Introduction

Transgenic animals currently play an important role in biological research and commercial developments. Pronuclear microinjection (PI), viral infection, receptor mediated gene transfer, sperm vector, the combination of sperm vector and intracytoplasmic sperm injection and nuclear transplantation have been successfully used to produce transgenic animals ***(1–8)***. However, low efficiency and high mosaic rates are still major problems for most gene delivery methods ***(9–10)***. The selection of transgenic embryos, before transfer into recipient females is believed to be one solution for high production costs caused by low transgenic rate ***(11)***. Several methods have been suggested for transgenic embryo selection, including the detection of transgene in embryos prior to implantation (preimplantation) using polymerase chain reaction, fluorescent *in-situ* hybridization, and the expression of a reporter gene ***(12–19)***. However, the presence of nonintegrated exogenous DNA in polymerase chain reaction analysis and the technically complicated process of fluorescent *in-situ* hybridization have limited their application ***(11)***.

Expression of firefly luciferase, β-galactosidase (*lacZ*), secreted placental alkaline phosphate, and green fluorescent protein (GFP) has been demonstrated in mammalian embryos ***(8,17,19–21)***. Among these transgenic reporters, GFP is the most commonly used reporter because of the visibility by conventional fluorescent microscopy, its autoactivation, and the fact that it is nondetrimental to embryo and fetal development ***(8,17,22,23)***. Several reports demonstrate the use of GFP as a marker for the selection of transgenic embryos prior to implantation ***(8,17,24,25)***. Despite the advantages of using GFP as a transgenic marker, a suitable mechanism of regulating expression must also be available

From: *Methods in Molecular Biology, vol. 183: Green Fluorescent Protein: Applications and Protocols*
Edited by: B. W. Hicks © Humana Press Inc., Totowa, NJ

for accurate selection. As a transgenic reporter, constitutive expression is not necessary throughout development but merely at a specific stage and time, or in a controllable manner ***(11,24)***. Stage specific promoters such as Oct/4 promoter, and inducible promoter systems such as tetracycline switch, will be an alternative regulatory mechanism for transgenic selection using GFP. Here we present a protocol for the application of GFP as a transgenic reporter in bovine embryos prior to implantation (preimplantation selection).

2. Materials

1. Maturation medium ***(26)***: TC-199 medium (GIBCO) supplemented with 10% heat inactivated fetal bovine serum, 0.2 m*M* sodium pyruvate, 5 μg/mL ovine-leuteinizing hormone, 25 μg/mL gentamycin sulfate, and 1 μg/mL estradiol-17β.
2. Modified Tyrode's-lactate medium (TL-stock/fertilization medium) ***(27)***: 114 m*M* NaCl, 3.2 m*M* KCl, 25 m*M* $NaHCO_3$, 0.4 m*M* NaH_2PO_4, 10 m*M* Na-lactate (60% syrup), 2 m*M* $CaCl_2$, 0.5 m*M* $MgCl_2$, 100 IU/mL penicillin, 50 μg/mL phenol red, 6 mg/mL bovine serum albumin (BSA) (fatty-acid-free [FAF]), 0.2 m*M* pyruvate, and 25 μg/mL gentamycin sulfate (mOsm 280–300).
3. Sperm-tyrodes-lactate (Sp-TL) ***(28)***: 100 m*M* NaCl, 3.1 m*M* KCl, 25 m*M* $NaHCO_3$, 0.29 m*M* NaH_2PO_4, 21.6 m*M* Na-lactate (60% syrup), 2.1 m*M* $CaCl_2$, 0.4 m*M* $MgCl_2$, 10 m*M* HEPES, 50 μg/mL phenol red, 6 mg/mL BSA-fraction V, 1.0 m*M* pyruvate, and 25 μg/mL gentamycin sulfate (pH 7.4; mOsm 290–300).
4. Percoll gradient ***(28)***: 90% percoll was prepared by mixing one part 10X Percoll salt solution and nine parts Percoll solution. 45% Percoll was prepared by mixing one part of 90% Percoll and one part Sp-TL. 2 mL 45% Percoll solution was laid on top of 2 mL 90% Percoll solution in a 15-mL conical tube and pre-equilibrated at 39°C with 5% CO_2 for at least 2 h. 10X Percoll salt: 32 m*M* KCl, 2.9 m*M* $NaHPO_4$, 800 m*M* NaCl, and 100 m*M* HEPES. 90% Percoll: Add one part 10X Percoll salt with 9 parts Percoll and supplement with 2.1 m*M* $CaCl_2$, 0.4 m*M* $MgCl_2$, 10 m*M* Na-lactate (60% syrup), and 25 m*M* $NaHCO_3$.
5. CR1-aa embryo culture medium ***(29)***: 114 m*M* NaCl, 3.2 m*M* KCl, 25 m*M* $NaHCO_3$, 0.4 m*M* NaH_2PO_4, 10 m*M* Na-lactate (60% syrup), 2 m*M* $CaCl_2$, 0.5 m*M* $MgCl_2$, 6 mg/mL BSA-FAF, 0.2 m*M* pyruvate, 5 m*M* hemicalcium lactate, 1 m*M* L-glutamine, and 25 μg/mL gentamycin sulfate (pH 7.4; mOsm 270–280).
6. TL-HEPES ***(27)***: 114 m*M* NaCl, 3.2 m*M* KCl, 2.0 m*M* $NaHCO_3$, 0.4 m*M* NaH_2PO_4, 10 m*M* Na-lactate (60% syrup), 2 m*M* $CaCl_2$, 0.5 m*M* $MgCl_2$, 10 m*M* HEPES, 100 IU/mL penicillin, 50 μg/mL phenol red, 1 mg/mL BSA-fractionV, 0.2 m*M* pyruvate, and 25 μg/mL gentamycin sulfate (pH 7.4; mOsm 255–270).
7. PHE ***(30)***: 0.5 m*M* penicillamine, 0.25 m*M* hypotaurine, 25 m*M* epinephrine. Hypotaurine (1 m*M*) and penicillamine (2 m*M*) stocks are prepared in 0.9% NaCl, aliquoted, and kept frozen. For 250 mM epinephine stock: 50 mL double-distilled water with 165 mg Na-lactate (60% syrup) and 50 mg sodium metabisulfite, pH 7.4, is prepared, and epinephrine (1.83 mg) is added to 40 mL of this solution. PHE is prepared by addition of 2.5 parts of hypotaurine (1 m*M*) and 2.5 parts

penicillamine (2 m*M*) stocks, 1 part 250 m*M* epinephine stock, and 4 parts of 0.9% NaCl. The mixture was filtered, aliquoted, and kept frozen at –20°C.

8. pEGFP expression vector (Clontech, Palo Alto, CA). 1% agarose gel containing 1 µg/mL ethidium bromide.
9. 10X TBE (1 L): 108 g Tris, 55 g boric acid, 9.3 g EDTA.
10. TE: 5–10 m*M* Tris, pH 7.4, 0.1–0.25 m*M* EDTA.
11. Gel purification kit (Qiagen™). DNA can also be visualized, using DNA stain, such as Syber Green, rather than ethidium bromide.
12. Inverted fluorescent microscope (e.g., Nikon TEM-300) equipped with fluorescein isothiocyanate filter set and Hoffmann Modulation Contrast optics. Inverted microscope, equipped with optics such as differential interference contrast, could also be used.
13. Narishigi micromanipulator.
14. Micropipet puller Model P-97 (Sutter, Novato, CA). Borosilicate micropipet with filament (OD 1.0 mm, ID 0.75 mm) (Sutter).
15. Micro-forge MF-1 (Technical Products, St. Louis, MO).
16. Microsyringe (Stoeling, Wood Dale, IL).
17. Eppendorf transjector 5246: The parameters for the transjector were set with an injection pressure of 300–500 hpa, compensation pressure of 15–25 hpa, and the time of injection was adjusted by observing the swelling of the pronuclei (Eppendorf, Westbury, NY).
18. Waterjacketed CO_2 incubator (Forma Scientific).
19. Hamilton microsyringe: 500 µL.
20. Heparin (2 µg/mL).
21. All chemicals were purchased from Sigma unless otherwise noted. All buffers and media are sterilized using 0.2-µm filter. Oocytes and embryos are handled in a warm room (30–32°C) or on a warm plate setting at 37–39°C, to minimize temperature fluctuation.

3. Methods

3.1. Oocyte Maturation, Fertilization, and Culture (26,31)

3.1.1. In Vitro Maturation (IVM) of Oocytes

1. Bovine ovaries are collected from a slaughterhouse and transported in normal saline at ~34–37°C in a thermos.
2. Ovaries are washed with running warm water and kept at 39°C in a water bath.
3. Follicles (2–6 mm in diameter) are aspirated with an 18 G needle and 10 mL syringe. Aspirate the follicles underneath the surface (cortex) by maintaining negative pressure in the syringe while the needle is passed through the cortex. 20–30 oocytes could be able to recover from each ovary, which depend on the age of the donor animals.
4. Transfer the oocytes into a 50-mL conical tube and keep at 39°C in a water bath.
5. After aspiration, allow the oocytes to settle down at the bottom of the conical tube.

6. Transfer the oocytes from the bottom of the tube using a 9-inch sterile Pasteur pipet into a 10-cm Petri dish containing 10 mL TL-HEPES.
7. Allow the oocytes to settle at the bottom of the dish.
8. Examine the oocytes under a dissecting microscope, and only those with multiple layers of compact cumulus cells (cumulus oocytes complex [COC]) and evenly granulated cytoplasm are selected for in vitro maturation.
9. Pick up COC surrounded by more than 2–3 layers of cumulus cells and transfer them into a 35-mm Petri dish with 4 mL TL-HEPES. A 500-μL Hamilton gas tight syringe, attached with a glass capillary, is used for oocyte transfer.
10. Wash the oocytes 3×, using fresh TL-HEPES, in a 35-mm Petri dish.
11. Transfer 10 COCs per group into a 50-μL drop of maturation medium pre-equilibrated at 39°C in a 60-mm Petri dish and covered with mineral oil.
12. Culture the oocytes at 39°C with 5% CO_2 for 24 h before insemination. One of the signs of maturation is the expansion of the cumulus cells (**Fig. 1**).

3.1.2. In Vitro Fertilization (IVF) of Oocytes

1. Mature oocytes (24–26 h post-in vitro maturation) are washed twice with 2–3 mL Sp-TL in a 35-mm Petri dish.
2. 10 oocytes per group are transferred into 44–μL drops of fertilization medium and returned to 39°C with 5% CO_2.
3. A straw of frozen semen is taken from liquid nitrogen, and plunged into a 37°C water bath for ~1 min, or until completely thawed.
4. Cut one end of the straw with scissors and place the open end into a 1.5-mL vial before cutting the other end.
5. Transfer the semen onto the top of a 45/90% Percoll gradient in a 15-mL conical tube.
6. Centrifuge at 700*g* using conventional benchtop serological centrifuge for 10–15 min.
7. Remove and discard the supernatant, taking care to avoid disturbing the pellet.
8. Determine sperm concentration in the pellet by hemocytometer and adjustd to 25,000 sperm/1 μL, using Sp-TL.
9. Transfer 2 μL sperm into the fertilization drop with oocytes followed by 2 μL heparin (2 μg/mL) and 2 μL PHE. Final concentration of sperm in a 50-μL fertilization drop is 1×10^6/mL.
10. Verify the sperm motility then return the oocytes to the incubator. One of the signs of sperm motility is the attachment of sperm on oocyte surface and the rotation of oocyte (**Fig. 2A**). Pronuclei are well-expanded at 18 h postinsemination (**Fig. 2B**).
11. The day of in vitro fertilization is d 0 of embryo development.

3.1.3. In Vitro Culture (IVC)

1. Cumulus cells are removed at 20–24 h postinsemination by vortexing for 3 min in a 1.5-mL Eppendorf vial with 50 μL TL-HEPES.
2. Flush the vial with 1–2 mL TL-HEPES and transfer the contents to a 35-mm Petri dish.

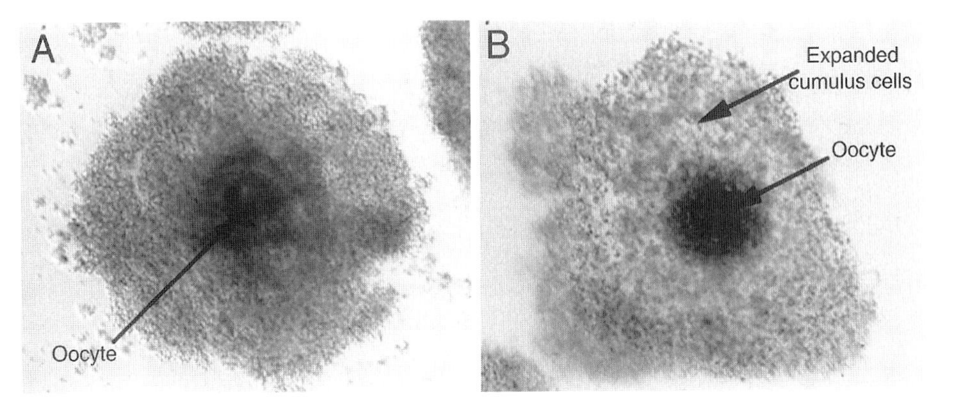

Fig. 1. Bovine oocyte maturation. **(A)** Oocyte freshly retrieved from ovary surrounded by compacted cumulus cells. **(B)** 24 h after culturing in maturation medium, cumulus cells were expanded and became fluffy. (Also on CD-ROM.)

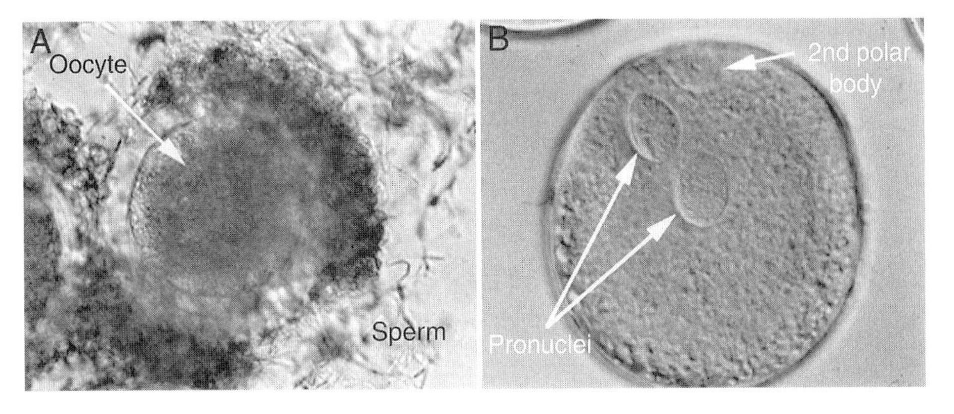

Fig. 2. Bovine oocyte fertilization. **(A)** Bovine oocyte bound with sperm on the surface at 16 h post-insemination. **(B)** Bovine zygote fixed in 1:3 (Glacial acetic acid: ethanol) at 18 h postinsemination. (Also on CD-ROM.)

3. Pick up the oocytes and wash them 3× in CR1-aa culture medium by transferring from drop to drop (50 μL drop of CR1-aa medium in a 60-mm dish covered with mineral oil and pre-equilibrated at 39°C with 5% CO_2).
4. 20–25 oocytes are transferred into a 50-μL drop of CR1-aa in a different 60-mm Petri dish. Be sure to pre-equilibrate the drop at 39°C with 5% CO_2 for at least 2 h.
5. On d 4, a fresh CR1-aa medium is prepared with 10% fetal bovine serum, to replace BSA-FAF. The culture plates are prepared as described, and pre-equilibrated before used for embryo culture.
6. The first cleavage takes place at ~32 h postinsemination (hpi), the second cleavage occurs at ~46 hpi, and the third cleavage at ~66 hpi (*32;* **Fig. 3**).

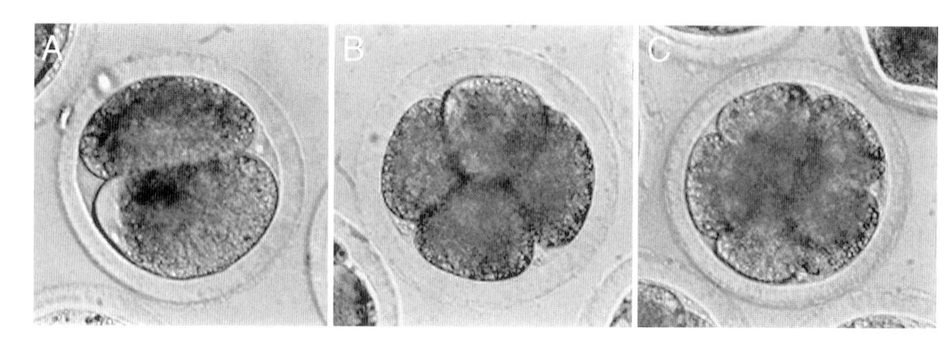

Fig. 3. Preimplantation stage bovine embryos. **(A)** 2-cell stage, **(B)** 4-cell stage; and **(C)** 8-cell stage bovine embryos. Two of the blastomeres of the 8-cell stage embryos are out of the focal plane. (Also on CD-ROM.)

3.2. EGFP Expression Vector

1. Expression vector pEGFP-N1 (Clontech) encodes for the EGFP under the control of a human cytomegaloviral promoter (hCMV) (*see* **Note 3**).
2. The pEGFP-N1 expression vector was digested by *ApaL*I and *Dra*III. A DNA fragment (~2.25 kb in length), which contains the CMV promoter and entire GFP coding region, is separated by gel electrophoresis on a 1% agarose gel in TBE. The DNA fragment with the appropriate size is excised from the gel and the DNA is purified using the gel purification kit.
3. The DNA fragment is diluted to 4 ng/µL in TE. The DNA solution can be divided into aliquots and kept frozen until pronuclear microinjection.

3.3. Pronuclear Microinjection (33,34)

1. After removing the cumulus cells from the oocytes at 16–18 h postinsemination (as described previously), 20–30 cumulus-free oocytes are transferred into a 1.5-mL vial with 50 µL TL-HEPES wash buffer, and centrifuged at 12,000*g* for 6 min at room temperature before proceeding with microinjection.
2. Remove the supernatant carefully and resuspend the oocyte pellet with 0.5–1 mL of TL-HEPES.
3. Transfer the oocytes into a 35-mm Petri dish with TL-HEPES.
4. 20–30 oocytes are placed at the bottom of a 100-µL drop of TL-HEPES in a 100-mm Petri dish covered with mineral oil, and placed on an inverted microscope for injection.
5. The rest of the oocytes are transferred into CR1aa-culture medium in the incubator, until pronuclear injection.
6. An injection needle with an internal filament and tip opening of less than 1 µm, was back-filled with DNA solution to the tip by capillary action. DNA solution can be loaded by placing the end of the needle into an Eppendorf vial with 5–10 µL DNA solution or place a small drop (~1 µL) of DNA solution at the opening end

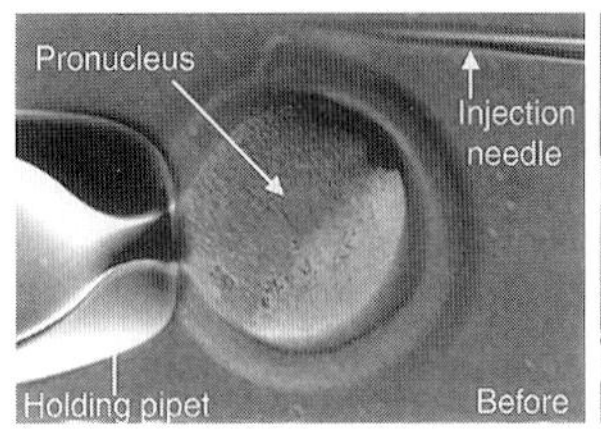

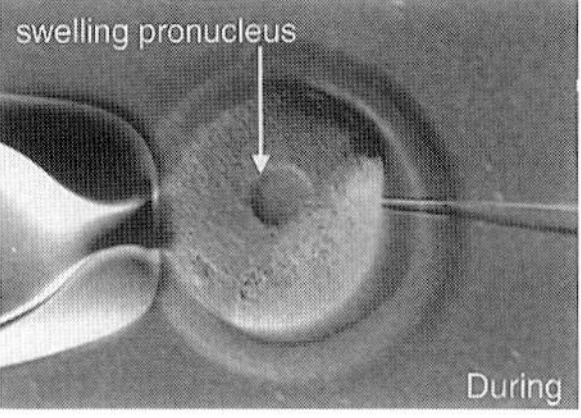

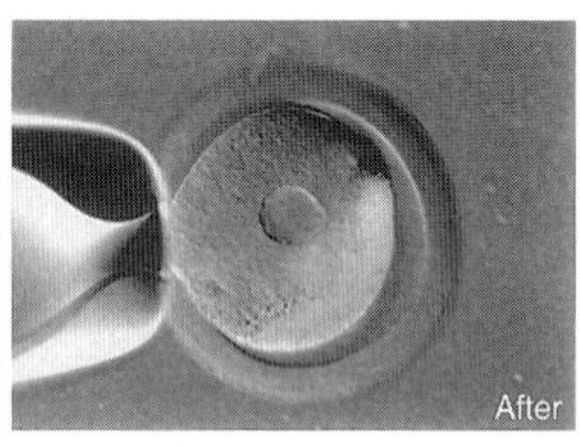

Fig. 4. Pronucleus microinjection. Bovine oocyte at 18 h postinsemination was centrifuged at 12,000*g* for 6 min to expose the pronuclei. DNA solution was injected into one of the nuclei, and the amount of injected DNA was determined by the size of the swollen pronucleus. (Also on CD-ROM.)

of the needle while the pipet is placed horizontally. The opening of the tip can be tested by placing an oocyte in front of the needle and expelling solution toward the oocyte. The movement of the oocyte indicates the opening of the tip. If the injection tip is not open, blockage may be eliminated by carefully sliding the tip of the injection needle over the end of the holding pipet.

7. The injection needle is connected with tygon tubing to a 50-mL glass syringe or automatic injector.
8. Oocytes are held in place by a holding pipet with an internal diameter of 30–40 µm and connected to a microsyringe filled with mineral oil. The holding pipet is back-filled with fluorinet, an inert solution that acts as a barrier to prevent the mixture of mineral oil and the buffer in the manipulation drop. A 1-mL syringe connected with a tygon tube (ID 1 mm) can be used for fluorinet filling.
9. The DNA solution is injected into the pronucleus, and the amount of DNA solution injected into the promuclei is determined by the size of the swollen pronuclei (**Fig. 4**; *see* **Note 4**).
10. After injection, zygotes are transferred to CR1-aa culture medium and cultured at 39°C with 5% CO_2, until analysis.

3.4. Analysis for Embryos Expressing GFP

1. Day 3–4 bovine embryos (8–16 cells) are placed individually in a 10-µL drop of TL-HEPES in a 100 mm Petri dish covered with mineral oil (**Fig. 5**).
2. Embryos are examined individually by fluorescence microscopy. A brief exposure (5–10 s) to excitation light is applied to visualize the green fluorescent (*see* **Notes 5** and **6**).
3. The total number of blastomeres expressing GFP in each embryo is counted and the ratio of GFP-expressing blastomeres to negative blastomeres is calculated in order to determine the mosaic rate of GFP expression.

4. Notes

1. Infection of oocytes before fertilization by injecting retroviral vector into the perivitelline space has been demonstrated as a powerful tool to create transgenic

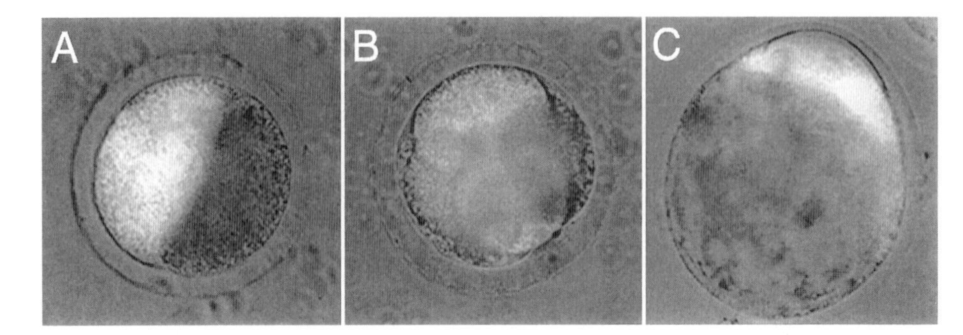

Fig. 5. Bovine embryos expressing GFP after PI at 18 h post-insemination. **(A)** One of the blastomeres in a two-cell stage embryo expressing the GFP. **(B)** Three of the blastomeres in a four-cell stage embryo expressing GFP. **(C)** GFP was expressed at the inner cell mass of a blastocyst. (For optimal, color representation please see CD-ROM.)

cattle *(7)*. Retroviral RNA encapsulated in viral core, is delivered into the cytoplasm followed by reverse transcription, transported into the nucleus, and inserted in the host cell genome. The expression of transgene is more likely to be derived from integrated transgene, rather than transient expression from free-form DNA. Therefore, GFP could be a reliable selection marker in retroviral-vector-produced embryos. However, the interpretation of GFP expression after perivitelline space injection needs to be done carefully because GFP-contained viral particles remain on the surface of blastomeres and inside the perivitelline space. The expression of GFP from inside the blastomeres can be determined with the aid of confocal microscopy (**Fig. 6**).

2. TransgenICSI is a newly developed gene transfer method with the combination of intracytoplasmic sperm injection (ICSI) and sperm as a DNA carrier (sperm vector). Rhesus embryos produced by ICSI using sperm bound with exogenous DNA have demonstrated the successful delivery of exogenous DNA into oocytes. The majority of the embryos express GFP at an early stage (1–4 cells); in some embryos, GFP expression seems to be related to maternal embryonic transition (MET), and expression is only initiated after the third division (unpublished data; **Fig. 7**). This suggests that GFP expression at a later stage may indicate gene integration if the expression is initiated after MET. Nevertheless, further studies need to be done to confirm this phenomenon. A similar procedure has been successfully used to produce transgenic mice *(8)*. Sperm after freeze–thaw, or treated with detergent followed by incubation with DNA solution are used for ICSI. Although most of the preimplantation embryo express GFP, only about 21% of the newborn were transgenic, which indicates a high transient expression rate in preimplantation embryos, and perhaps high mosaic rate in the pups. This phenomenon is similar to that of PI.
3. Promoter selection is critical for transgenic reporters. Several promoters have been successfully used to express GFP in early-stage embryos. Some of the stud-

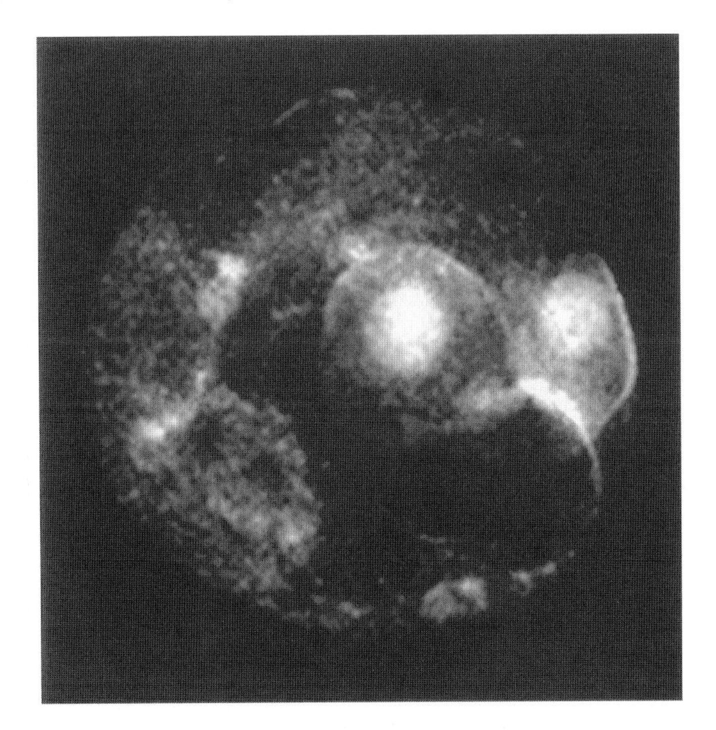

Fig. 6. Confocal microscopy of bovine morula-stage embryo, after infection by vesicular stomatitis virus envelope glycoprotein-G (VSV-G) pseudotyped retroviral vector before fertilization. Most of the blastomeres express GFP in the nucleus and the cytoplasm. It is very distinctive between GFP- positive and -negative cells in individual embryo. (For optimal, color representation please see CD-ROM.)

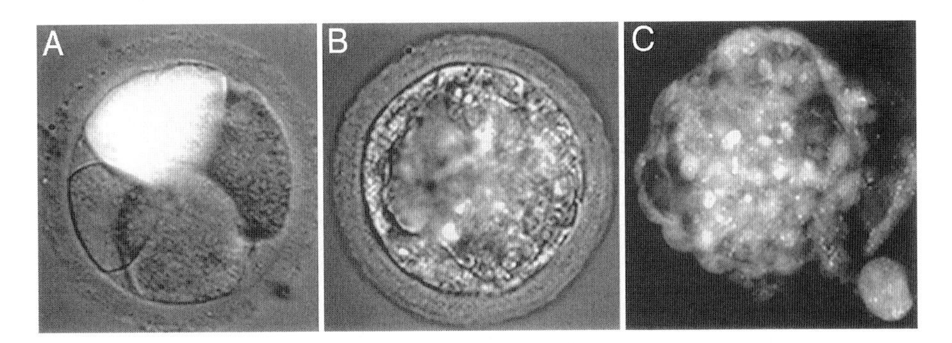

Fig. 7. Rhesus embryos expressing GFP after intracytoplasmic sperm injection using sperm bound with GFP encoded DNA construct. **(A)** Four-cell-stage embryo with one of the blastomeres expressing GFP. **(B)** Blastocyst stage embryo expressing GFP in all of the blastomeres. **(C)** Confocal microscopy of a blastocyst expressing GFP. (For optimal, color representation please see CD-ROM.)

ies indicated that a reliable selection is also related to the promoter that regulates the expression of GFP in preimplantation-stage embryos. The polypeptide chain elongation factor 1α (EF-1α) promoter expresses ubiquitously. In EF-wt-GFP transgenic mice, a very weak expression was found in the liver *(23)*. Transgenic mice that carried EGFP gene under the control of a combination of cytomegalovirus immediate-early enhancer chicken beta-actin promoter, and 3' splice sequence derived from the rabit beta-globin gene (CAG) have shown expression in all tissues except erythrocytes and hair. The expression of CAG-EGFP was first detected in preimplantation stage-embryos and throughout pregnancy *(39)*. Stage-specific promoters such as Oct4 are highly active in undifferentiated cells and preimplantation-stage embryos *(37)*. In Oct4/GFP transgenic mice, the expression of GFP in primordial germ cells was first detected at d 8 postcoitum, and subsequently found in both male and female germ cells during development *(24)*. The CMV promoter expresses constitutively but lacks specificity *(25)*. However, CMV is one of the strongest viral promoters that express in a wide range of cell types. It does have advantages over the other promoters in most preliminary studies. However, the concerns of viral promoters in undifferentiated cell types need to be considered *(7,11)*.

4. The timing of the injection is critical to the expression pattern of GFP after pronuclear injection (PI) at the zygotic stage *(19)*. High mosaic rates have been demonstrated in bovine. In bovine, DNA replication begins at ~10 h post-insemination, and ends at ~20–24 h *(35,36)*. However, PI in bovine mostly takes place at 18–22 h postinsemination, because pronuclei are fully expanded, which allows easier penetration, compared to earlier stages; pronuclei are small and less turgid, and it is at the end of replication. Therefore, exogenous DNA integration mostly takes place after DNA replication at the first cell cycle and mosaic embryo is produced. Therefore, uneven distribution of transgene and expression pattern is predictable *(19)*. Only if DNA integration occurred before DNA replication, at the first cell cycle, will nonmosaic embryos be produced (**Table 1**). On the other hand, the presence of nonintegrated exogenous DNA in preimplantation embryo indicates the possibility of subsequent integration of the transgene during embryonic development (**Table 1**; *19*).
5. Transient expression of exogenous DNA after PI is one of the major barriers for transgenic reporter application. The expression of reporter does not indicate successful integration of the transgene. The use of stage-specific promoters such as Oct4 promoter has demonstrated a regulated expression pattern of GFP gene and is believed to be a reliable regulator for transgenic embryo selection *(24,37)*. However, further experiments need to be done to confirm this phenomenon.
6. High success rate in transgenic embryo selection, using GFP, has been demonstrated in mice. Approximately 77% of pups born from uniformly expressing embryos were transgenic and approx 21% of pups born from mosaic expressing embryos were transgenic *(38)*. However, similar results may not hold for bovine, since the progression of the cell cycle is different and the timing of gene integration may be altered. PI in mouse is at the peak of DNA replication

Table 1
Prediction of DNA Integration and GFP Expression Pattern Related to Cell Cycle in Embryos Produced by Pronuclei Microinjection

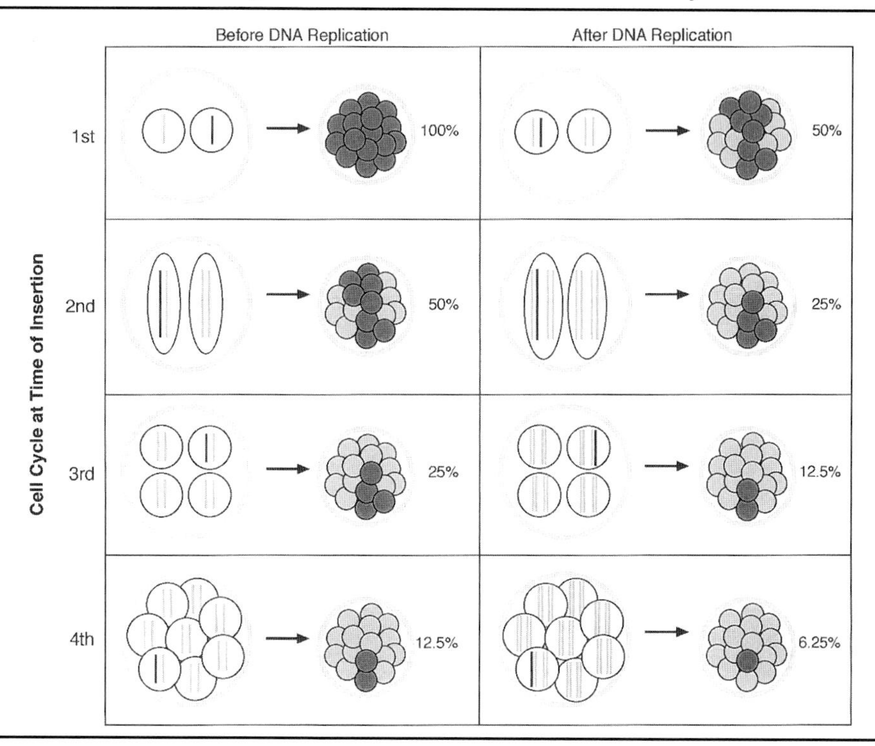

Green line represents transgene that encoded GFP, and when it was integrated at specific stages during embryonic development. (Also on CD-ROM.)

compared with bovine, which is at the end of replication ***(11)***. Therefore, high mosaic rate is expected in bovine embryos based on the assumption of subsequent integration in the previous subheading.

References

1. Brackett, B. G., Baranska, W., Sawicki, W., and Koprowski, H. (1971) Uptake of heterologous genome by mammalian spermatozoa and its transfer to ova through fertilization. *Proc. Natl. Acad. Sci. USA* **68,** 353–357.
2. Jaenisch, R. (1974) Infection of mouse blastocysts with SV 40 DNA: Normal development of infected embryos and persistance SV 40-specific DNA sequences in the adult animals. *Cold Spring Harbor Symp.* **39,** 375–380.
3. Palmiter, R. D., Brinster, R. L., Hammer, R. E., et al. (1982) Dramatic growth of mice that develop from eggs microinjected with metallothionein-growth hormone fusion genes. *Nature* **300,** 611–615.

4. Lavitrano, M., Camaioni, A., Frati, V. M., et al. (1989) Sperm cells as vectors for introducing foreign DNA into eggs: genetic transformation of mice. *Cell* **57,** 717–723.
5. Krimpenfort, P., Rademakers, A., Eyestone, W., et al. (1991) Generation of transgenic dairy cattle using in vitro embryo production. *Biotechnology* **9,** 844–847.
6. Schnieke, A. E., Kind, A. J., Ritchie, W. A., et al. (1997) Human factor IX transgenic sheep produced by transfer of nuclei from transfected fetal fibroblasts. *Science* **278,** 2130–2133.
7. Chan, A. W. S., Homan, E. J., Ballou, L. U., Burns, J. C., and Bremel, R. D. (1998) Transgenic cattle produced by reverse-transcribed gene transfer in oocytes. *Proc. Natl. Acad. Sci. USA* **95,** 14,028–14,033.
8. Perry, A. C., Wakayama, T., Kishikawa, H., et al. (1999) Mammalian transgenesis by intracytoplasmic sperm injection. *Science* **284,** 1180–1183.
9. Wall, R. J. and Seidel, G. E. J. (1992) Transgenic farm animals-a critical analysis. *Theriogenology* **38,** 337–357.
10. Brem, G. and Muller, M. (1994) Large transgenic mammals, in *Animals with Novel Genes* (Maclean, N., ed.), Cambridge University Press, Cambridge, pp. 179–245.
11. Chan, A. W. S. (1999) Transgenic animals: current and alternative strategies. *Cloning* **1,** 25–46.
12. Rexroad, C. E. J. (1992) Transgenic technology in animal agriculture. *Anim. Biotechnol.* **3,** 1–13.
13. Kim, T., Leibfried-Rutledge, M. L., and First, N. L. (1993) Gene transfer in bovine blastocysts using replication-defective retroviral vectors packaged with Gibbon ape leukemia virus envelopes. *Mol. Reprod. Dev.* **35,** 105–113.
14. Bowen, R. A., Reed, M. L., Schrieke, A., et al. (1994) Transgenic cattle resulting from biopsied embryos expression of c-ski in a transgenic calf. *Biol. Reprod.* **50,** 664–668.
15. Cousens, C., Carver, A. S., Wilmut, I., et al. (1994) Use of PCR-based methods for selection of integrated transgenes in preimplantation embryos. *Mol. Reprod. Dev.* **39,** 384–391.
16. Williams, J. L., Harvey, M., Wilburn, et al. (1996) Analysis of transgenic mosacism in microinjected mouse embryos using fluorescent in situ hybridization at various development time points. *Theriogenology* **45,** 335.
17. Takada, T., Iida, K., Awaji, T., et al. (1997) Selective production of transgenic mice using green fluorescent protein as a marker. *Nature Biotechnol.* **15,** 458–461.
18. Chan, P. J., Kalugdan, T., Su, B. C., Whitney, E. A., Perrott, W., Tredway, D. R., and King, A. (1995) Sperm as a noninvasive gene delivery system for preimplantation embryos. *Fertil. Steril.* **63,** 1121–1124.
19. Chan, A. W. S., Kukolj, G., Skalka, A. M., and Bremel, R. D. (1999) Timing of DNA integration, transgenic mosaicism, and pronuclear microinjection. *Mol. Reprod. Dev.* **52,** 406–413.
20. Menck, M., Mercier, Y., Campion, E., et al. (1998) Prediction of transgene integration by non-invasive bioluminescent screening of microinjected bovine embryos. *Transgenic Res.* **7,** 331–341.

21. Saeki, K., Matsumoto, K., Kaneko, T., et al. (1998) Onset of gene activation in early bovine embryos detected in a luminescent system. *Theriogenology* **49,** 277.
22. Heim, R. and Tsien, R. Y. (1996) Engineering green fluorescent protein for improved brightness, longer wavelengths and fluorescence resonance energy transfer. *Curr. Biol.* **6,** 178–182.
23. Ikawa, M., Yamada, S., Nakanishi, T., and Okabe, M. (1999) Green fluorescent protein (GFP) as a vital marker in mammals, in *Current Topics in Developmental Biology* (Schatten, G. P. and Pedersen, R. A. eds.), Academic Press, San Diego, pp. 1–20.
24. Yoshimizu, T., Sugiyama, N., De Felice, M., et al. (1999) Germline-specific expression of the Oct-4/green fluorescent protein (GFP) transgene in mice. *Dev. Growth Differ.* **41,** 675–684.
25. Chan, A. W. S., Luetjens, C. M., Dominko, T., et al. (2000) Foreign DNA transmission by ICSI: injection of spermatozoa bound with exogenous DNA results in embryonic GFP expression and live Rhesus monkey births. *Mol. Hum. Reprod.* **6,** 26–33.
26. Leibfried-Rutledge, M. L., Critser, E. S., Eyestone, W. H., Northey, D. L., and First, N. L. (1987) Development potential of bovine oocytes matured in vitro or in vivo. *Biol. Reprod.* **36,** 376–383.
27. Bavister, B. D., Leibfried, M. L., and Lieberman, G. (1983) Development of preimplantation embryos of the golden hamster in a defined culture medium. *Biol. Reprod.* **28,** 235–47.
28. Parrish, J. J., Susko-Parrish, J., Winer, M. A., et al. (1986) Bovine in vitro fertilization with frozen-thawed semen. *Theriogenology* **25,** 591–600.
29. Rosenkrans, C. F. J. and First, N. L. (1994) Effect of free amino acids and vitamins on cleavage and development rate of bovine zygotes in vitro. *J. Anim. Sci.* **72,** 434–437.
30. Leibfried, M. L. and Bavister, B. D. (1982) Effects of epinephrine and hypotaurine on in-vitro fertilization in the golden hamster. *J. Reprod. Fertil.* **66,** 87–93.
31. Critser, E. S., Leibfried-Rutledge, M. L., and First, N. L. (1986) Influence of cumulus cells association during in vitro maturation of bovine oocytes on embryonic development. *Biol. Reprod.* **34,** 192.
32. Memili, E. and First, N. L. (1998) Developmental changes in RNA polymerase II in bovine oocytes, early embryos, and effect of alpha-amanitin on embryo development. *Mol. Reprod. Dev.* **51,** 381–389.
33. Hogan, B., Beddington, R., Costantini, F., and Lacy, E., eds. (1994) *Manipulating the Mouse Emrbyo.* Cold Spring Harbor Laboratory, Cold Spring Harbor.
34. Chan, A. W. S. (1997) *Alternative Methods to Produce Transgenic Animals.* PhD Thesis, University of Wisconsin-Madison.
35. Eid, L. N., Lorton, S. P., and Parrish, J. J. (1994) Paternal influence on S-phase in the first cell cycle of the bovine embryos. *Biol. Reprod.* **51,** 1232–1237.
36. Gagne, M., Pothier, F., and Sirard, M. A. (1995) Effect of microinjection time during postfertilization S-phase in bovine embryonic development. *Mol. Reprod. Dev.* **41,** 184–194.

37. Keiser, J. T., Jobst, P. M., Garst, A. S., et al. (2000) Preimplantation screening for transgenesis using an embryonic specific promoter and green fluorescent protein. *Theriogenology* **53,** 517.
38. Kato, M., Yamanouchi, K., Ikawa, M., et al. (1999) Efficient selection of transgenic mouse embryos using EGFP as a marker gene. *Mol. Reprod. Dev.* **54,** 43–48.
39. Okabe, M., Ikawa, M., Kominami, K., Nakanishi, T., and Nishimune, Y. (1997) 'Green mice' as a source of ubiquitous green cells. *FEBS Lett.* **407,** 313–319.

16

Development of Glycosylphosphatidylinositol-Anchored Enhanced Green Fluorescent Protein

One-Step Visualization of GPI Fate in Global Tissues and Ubiquitous Cell Surface Marking

Gen Kondoh

1. Introduction

This chapter describes the development of enhanced green fluorescent protein-glycosylphosphatidylinositol (EGFP-GPI), an EGFP version specifically targeted to the cell surface, which contains a GPI anchor. Many membranous proteins in eukaryotes are transported and attached to the plasma membrane via a GPI anchor *(1)*. In mammals, such proteins are highly tissue-specific, and seem to play critical roles *in situ (2)*. The distribution and kinetics of formation of the GPI anchor are still under investigation. To elucidate the fate and timing of GPI anchor expression in the whole body, the author took a transgenic approach. Because many GPI-anchored proteins are highly tissue-specific, and turnover of GPI-anchored proteins would be extremely difficult to evaluate equally among all tissues in an organism using endogenous proteins, the use of GFP in transgenic animals was considered. The GFP and its more efficient mutant, EGFP, have been used widely as a marker in many organisms *(3)*, because GFP requires no additional reagents for its operation, is harmless to many living organisms, and is easily detectable.

Generally, GPI-anchored proteins are synthesized as precursors that contain two signal sequences: an N-terminal signal sequence for membrane translocation into the endoplasmic reticulum (ER), and a C-terminal GPI-anchor signal polypeptide. Both signal peptides are cleaved soon after translocation into the ER lumen. The C-terminus of the protein is then covalently modified with the GPI anchor that is presynthesized *en bloc* in the ER. In the use of cell

From: *Methods in Molecular Biology, vol. 183: Green Fluorescent Protein: Applications and Protocols*
Edited by: B. W. Hicks © Humana Press Inc., Totowa, NJ

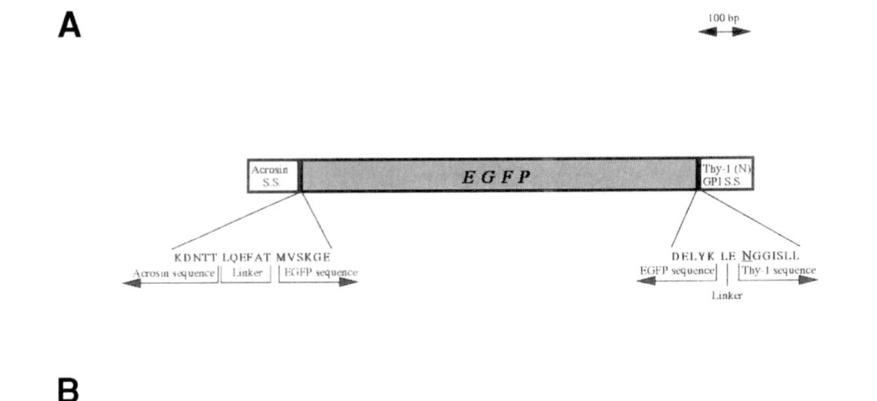

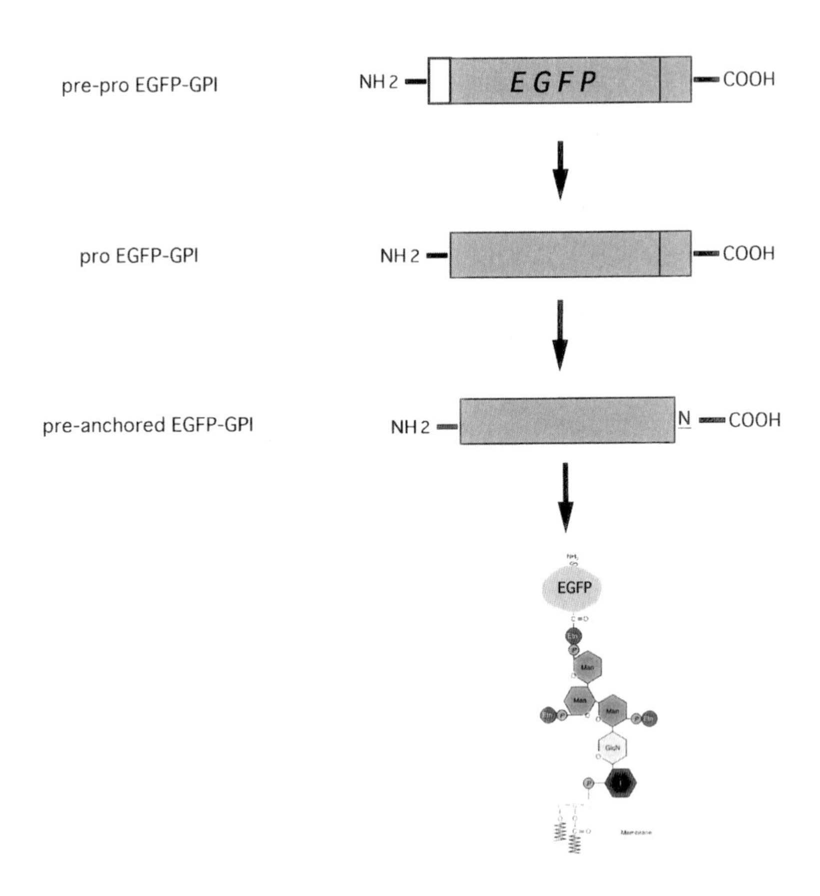

Fig. 1. **(A)** Structure of EGFP-GPI fusion gene. **(B)** Putative scheme of EGFP-GPI processing. (For optimal, color representation please see accompanying CD-ROM.)

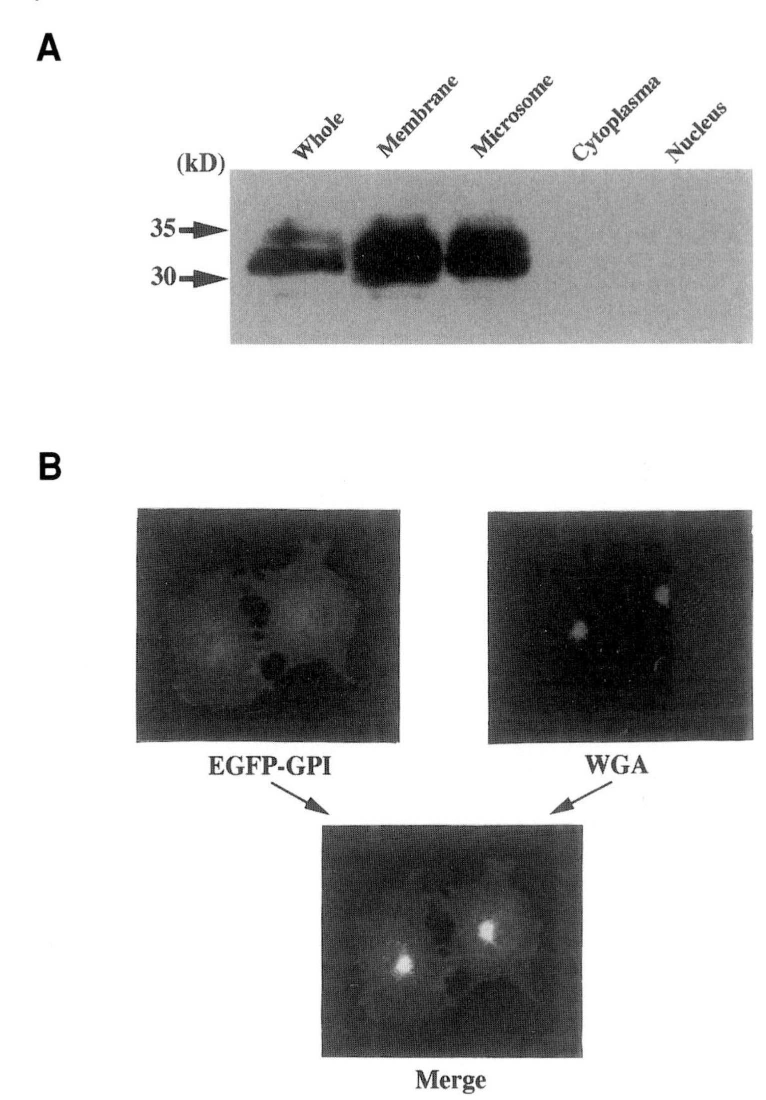

Fig. 2. (**A**) Cellular distribution of EGFP-GPI by Western blotting. (**B**) Accumulation of EGFP-GPI on the Golgi complex. (For optimal, color representation please see accompanying CD-ROM.)

surface marking, EGFP-GPI is applicable to many eukaryote systems, including primitive protozoa and yeast, because it is processed and targeted to the membrane via a ubiquitous GPI anchoring system.

Transgenic mice expressing EGFP-GPI had normal birth weights, no morphological abnormalities through at least the first 2 yr after birth, remain

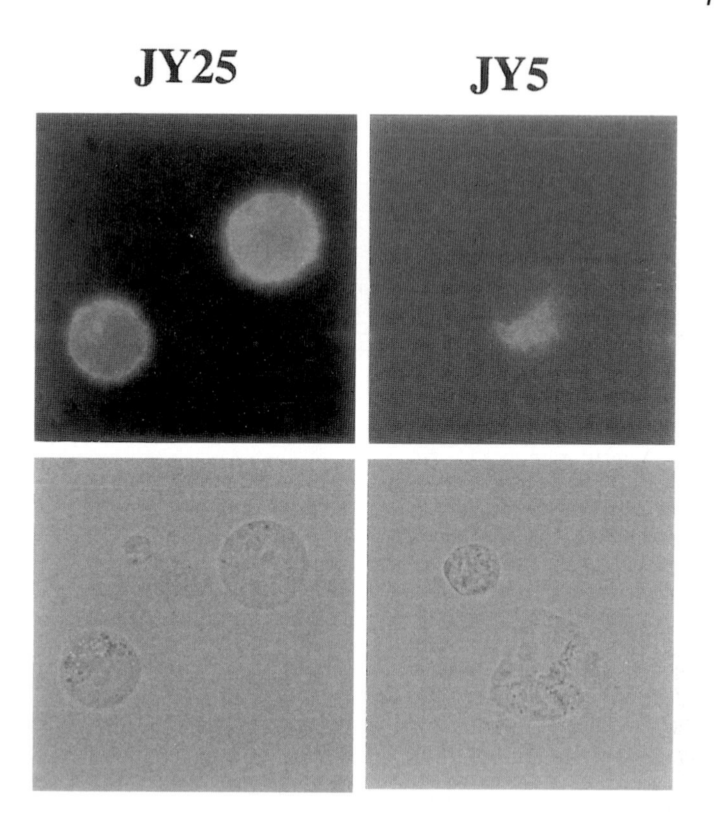

Fig. 3. Appearance of EGFP-GPI in GPI-positive (JY25) and GPI-negative (JY5) cells. Lower panel, phase contrast view. (For optimal, color representation please see accompanying CD-ROM.)

fertile, and display a broad expression of the transgene. Histologically, localization of EGFP-GPI protein displays tissue-to-tissue variations, with two major classifications. It is distributed in either a polarized and localized fashion, or in a homogeneous and nonpolarized fashion *(4)*. This evidence demonstrates that the EGFP-GPI efficiently labels the surface of culture cells in transgenic mice, and that these mice are effective for investigating the fate of GPI in vivo.

One noteworthy benefit of EGFP-GPI is that it is fully exposed to the cell surface. Usually, GPI-anchored proteins are more stable than transmembrane proteins, which means that EGFP-GPI could be used as an immunotarget for antibody-mediated collection or elimination of cells of interest. Possibly EGFP-GPI could be used as a tool for cancer gene therapy. Because GFP is originally a jellyfish protein, it should be antigenic in mammals. If EGFP-GPI expression could be virally delivered specifically to tumor lesions, some degree of inflammation might occur and induce a series of tumor-rejecting reactions.

Table 1
Category of EGFP-GPI Localization in the Transgenic Mice

1. Cells showing polarization/localization of EGFP-GPI:
Apical polarization
Epithelial cells
Specific localization
Neuron Nerve fiber
Liver parenchymal cells bile canaliculi
2. Cells showing non-polarized localization of EGFP-GPI:
Mesenchymal cells
Muscle cells
Hematopoietic cells
Vascular endothelialcells
Endocrine cells

Polarized cell

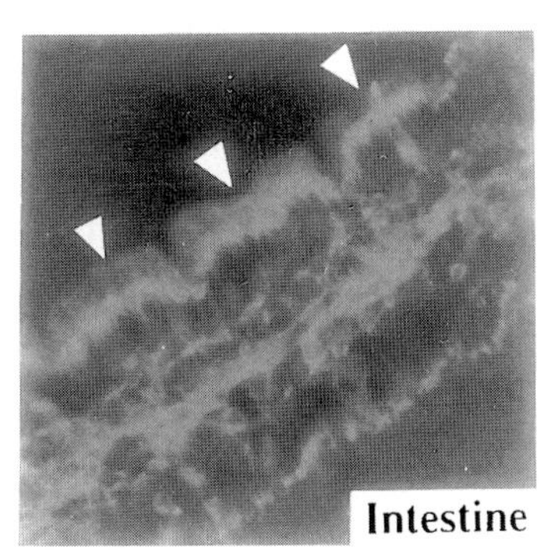

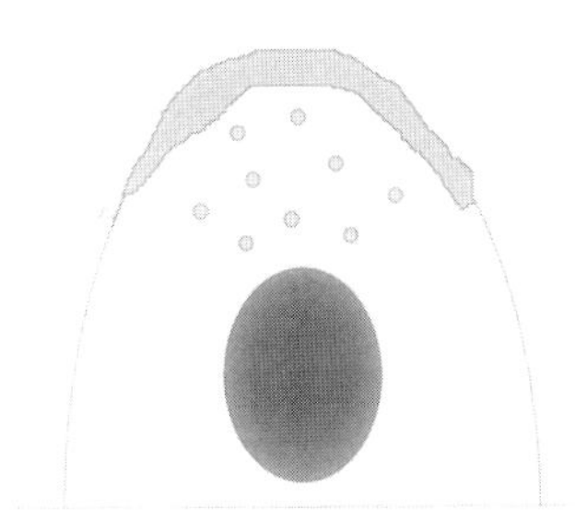

Non polarized cell

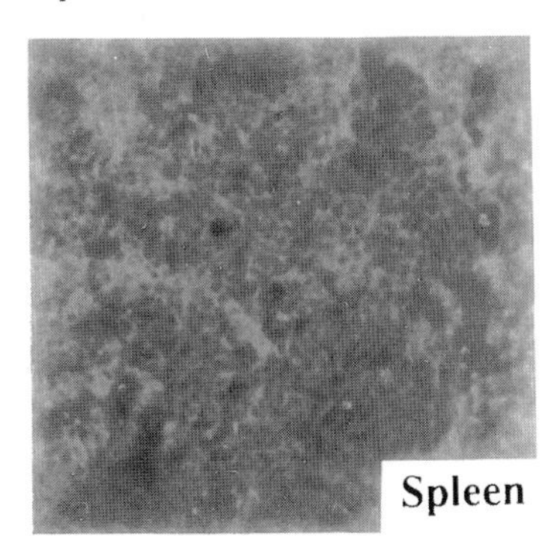

Fig. 4. Localization of EGFP-GPI in polarized and nonpolarized cells of transgenic mice. (For optimal, color representation please see accompanying CD-ROM.)

2. Materials

1. pEGFP: the cDNA vector containing EGFP (Clontech, Palo Alto, CA).
2. Plasmid DNA with the EGFP fused to the mouse acrosin signal sequence was a generous gift from Dr. M. Okabe, Osaka University, Japan *(5)*.
3. Plasmid DNA, pThy-1(N), with the mouse Thy-1 GPI anchor signal sequence (with a cysteine-to-asparagine mutation at the ω site of mouse Thy-1 GPI anchoring sequence), was a generous gift from Dr. K. Ohishi and Dr. T. Kinoshita, Osaka University *(6)*.
4. A pBluescriptII plasmid containing the CAAG promoter and rabbit β-globin poly(A) signal were generous gifts from Dr. J. Miyazaki, Osaka University *(7)*.
5. pCRII vector (Invitrogen, Carlsbad, CA).
6. Restriction endonucleases and buffers: *Xho*I, *Kpn*I, *Not*I, *Eco*RI, *Sal*I, and *Hind*III (Takara shuzo, Shiga, Japan).
7. Expand high-fidelity polymerase chain reaction (PCR) system (Boehringer Mannheim, Mannheim, Germany).
8. PCR primers:
   5-prime primer: 5'-TTAGGGCAGGAGTATGGTAGAGATGC-3' and,
   3-prime primer: 5'-CTCGAGT TTGTATAGTTCATCCATGC-3'.
9. TA-cloning kit (Invitrogen, CA).
10. DNA ligation kit ver. 1 (Takara shuzo).
11. DNA blunting kit (Takara shuzo).
12. Complete™ protease-inhibitor (Boehringer Mannheim): 1 tablet in 10 mL solution.
13. Polytron homogenizer.
14. TNE solution: 10 m*M* Tris-HCl, pH 7.8, 150 m*M* NaCl, 1 m*M* EDTA.
15. Triton-X-114–TNE solution: 1% (v/v) Triton-X-114 in TNE.
16. TBS-T solution: 20 m*M* Tris-HCl, pH 7.6, 138 m*M* NaCl, 0.05% Tween-20.
17. Blocking solution: Boehringer Mannheim 10% (v/v) in TBS-T.
18. Rabbit polyclonal antibody against GFP (MBL, Nagoya, Japan).
19. Enhanced chemuluminescence (ECL)-system (Amersham Pharmacia Biotech, Buckinghamshire, UK).
20. Fix solution: 4% (w/v) paraformaldehyde in phosphate-buffered saline (PBS).
21. Fluorescence microscope with GFP-specified filters (Olimpus, Tokyo, Japan).
22. Phenol–chloroform.
23. Ethanol.
24. Low-melting-temperature agarose gels and agarose gel apparatus.
25. 15% Sodium dodecyl sulfate-polyacrylamide gel electrophoresis (SDS-PAGE) gels and electrophoresis apparatus.
26. Nitrocellulose membranes and electrophoretic transfer apparatus.
27. Anesthetics, 10% Nembutal (Dainihon-seiyaku, Osaka, Japan)
28. PBS containing 20% sucrose.
29. Cryostat.
30. Tissue-Tek O. C. T. compound (Sakura Finetek, Torrance, CA).

3. Methods

3.1. Generation of EGFP-GPI Fusion Gene (see *Note 1*)

The EGFP was fused with the membrane translocation signal sequence of acrosin at N-terminal as described *(5)*. This plasmid was the generous gift of Dr. M. Okabe.

1. Amplify the plasmid containing the acrosin-EGFP construct with PCR primers to generate *Xho*I sequence on the 3' end.
2. PCR reagents: 1 µL of 10 ng/µL template DNA, 5 µL of 10X enzyme-attached reaction buffer, 1.25 µL 10 m*M* dNTP, 28.75 µL sterile water, 2 µL 20 µ*M* 5-prime PCR primer, 2 µL 20 µ*M* 3-prime PCR primer, 0.5 µL PCR enzyme. PCR condition: 93°C for 30 s, 60°C for 60 s, 68°C for 120 s, for 30 cycles.
3. The acrosin-EGFP encoding fragment was isolated by agarose gel electrophoresis.
4. Clone the acrosin-EGFP fragment into the pCRII vector by TA-cloning kit according to the manufacturer's procedure.
5. The resulting pCRII vector containing acrosin-EGFP was isolated by agarose gel electrophoresis and sequenced.
6. Digest the pCRII vector with *Kpn*I and *Xho*I to make the EGFP insert. Purify the fragment by agarose gel electrophoresis using low-melting-temperature agarose and isolate the fragment.
7. Digest the pThy-1(N) vector (6) with *Kpn*I and *Xho*I using buffer and directions supplied by the manufacturer. Purify the fragment by agarose gel electrophoresis using low-melting-temperature agarose and isolate the fragment.
8. Ligate the EGFP insert into the Thy-1 vector by DNA ligation kit according to the manufacturer's procedure.
9. Digest the plasmid containing the acrosin-EGFP-Thy-1(N) fusion gene with *Kpn*I- *Not*I. Separate the digest by agarose gel electrophoresis using low-melting-temperature agarose and isolate the 0.9-kb fragment.
10. Blunt the ends of the isolated fragment by DNA blunting kit according to the manufacturer's directions.
11. Digest the pBluescriptII vector containing CAAG promoter and rabbit β-globin poly(A) signal with *Eco*RI using buffer and directions supplied by the manufacturer.
12. Ligate the blunt-ended EGFP-Thy1(N) insert into the *Eco*RI-digested pBluescript vector.
13. Isolate the product by gel electrophoresis and sequence to verify that the construct is correct. This construct is designated as "pCAAG-EGFP-GPI" (*see* **Note 2**).

3.2. Transgenic Mouse Production (see *Note 2*)

1. Digest the pCAAG-EGFP-GPI plasmid with *Sal*I and *Hin*dIII to remove the vector sequence.

2. Separate the digest by agarose gel electrophoresis using low-melting-temperature agarose and isolate a 3.2-kb transgene fragment.
3. Purify the fragment by serial phenol–chloroform extractions, and precipitate with ethanol.
4. Resuspend the DNA in sterile saline.
5. Inject DNA into one-cell embryos of B6C3F1 mice.
6. Transplant DNA-injected embryos into a pseudopregnant ICR mouse.

3.3. Western Blotting for EGFP-GPI (see Note 3)

1. Homogenize cells and tissues using a polytron homogenizer for 1 min in ice-cold TNE solution containing complete protease inhibitor.
2. Centrifuge the homogenates for 30 min at 100,000*g* at 4°C.
3. Collected the supernatants and store at –20°C (the water-soluble fraction).
4. Wash the pellets with plenty of TNE buffer and centrifuge at 10,000*g*.
5. Homogenize the pellet in 1% Triton-X-114–TNE solution containing complete protease inhibitor and centrifuge at 100,000*g* for 30 min at 4°C. Collect the supernatants and store at –20°C (the detergent soluble fraction).
6. Subject 10 µg samples of water-soluble and detergent-soluble fractions to 15% SDS-PAGE.
7. Electrophoretically transfer the proteins from the gel to a nitrocellulose membrane.
8. Immerse the membrane in blocking solution for 1 h at room temperature with rocking.
9. Probe the membrane with anti-GFP antibody diluted 200-fold into TBS-T, wash the blot, and detect with the ECL-system according to the manufacturer's procedures.

3.4. Fluorescence Histology (see Notes 4, 5)

1. Deeply anesthetize the mice by 10% Nembutal injection and fix by perfusing the fix solution into the left ventricle of the heart.
2. Excise tissues, and place them in the fix solution.
3. Incubate fixed tissue in PBS containing 20% sucrose for 48 h at 4°C.
4. Embed pieces of tissue in cryostat compound.
5. Quickly freeze embedded tissue with dry ice and make 5–10-µm-thick sections, using a cryostat.
6. Examine sections under a fluorescence microscope with GFP-specified or fluorescein isothiocyanate-matched filters.

4. Notes

1. **Figure 1A** indicates the structure of the DNA containing the EGFP-GPI fragment. The acrosin signal sequence and mutant Thy-1 GPI-anchoring signal sequence were fused to the coding region of EGFP at N-terminus and C-terminus, respectively. The amino acid sequences of both boundaries were indicated. The ω residue asparagine was underlined. In **Fig. 1B** (*see* page 216), putative EGFP-GPI processing was schematized. Both signal sequences were serially cleaved

off, and EGFP with additional amino acids was finally connected to the phosphoethanolamine residue of GPI anchor via an amide bond.

2. The plasmid DNA construct and the transgenic mice expressing EGFP-GPI are both available from the author upon request.
3. The subcellular localization of EGFP-GPI was examined by Western blotting. The lysates of EGFP-GPI-transfected COS7 cells were fractionated by sucrose density-gradient centrifugation and transferred to nitrocellulose for Western blots. As shown in **Fig. 2A**, EGFP-GPI was strictly fractionated to the membrane and microsomal fraction where GPI-anchored proteins harbor. EGFP-GPI was also detected on an intracellular structure in all kinds of cells examined, as well as on the plasma membrane. Double staining with rhodamine-conjugated wheat germ agglutinin, a marker for Golgi complex *(9)*, was performed, and the intracelluar fluorescence of EGFP-GPI was precisely co-localized to the wheat germ agglutinin staining, indicating that EGFP-GPI is highly accumulated in the Golgi complex (notice yellow signal of Merge picture of **Fig. 2B**). (*See* page 217.)
4. **Figure 3** (*see* page 218) shows appearances of EGFP-GPI when pCAAC-EGFP-GPI was transfected into culture cells. To elucidate whether membranous and intracellular expressions of EGFP-GPI depend on GPI biosynthesis, the fusion gene was introduced to GPI-anchor-deficient JY5 cell, compared with parental GPI-anchor-positive JY25 cells *(8)*. Remarkable membranous and intracellular fluorescence were observed in EGFP-GPI-transfected JY25 cells; only weak fluorescence was detected in the cytoplasmic region of JY5 transfectants.
5. Results from a histological survey of EGFP-GPI transgenic mice are summarized in **Table 1** (*see* page 219.) Localization of EGFP-GPI protein was different among polarized and nonpolarized cells. Generally, epithelial cells have shown apical/luminal polarization of EGFP-GPI, as expected for endogenous GPI-anchored proteins (**Fig. 4**, *upper*) (*see* page 219.) The EGFP-GPI was also localized to nerve fibers in the nervous system. The expression pattern in the forebrain and cerebellum was similar to that of Thy-1, a GPI-anchored protein expressed in the brain *(10)*. In the liver parenchym, EGFP-GPI was localized along the network of bile canaliculi. The pattern of expression was similar to that of alkaline phosphatase, a well-known GPI-anchored protein distributed on bile canaliculi of the liver *(11)*. Tissues of mesodermal origin, such as mesenchymal cells, muscle cells, vascular endothelial cells, and hematopoietic cells, could be categorized as homogeneous/nonpolarized types (**Fig. 4**, *lower*).

References

1. Englund, P. T. (1993) The structure and biosynthesis of glycosyl phosphatidylinositol protein anchors. *Annu. Rev. Biochem.* **62,** 121–138.
2. Kinoshita, T., Inoue, M., and Takeda, J. (1995) Defective glycosyl phosphatidylinositol anchor synthesis and paroxymal nocturnal hemoglobinuria. *Adv. Immunol.* **60,** 57–103.
3. Chalfie, M. and Kain, S., eds. (1998) *Green Fluoresscent Protein* Wiley-Liss, New York.

4. Kondoh, G., Gao, X.-H., Nakano, Y., et al. (1999) Tissue-inherent fate of GPI revealed by GPI-anchored GFP transgenesis. *FEBS Lett.* **458,** 299–303.5. Nakanishi, T., Ikawa, M., Yamada, S., et al. (1999) Real-time observation of acrosomal dispersal from mouse sperm using GFP as a marker protein. *FEBS Lett.* **449,** 277–283.
6. Ohishi, K, Inoue, N., Maeda, Y., et al. (2000) Gaa1p and Gpi8p are components of a glycosyl phosphatidylinositol (GPI) transamidase that mediates attachment of GPI to proteins. *Mol. Biol. Cell* **11,** 1523–1533.
7. Niwa, T., Yamamura, K., and Miyazaki, J. (1991) Efficient selection for high-expression transfectants with a novel eukaryotic vector. *Gene* **108,** 193–199.
8. Miyata, T., Takeda, J., Iida, Y., et al. (1993) The cloning of PIG-A, a component in the early step of GPI-anchor biosynthesis. *Science* **259,** 1318–1320.
9. Colley, K. J., Lee, E. U., and Paulson, J. C. (1992) The signal anchor and stem regions of the galactoside 2, 6-sialyltransferase may each act to localize the enzyme to the Golgi apparatus. *J. Biol. Chem.* **267,** 7784–7793.
10. Moore, M. J., Dikkes, P., Reif, A. E., et al. (1971) Localization of theta alloantigens in mouse brain by immunofluorescence and cytotoxic inhibition. *Brain Res.* **28,** 283–293.
11. De Wolf-Peeters, C., De Vos, R., and Desmet, V. (1972) Electron microscopy and histochemistry of canalicular differentiation in fetal and neonatal rat liver. *Tissue Cell* **4,** 379–388.

17

Transgenic Zebrafish Expressing Green Fluorescent Protein

Ebrahim Shafizadeh, Haigen Huang, and Shuo Lin

1. Introduction

The zebrafish has recently emerged as a powerful model for study of the genetics and molecular biology of vertebrate development. Zebrafish reach sexual maturity in ~2–3 mo, and produce hundreds of embryos in weekly intervals. The embryos develop rapidly and organogenesis occurs within the first 2 d of development in transparent bodies. The availability of thousands of chemically induced mutations *(1,2)*, as well as the results of insertional mutagenesis studies *(3–6)*, provide valuable resources to study genetics of vertebrate development.

Another factor that has made zebrafish a powerful model is the ability to utilize transgenic technology to introduce foreign genes into the zebrafish embryos *(7–9)*. Experiments have shown that germline transmission of a foreign gene in zebrafish results in a stable expression of the gene in subsequent generations *(10–13)*. Transgenic technology has been extensively used in zebrafish to study developmental processes. It has been used to rescue mutant phenotypes, by injecting the candidate genes that have been mapped to the critical region during positional cloning experiments *(14)*. Transgenic techniques are used in insertional mutagenesis experiments to disrupt endogenous genes. Promoter analysis to study *cis*-acting elements responsible for tissue-specific gene regulation *(15,16)*, cell lineage analysis *(17)*, and transplantation experiments, is also among the many applications of transgenic technology that have been used in zebrafish.

In order for transgenic techniques to be useful in studying promoter function, cell lineage tracing and transplantation, a reporter gene is required. Although transgene expression can be detected by *in situ* hybridization or his-

From: *Methods in Molecular Biology, vol. 183: Green Fluorescent Protein: Applications and Protocols*
Edited by: B. W. Hicks © Humana Press Inc., Totowa, NJ

tochemical analysis, it would be useful to have a reporter gene to monitor gene expression in living embryos in real time. The expression of *lacZ* reporter gene under control of the promoter and enhancer of the *Xenopus* elongation factor1a gene has been used as such a marker; however, this technique requires substrates and often has high background ***(18)***.

Introduction of the green fluorescent protein (GFP) as a marker of gene expression ***(19)*** has profoundly improved transgenic zebrafish technology ***(20–24)***. In transparent zebrafish embryos, GFP appears highly stable, shows very little photobleaching, and its fluorescence can be easily detected by various microscopic techniques. Transgenic techniques using GFP as a reporter gene have been instrumental in studying mechanisms of development, gene regulation, cell migrations, and lineage. This chapter describes protocols for generating and analyzing transgenic zebrafish expressing GFP as a reporter gene.

2. Materials

2.1. Equipment

1. Micropipet puller: Model P-91, Sutter.
2. Microinjection capillaries with 1-mm outside diameter.
3. Micromanipulator with stand and tubing (*see* **Note 1**): model MN-151, Narishige, Japan.
4. Microscopes: SV6 or SV11 stereomicroscope; Axioplan 2 microscope with FITC filter set 09 (excitation BP 450–490, beamsplitter FT510, emission LP520); AttoArc microscope illuminator with HBO100w; Achoplan X40/0,75W, X63/0,90W, plan-Neofluor X10/0,30, X20/0,50; Color video camera: ZVS-3C75DE, Carl Zeiss; Zeiss Axioplan-2 mot upright microscope equipped with a Zeiss 510 NLO scan head for multiple photon microscopic analysis.
5. Mouse cage (24 × 14 × 13 cm, cat. no. 6601177, Nalgene). Breeding traps and dividers (CD-203) (Aquaculture, Dade City, FL).
6. Glass slides with depression wells.
7. Microinjection plates (*see* **Note 2**).
8. Petri dishes: 100 × 200 mm and 60 × 15 mm.

2.2. Reagents

1. Holtfreter's solution (freshly prepared before each injection): 3.5 g/L NaCl, 0.2 g/L sodium bicarbonate, 0.1 g/L $CaCl_2$, 0.05 g/L KCl; adjust the pH to 6.5–7.0 with HCl.
2. Pronase (Sigma): 30 mg/mL in water. Self-digest at 37°C for 1 h. Aliquot, and store at –20°C.
3. 1 *M* KCl: Autoclave, store 1-mL aliquots at –20°C, and use as 10X stock.
4. DNA extraction buffer: 10 m*M* Tris-HCl, pH 8.2, 10 m*M* EDTA, 200 m*M* NaCl, 0.5% sodium dodecyl sulfate, 200 mg/mL pronase.
5. Tricaine (3-aminobenzoic acid ethyl ester)(Sigma): Stock solution is 0.16% in water (should have a pH of 7.6). Store at 4°C.

6. Light mineral oil (Fisher).
7. Plasmid DNA preparation kit (Qiagen).
8. Geneclean II kit (Bio101).
9. Plasmid constructs containing GFP reporter gene.
10. Restriction enzymes to excise GFP from plasmid.
11. *Taq* polymerase, 10X polymerase chasin reaction (PCR) buffer: 100 m*M* Tris HCl, pH 8.3, 500 m*M* KCl, 15 m*M* $MgCl_2$, 0.01% (w/v) gelatin. Deoxyribonucleoside triphosphate mix (Promega): 200 m*M* each of deoxycytidine, deoxyadenosine, deoxyguanosine, and deoxythymidine triphosphophates.
12. Agarose and gel apparatus.
13. TE: 10 m*M* Tris-HCl, pH 8.0, 1 m*M* EDTA.
14. 70% Ethanol and 100% ethanol.
15. 1% Methyl cellulose in Holtfreter's solution.
16. Primers:
    a. GFP (sense) 5'-AATGTATCAATCATGGCAGAC-3'
    b. GFP (antisense) 5'-TGTATAGTTCATCCATGCCATGTG-3'
    c. Elongation Factor-1α (EF1α) (sense) 5'-TACGCCTGGGTGTTGGACAAA-3'
    d. EF1α (antisense) 5'-TCTTCTTGATGTATCCGCTGAC-3'

3. Methods

3.1. Preparation of Plasmid DNA (see Note 3)

1. Use a Qiagen or other plasmid DNA preparation kit to prepare plasmid DNA.
2. Linearize the plasmid with appropriate restriction enzyme (*see* **Note 4**).
3. Run the digested plasmid on an agarose gel. Excise and purify the desired fragment using a Geneclean II kit.
4. Dissolve the purified DNA in 1X TE at a concentration of 200–500 μg/mL (*see* **Note 5**).
5. Dilute the DNA in 0.1 *M* KCl to 50–100 μg/mL. Store at –20°C.

3.2. Collection of Eggs (see Note 6)

1. Select 20–50 healthy adult zebrafish (4–12 mo old) and keep males and females in separate tanks as mating stocks.
2. Feed these mating stocks with an extra meal of food daily to assure healthy condition and only use them for mating once a week.
3. Set up male and female fish in the afternoon before injection. Transfer a male and 1–2 females to the mating cage, situated in a mouse cage. Separate them by placing a divider between the females and the male.
4. On the morning of injection, join the fish together. They usually lay eggs within 10 min.
5. Collect eggs in a small beaker ~10 min after being laid.
6. Dechorionate eggs by replacing the water in the beaker with 1 mg/mL pronase in Holtfreter's solution. Place the beaker at room temperature for several min.
7. As soon as several chorions drop off the embryos and start to float, wash the eggs 5–10× with Holtfreter's solution (*see* **Note 7**).

3.3. Preparation and Loading the Microinjection Needles

1. Using a micropipet puller, prepare microinjection needles (*see* **Note 7**).
2. Under a stereoscope, open the needle by breaking off the tip with a sharp blade or forceps. The opening of the needle should be 0.05–0.15 mm in diameter. An opening that is too large or too small will result in damaging the embryos or difficulty loading the DNA.
3. Attach the needle to the needle holder.
4. Spin the DNA in 0.1 *M* KCl solution at high speed for 2–5 min to precipitate any debris that may clog the needle.
5. Depending on how many embryos are to be injected, transfer 2–5 μL DNA into the well of a clean depression glass slide. Cover the solution with a couple of drops of light mineral oil to prevent evaporation during the loading of the needle.
6. Place the tip of the needle into the DNA solution and draw the DNA into the needle (*see* **Note 8**).

3.4. Microinjection

1. Fill the injection ramp with Holtfreter's solution.
2. Transfer 100–200 dechorionated eggs into the microinjection plates so that they line up in the groove.
3. Using the micromanipulator, position the needle so that it penetrates the cytoplasm of the single cell embryo from the top. Moving the eggs around is often required, to be able to position them correctly for injection.
4. Microinject the DNA into the cytoplasm of the one-cell-stage embryos. The volume of injected DNA can be seen under the stereoscope and should be about one-fifth of the volume of the cytoplasm (*see* **Note 9**).
5. After all embryos are injected, very gently transfer them into 100 × 200 mm Petri dishes containing Holtfreter's solution. Incubate at 28–30°C.
6. After ~6 h, remove unfertilized and deformed eggs. Continue the incubation overnight.
7. Next day, remove any dead embryos. Transfer healthy embryos into 0.5X Holtfreter's solution. At 48 h postinjection, transfer the embryos into regular fish water.
8. Grow the injected embryos into mature transgenic fish. This usually requires ~3 mo.

3.5. Identification of Germline Transgenic Founder Fish (see Note 10)

1. Set up injected adult fish for mating as described above. Collect the embryos and grow them at 28–30°C for 24 h. Keep the injected fish in a separate labeled tank.
2. Dechorionate the F1 embryos. Transfer 100–200 embryos from each pair into a labeled tube.
3. Add 1 mL DNA extraction buffer. Vortex briefly and incubate at 55°C for 2–4 h, with constant rocking (or vortex every h).
4. Spin at maximum speed in a microcentrifuge for 10 min at room temperature. Collect the supernatants and transfer to a clean tube.

5. Precipitate DNA with 3 vol 95% ethanol. Spin and wash the pellet with 70% ethanol. Resuspend the pellet in 100 µL 1X TE.
6. Set up PCR reactions with GFP and EF1α primers as internal control. Each 50 µL PCR reaction contains 1 µL DNA template, 400 µ*M* dNTP, 1 µ*M* GFP primers, 0.5 µ*M* EF1α primers, 5 µL 10X buffer, and 1 U *Taq* DNA polymerase. After an initial 2-min denaturation at 94°C, perform 30 cycles at 94°C for 30 s, at 50°C for 30 s, and at 72°C for 30 s. Finish with a 5-min step at 72°C. Positive and negative PCR controls should be included (*see* **Note 11**).
7. Run 10–15 µL of PCR products on a 2% agarose gel. GFP primers produce a 267-bp product and the PCR product of EF1α primers is 450 bp.
8. A positive pool is indicative of transgenic founder fish.

3.6. Fluorescence Microscopic Analysis of GFP Expression

1. Embryos that are older than 20 somites are dechorionated as described in the previous subheading, and anesthetized in Holtfreter's solution containing 0.003% tricaine.
2. Several embryos are transferred onto a chamber slide containing 1% methyl cellulose dissolved in Holtfreter's/tricaine solution and are oriented under a dissection microscope using a small needle to obtain a desirable position.
3. The slide is then immediately placed on a Zeiss AxioPlan 2 fluorescence microscope and examined under an FITC filter. A color video camera or a digital camera is used to take images of fluorescent embryos. In order to obtain an optimum GFP signal, a Plan-NeoFluor lens with a high numerical aperture is preferred.
4. Confocal microscopy allows one to observe three-dimensional structures of GFP expression patterns. Two-photon excitation microscopy appears to have many advantages, compared to the conventional confocal microscope, when analyzing GFP expression in living transgenic embryos. The authors have used a Zeiss Axioplan-2 mot upright microscope equipped with a Zeiss 510 NLO scan head. Live zebrafish embryos are anesthetized and imaged, using a 40× IR (NA 0.8) or a 63× (NA 1.2) enhanced water-immersion objective. GFP fluorescence is excited using ultrashort (~100 fs) pulses of 850 nm light generated by directly coupling a Coherent Ti-sapphire laser (Mira) that is pumped with a 10 W solid-state laser (Verdi). The intensity of the excitation beam is regulated by an acousto-optic modulator, and set to the minimum power required to generate two-photon fluorescence (typically under 15 mW). The acousto-optic modulator is also used to minimize infrared exposure by blocking the beam during the back-scan. GFP emission is collected through a 500–550-nm band pass filter. Multiphoton fluorescence can be confirmed by the disappearance of the fluorescence image when mode-locking of the Ti-sapphire laser is disconnected.
5. Adobe PhotoShop software is often used to process digital images acquired using a Zeiss AxioPlan 2 fluorescence microscope or a two-photon excitation microscope. Sample images of transgenic zebrafish embryos are shown in **Fig. 1**.

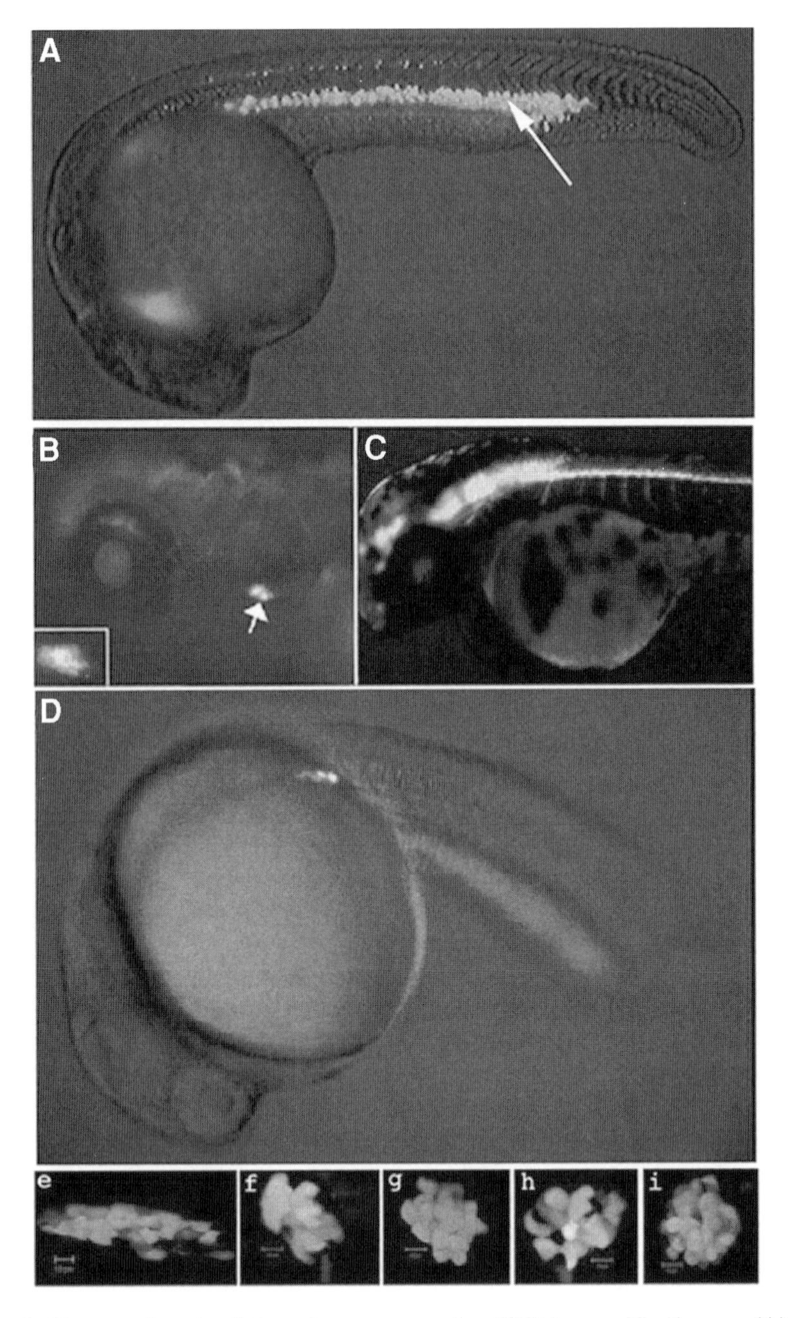

Fig. 1. Transgenic zebrafish embryos expressing GFP in specific tissues. **(A)** GFP expression in hematopoietic progenitor cells driven by the GATA-1 promoter. **(B)** GFP expression in T-cells located in thymus driven by the Rag-1 promoter. **(C)** GFP expression in central nervous system driven by the neuron-specific enhancer of the *GATA-2* gene. **(D)** GFP expression in pancreatic β cells driven by the insulin

4. Notes

1. There are many different ways to set up a microinjection system, depending on the budget of the lab. A typical microinjection system includes a pipet holder, Teflon tubing, a magnetic stand and a micromanipulator to hold the pipet holder, and a microinjector that works either with oil or air. The authors use a 10-mL glass syringe filled with 3–4 mL light mineral oil, to load the needle and inject the DNA. A complete system can be very expensive, but is not required for a successful injection.
2. The authors prepare injection plates by using the lid of a 60-mm Petri dish ***(12)***. Prepare 15 mL 1% agarose in water, and pour it into the lid. Place a wide glass slide (5 × 7 cm) onto the agarose at 10–20 degree angle. Removing the glass slide after the agarose is set will leave a ramp ending in a groove. Clean the ramp after injection with ethanol and store them at 4°C, covered in plastic wrap.
3. For RNA injection, in vitro transcribed mRNA should be prepared, using commercially available kits. Capped mRNA with a poly(A)tail should be designed.
4. The authors isolate the DNA insert from the vector when possible.
5. The quality and purity of the DNA is one the most important factors in a successful injection. Run 1–2 μL DNA on an agarose gel to make sure of the integrity of the DNA. Spin the stock solution of the DNA before transferring it to KCl solution.
6. Another important factor that assures a good post-injection survival rate is the quality of the eggs. To obtain good eggs, the authors have dedicated "for injection only" male and female fish that are mated once a week. If not mated on a regular basis, fish will give bad eggs.
7. It is possible to inject eggs in their chorions. However, the needles must be pulled so that instead of having a long, fine tip, they have a thick stem that quickly tapers to a narrow tip. This way, the needle can penetrate the chorion.
8. Depending on what apparatus is used for microinjection, loading procedures may vary. The authors use a syringe filled with 3–4 mL mineral oil, and pulling the plunger creates enough suction to draw the DNA into the needle. Filling the needle is also possible by using a microloader tip (Eppendorf).
9. A solution containing mRNA can be injected in the yolk or the cytoplasm of the egg. When working with RNA, keep all reagents and materials free from RNase contamination.
10. Expression of GFP reporter gene can be used to identify the founder fish too. Using a fluorescent microscope, screen the embryos of injected fish mated to a wild. Because of the mosaic nature of transmission of the transgene to the F1 generation, at least 100 embryos should be screened.
11. PCR reaction should be sensitive enough to detect one transgenic GFP-expressing embryo in 300 nontransgenic embryos. To make a positive control, add one

---

(*Fig. 1. continued*) promoter. **(e–i)** Reconstruction of pancreatic structure from d 1–d 5 of zebrafish embryonic development using two-photon excitation microscopy. All promoter sequences are from zebrafish. Anterior is to the left and dorsal is to the top. (For optimal, color representation please see accompanying CD-ROM.)

GFP-positive embryo to 300 nontransgenic embryos. Extract DNA and use it as template for positive control. If no transgenic embryo is available, dilute the GFP-containing transgene with embryonic DNA at an appropriate ratio and use this as a positive control.

References

1. Driever, W., Solnica-Krezel, L., Schier, A. F., et al. (1996) A genetic screen for mutations affecting embryogenesis in zebrafish. *Development* **123,** 37–46.
2. Haffter, P., Granato, M., Brand, M., et al. (1996) The identification of genes with unique and essential functions in the development of the zebrafish, Danio rerio. *Development* **123,** 1–36.
3. Amsterdam, A. and Hopkins, N. (1999) Retrovirus-mediated insertional mutagenesis in zebrafish. *Meth. Cell Biol.* **60,** 87–98.
4. Allende, M. L., Amsterdam, A., Becker, T., et al. (1996) Insertional mutagenesis in zebrafish identifies two novel genes, pescadillo and dead eye, essential for embryonic development. *Genes Dev.* **10,** 3141–3155.
5. Gaiano, N., Amsterdam, A., Kawakami, K., et al. (1996) Insertional mutagenesis and rapid cloning of essential genes in zebrafish. *Nature* **383,** 829–832.
6. Amsterdam, A., Yoon, C., Allende, M., et al. (1997) Retrovirus-mediated insertional mutagenesis in zebrafish and identification of a molecular marker for embryonic germ cells. *Cold Spring Harbor Symp. Quant. Biol.* **62,** 37–450.
7. Jaenisch, R. (1988) Transgenic animals. *Science* **240,** 1468–1474.
8. Stuart, G. W., Vielkind, J. R., McMurray, J. V., et al. (1990) Stable lines of transgenic zebrafish exhibit reproducible patterns of transgene expression. *Development* **109,** 577–584.
9. Stuart, G. W., McMurray, J. V., and Westerfield, M. (1988) Replication, integration and stable germ-line transmission of foreign sequences injected into early zebrafish embryos. *Development* **103,** 403–412.
10. Lin, S., Gaiano, N., Culp, P., et al. (1994) Integration and germ-line transmission of a pseudotyped retroviral vector in zebrafish. *Science* **265,** 666–669.
11. Gaiano, N., Allende, M., Amsterdam, A., et al. (1996) Highly efficient germ-line transmission of proviral insertions in zebrafish. *Proc. Natl. Acad. Sci. USA* **93,** 7777–7782.
12. Culp, P., Nusslein-Volhard, C., and Hopkins, N. (1991) High-frequency germ-line transmission of plasmid DNA sequences injected into fertilized zebrafish eggs. *Proc. Natl. Acad. Sci. USA* **88,** 7953–7957.
13. Collas, P. and Alestrom, P. (1998) Nuclear localization signals enhance germline transmission of a transgene in zebrafish. *Transgenic Res.* **7,** 303–309.
14. Yan, Y. L., Talbot, W. S., Egan, E. S., et al. (1998) Mutant rescue by BAC clone injection in zebrafish. *Genomics* **50,** 287–289.
15. Meng, A., Tang, H., Yuan, B., et al. (1999) Positive and negative cis-acting elements are required for hematopoietic expression of zebrafish GATA-1. *Blood* **93,** 500–508.

16. Meng, A., Tang, H., Ong, B. A., et al. (1997) Promoter analysis in living zebrafish embryos identifies a cis-acting motif required for neuronal expression of GATA-2. *Proc. Natl. Acad. Sci. USA* **94,** 6267–6272.
17. Long, Q., Meng, A., Wang, H., et al. (1997) GATA-1 expression pattern can be recapitulated in living transgenic zebrafish using GFP reporter gene. *Development* **124,** 4105–4111.
18. Lin, S., Yang, S., and Hopkins, N. (1994) lacZ expression in germline transgenic zebrafish can be detected in living embryos. *Dev. Biol.* **161,** 77–83.
19. Chalfie, M., Tu, Y., Euskirchen, G., et al. (1994) Green fluorescent protein as a marker for gene expression. *Science* **263,** 802–805.
20. Amsterdam, A., Lin, S., and Hopkins, N. (1995) The Aequorea victoria green fluorescent protein can be used as a reporter in live zebrafish embryos. *Dev. Biol.* **171,** 123–129.
21. Amsterdam, A., Lin, S., Moss, L. G., et al. (1996) Requirements for green fluorescent protein detection in transgenic zebrafish embryos. *Gene* **173,** 99–103.
22. Ju, B., Xu, Y., He, J., et al. (1999) Faithful expression of green fluorescent protein (GFP) in transgenic zebrafish embryos under control of zebrafish gene promoters. *Dev. Genet.* **25,** 158–167.
23. Moss, J. B., Price, A. L., Raz, E., et al. (1996) Green fluorescent protein marks skeletal muscle in murine cell lines and zebrafish. *Gene* **173,** 89–98.
24. Peters, K. G., Rao, P. S., Bell, B. S., et al. (1995) Green fluorescent fusion proteins: powerful tools for monitoring protein expression in live zebrafish embryos. *Dev. Biol.* **171,** 252–257.

18

Transgenic Insects

Expressing Green Fluorescent Protein–Silk Fibroin Light Chain Fusion Protein in Transgenic Silkworms

Hajime Mori

1. Introduction

Lepidopteran insects and their cells are very useful as hosts for the production of heterologous proteins by baculovirus expression vectors ***(1,2)***. However, the gene expression is transient because the infected cells ultimately die from viral infection. Furthermore, the polyhedrin promoter is expressed only during the very late phase of infection.

In order to create stable transgenesis in the silkworm, *Bombyx mori*, using baculovirus, the author performed gene targeting by homologous recombination ***(3)***. The *B. mori fibroin light* (*L*)*-chain gene* was cloned and a *green fluorescent protein* (*GFP*) gene was inserted into exon 7 of the *L-chain* gene. The *L-chain–GFP chimeric* gene was used to replace the polyhedrin gene of *Autographa californica* nucleopolyhedrovirus (AcNPV). This recombinant virus was used to target the *L-chain–GFP* gene to the endogenous L-chain region of the silkworm genome. Female moths were infected with the recombinant virus, then mated with normal male moths. Genomic DNA from their progeny was screened for the desired targeting event. The chimeric gene integrated into the *L-chain* gene on the genome by homologous recombination and stably transmitted over multiple generations. Homologous recombination was expected to occur between the long- and short-arm sequences common to both the chimeric *L-chain* gene in the recombinant AcNPV and the *L-chain* gene in the *B. mori* genome. The chimeric gene was expressed as an L-chain–GFP fusion protein under the control of the endogenous L-chain promoter. The open reading frame of the *chimeric* gene terminates with the stop codon (TAG) of

From: *Methods in Molecular Biology, vol. 183: Green Fluorescent Protein: Applications and Protocols*
Edited by: B. W. Hicks © Humana Press Inc., Totowa, NJ

the *GFP* gene. The *chimeric* gene was expressed only in the posterior silk gland where the endogenous L-chain promoter is active, and the gene product is spun into the silk of the cocoon layer.

2. Materials

1. Polymerase chain reaction (PCR) primers:
   a. For PCR amplification of long-arm DNA fragment:
      1. Forward primer: 5'-AATTAGCTCTAGATGAGCTCCCGGCGTACC-3'.
      2. Reverse primer: 5'-CAACTAAGGGATCCGCGTCATTACCGTTGC-3'.
      In the reverse primer, the AAGCCGGTCGCG sequence included in exon 7 of *L-chain* gene is converted into AAGGGATCCGCG, creating a *Bam*HI site.
   b. For PCR amplification of the short-arm DNA fragment:
      3. Forward primer: 5'-TACCCACTGTCCAATCCACCG-3'.
      4. Reverse primer: 5'-CCGGCTTAGTTGCTAATGCTC-3'.
   c. Primers for screening recombinant virons:
      5. 5'-GGAGAAGAACTTTTCACTGGAG-3' and
      6. 5'-ATCCATGCCATGTGTAATCCC-3' to screen for the *GFP* gene.
      7. 5'-ACTACAAGACACGTGCTG-3' in the *GFP* gene and
      8. 5'-AGCATGACAACAGTACCG-3' in the intron of downstream sequence from exon 7 are used for screening of the targeting event (*see* **Fig. 1**).
2. Genomic DNA extraction kit (Stratagene).
3. LA PCR kit ver.2 (TaKaRa) for PCR amplification of the long arm. PCR mixture contains: 2.5 U LA *Taq*, 10X LA PCR buffer (supplied by kit), 400 m*M* deoxyribonucleoside triphosphate (dNTP) each, 2 m*M* $MgCl_2$ in 50 µL.
4. DNA polymerase for PCR amplification of the short arm: KOD-Plus*- (Toyobo). PCR mixture contains 1 U KOD-Plus-, 5 µL 10X PCR buffer (supplied by kit), 200 µ*M* dNTP each, 2 m*M* $MgCl_2$, 15 pmol of each primer in 50 µL.
5. DNA polymerase for screening: r*Taq* (Toyobo). PCR mixture contains 1 U r*Taq*, 10 m*M* Tris-HCl (pH 8.3), 200 µ*M* dNTP each, 1.5 m*M* $MgCl_2$, 50 m*M* KCl, 15 pmol of each primer in 50 µL.
6. MicroAmp reaction vials (PE Applied Biosystems). Gene AmpPCR System 2400 Thermocycler (PE Applied Biosystems).
7. Restriction enzymes (New England Biolabs).
8. Baculovirus transfer vector pVL1392 (PharMingen).
9. Lipofectin Reagent (Gibco-BRL).
10. pGFP vector (Clontech).
11. Linear AcNPV DNA (Baculogold Baculovirus DNA, BD PharMingen).
12. TE: 10 m*M* Tris-HCl, pH 8.0, 1 m*M* EDTA.
13. *Spodoptera frugiperda* IPLB-SF21-AE cell line (*Sf*21 cells) maintained in TC-100 (Gibco-BRL) with 10% fetal bovine serum.

---

* "KOD-plus-" was a trade name of a high fidelity DNA polymerase for PCR by a Japanese company. But "KOD" (not "KOD-plus-") or other DNA polymerase with a high fidelity is also useable in this experiment.

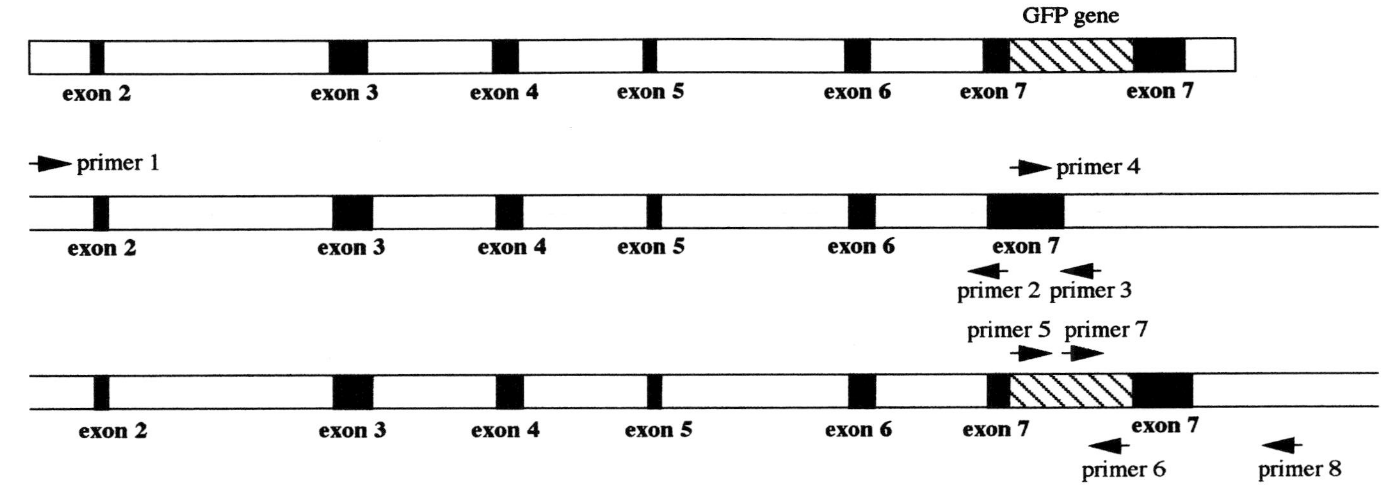

Fig. 1. Homologous recombination between the targeting vector and endogenous DNA. Targeting vector (*top*), intron and exon of *L-chain* gene (*middle*), and a product of the targeting event by homologous recombination (*bottom*). Reprinted with permission ***(3)***.

14. Ecdysteroid hormone: 1 mg 20-hydroxyecdysone (Sigma) are dissolved in 2.5 mL of 10% ethanol (0.4 mg/mL), and stored at –20°C.
15. 70% Lithium thiocyanate (LiSCN).
16. Denaturation buffer: 62.5 m*M* Tris-HCl buffer (pH 6.8) containing 2% sodium dodecyl sulfate (SDS) and 5% 2-mercaptoethanol. 12.5% resolving SDS-polyacrylamide gel electrophoresis (PAGE) gel.
17. Polyclonal rabbit anti-GFP antibody (Clontech). Goat antirabbit immunoglobulin (H+L) horseradish peroxidase conjugate (Bio-Rad).
18. TBS: 20 m*M* Tris-HCl, 500 m*M* NaCl (pH 7.5).
19. Blocking solution: 3% gelatin in TBS.
20. Antibody buffer: 1% gelatin in TBS.
21. Enhanced Chemiluminescence (ECL) Plus Western blotting detection reagents (Amersham Pharmacia Biotech).

3. Methods

3.1. Construction of Targeting Vector

1. Prepare two primers for PCR amplification of the 5.0-kbp long arm and the 0.5-kbp short-arm DNA fragments according to the published sequence data for the *L-chain* gene (***4***; **Fig. 1**). Use LA *Taq* and KOD-Plus- for PCR amplification of the long arm and the short arm, respectively.
2. Extract the genomic DNA of fifth instar larvae from hemocytes using a genomic DNA extraction kit.
3. Amplify the long-arm DNA fragment (*see* **Note 2**). PCR amplification is carried out in PCR reaction mixture with 0.4 m*M* of primers 1 and 2, 200 ng of the genomic DNA as template, and 2.5 U LA *Taq* DNA polymerase. After heating the reaction mixture (50 µL) in MicroAmp reaction vials at 94°C for 1 min in a Gene AmpPCR System 2400, 30 cycles of amplification are performed using 20 s at 98°C, followed by 15 min at 68°C. A final 10 min step at 72°C is performed at the completion of these cycles.
4. The long arm is cut out with *Xba*I and *Bam*HI, and the fragment is inserted into baculovirus transfer vector, pVL1392, at the *Xba*I and *Bam*HI site, to obtain pVLFL-5K (*see* **Note 1**).
5. Amplify the short-arm DNA fragment. PCR amplification is carried out in PCR reaction mixture with 0.4 m*M* of primers 3 and 4, 200 ng of the genomic DNA as template, and 1 U KOD-Plus- DNA polymerase. After heating the reaction mixture at 94°C for 2 min in a Gene AmpPCR System 2400, perform 35 cycles of (denaturation) 1 min at 94°C, (annealing) 1 min at 55°C, and (extension) 2 min at 72°C. Perform a final 7 min step at 72°C at the completion of the 35 cycles. Excise the short arm with *Eco*RI, and insert it into the GFP plasmid at the dephosphorylated *Eco*RI site to construct pGFP-0.5K. Separate *Bam*HI and *Stu*I fragments from pGFP-0.5K, and ligate with a *Bam*HI adaptor at the *Stu*I site. The resulting 1.2-kbp *Bam*HI fragment, composed of the *GFP* gene and short-arm DNA fragment is inserted into pVLFL-5K at the dephosphorylated *Bam*HI site,

and the recombinant baculovirus transfer vector, pAcFLGFP, is constructed (**Fig. 1**). The nucleotide sequence of the intron–exon boundary of the *L-chain* gene should also be determined by dideoxy termination sequencing (*see* **Note 1**).
6. Transfect *Sf*21 cells in a 60-mm culture plate with 5 μg recombinant transfer vector and 0.5 μg linearized Baculogold AcNPV DNA as described in the manufacturer's literature (*see* **Note 3**) using Lipofectin Reagent (Gibco-BRL).
7. Plaque purify the recombinant AcNPV that is rescued by the transfer vector.

3.2. Screening of Silkworm by PCR

To confirm that the targeting event occurs, PCR amplifications are performed.

1. Inject 1-d-old, fifth-instar, female and male larvae (F0) with 50 μL TC-100 medium containing 5×10^5 plaque-forming unit (pfu) of the recombinant virus by a hypodermic injection (**Fig. 2**). Approximately 50 insects are inoculated in each experiment. Although normal larval-pupal ecdysis is observed, pupal-adult metamorphosis is arrested (*see* **Note 4**). However, metamorphosis will resume after administration of ecdysteroid hormone *(5)*.
2. Mate virus-inoculated female and male moths with a normal moth.
3. Extract the genomic DNA from 100 embryos, and use it as the template for a PCR, using primers 5 and 6 to screen for the presence of the *GFP* gene. Use the r*Taq* polymerase for *GFP* gene screening. Rear the larvae from the PCR-positive siblings, and discard PCR-negative siblings.
4. At F1, collect 100–150 individuals. Separately harvest hemocytes from each fifth instar larva. Genomic DNA is extracted from the hemocytes, and screened by PCR analysis, using primers 5 and 6 to screen for the presence of the *GFP* gene.
5. Mate male or female moths containing *GFP* gene with normal moths, and their progenies (F2) are reared. About 150 offspring are produced from each cross and hemocytes are harvested separately from each larvae in fifth instar.
6. Assay genomic DNA from the hemocytes for transgenesis using a PCR screen that specifically detects a novel DNA junction created by the targeting event. Only homologous recombination can juxtapose the two sequences that create this junction, one from the *GFP* gene in the targeting vector, and the other within the downstream sequence beyond the end of the targeting vector. The PCR reaction with genomic DNA using primers 7 and 8, amplifies a 0.9-kbp DNA fragment, and verifies that the targeting vector has correctly recombined with the genomic DNA.

3.3. Screening of Silkworm by Western Blot

Western blot analysis for chimeric protein expression is performed on silk proteins from the cocoon layer (**Fig. 3**).

1. Add 150 mL 70% LiSCN to 5 mg of the cocoon layer, to dissolve the silk protein.
2. Add 40 mL denaturation buffer to 10 mL dissolved silk protein.
3, Analyze 20 μL of the denatured silk protein by SDS-PAGE on a 12.5% resolving gel.
4. Transfer the proteins to nitrocellulose.

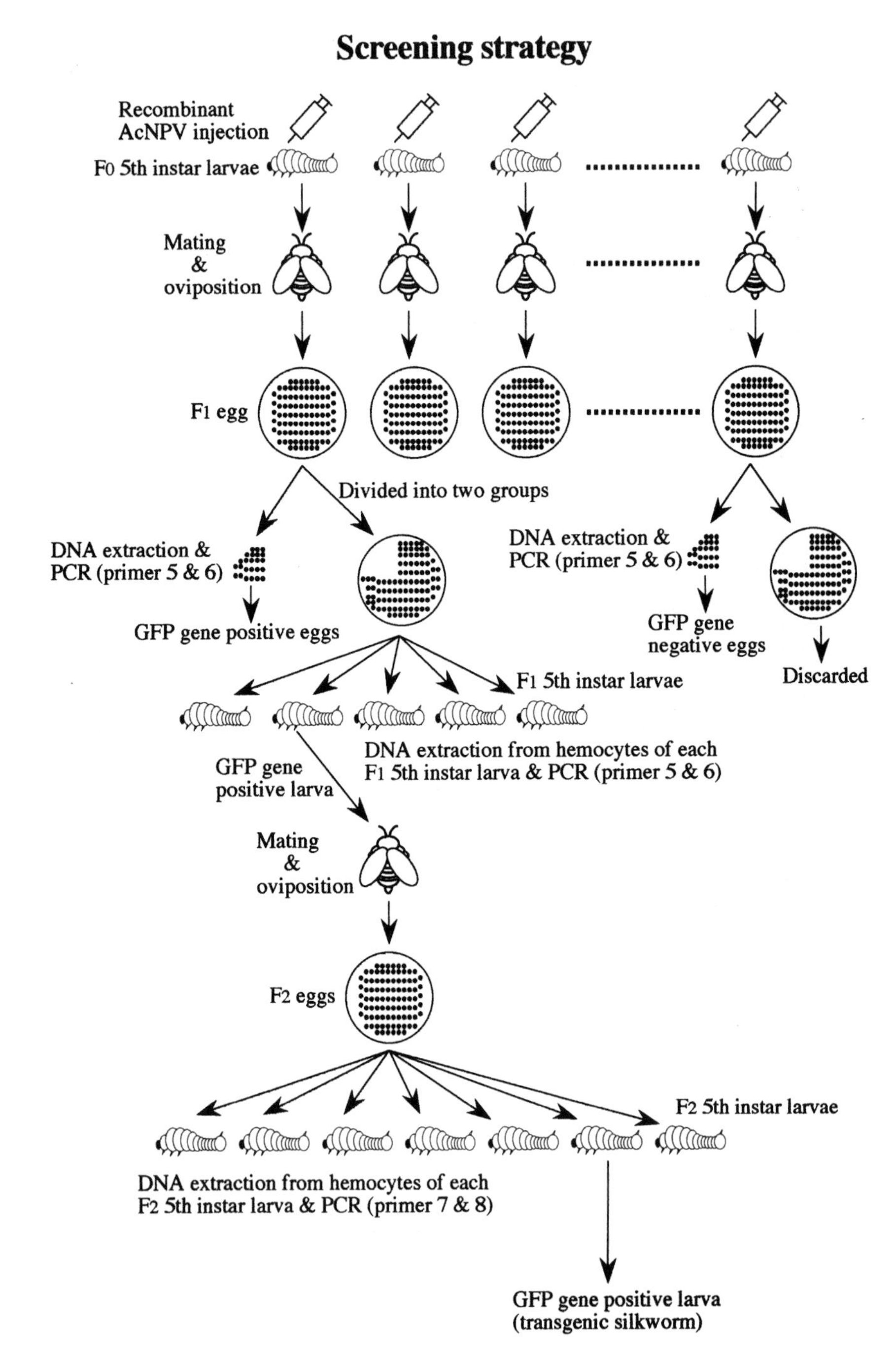

Fig. 2. Scheme of the screening method for analysis of gene targeting. Reprinted with permission *(3)*.

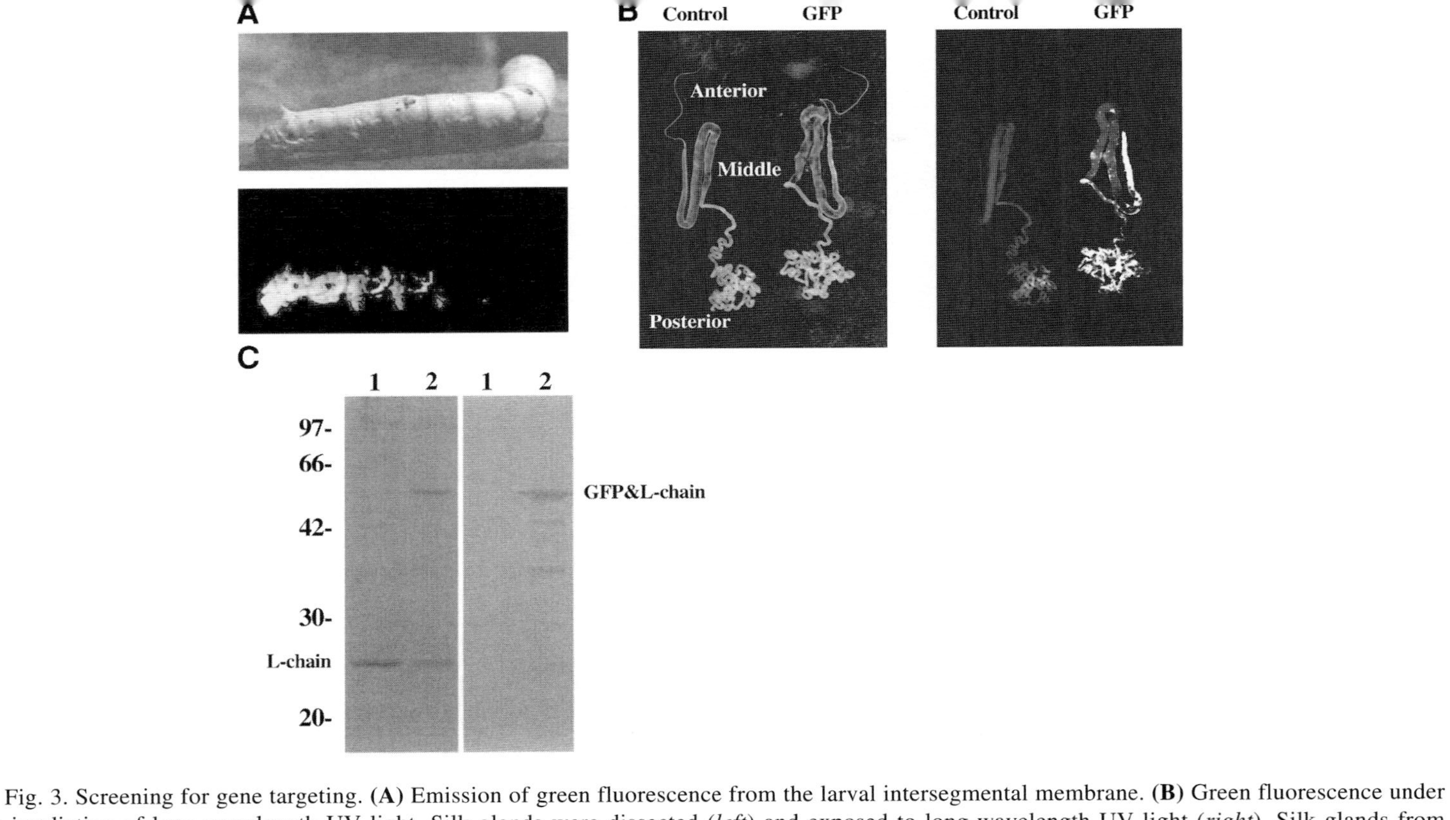

Fig. 3. Screening for gene targeting. **(A)** Emission of green fluorescence from the larval intersegmental membrane. **(B)** Green fluorescence under the irradiation of long-wavelength UV light. Silk glands were dissected (*left*) and exposed to long-wavelength UV light (*right*). Silk glands from control animal and the *GFP* gene-targeted animal are shown as control and GFP, respectively. **(C)** SDS-PAGE (*left*) and Western blot analysis (*right*) of silk protein. Lane 1, normal cocoon layer; lane 2, cocoon layer from the *GFP* gene-inserted animal. Antibody used in the Western blot analysis was specific for GFP. The size (kDa) of protein markers is shown at the left. (For optimal, color representation please see accompanying CD-ROM.). Reprinted with permission ***(3)***.

5. Detect the recombinant silk protein by Western blotting. The nitrocellulose blot is placed in blocking solution for 1 h, and the blots are incubated overnight with antibody buffer containing rabbit anti-GFP antibody at 1:1000 dilution. Wash the blots 3 × 15 min in TBS. Incubate the blot with goat antirabbit immunoglobulin (H+L) horseradish peroxidase conjugate (Bio-Rad) at 1:3000 dilution for 1 h. Wash the blots 3 × 15 min in TBS, and detect presence of GFP by ECL Plus HRP Detection Reagents.

3.4. Screening of Silkworm by Green Fluorescence

Emission of the green fluorescence is very useful for screening of transgenesis in the silkworm (**Fig. 3**).

1. Irradiate the suspected transgenic silkworms with a long-wavelength ultraviolet (UV) light. Green fluorescence can be observed from the larval intersegmental membrane because the intersegmental membrane forms flexible, translucent joints.
2. Dissect and excise the silk gland. Under the exposure of long-wavelength UV light, emission of green fluorescence will be observed from the *GFP* gene-targeted silkworm larvae, and no green fluorescence will be observed from control larvae. Two subunits of fibroin (H- and L-chains) are synthesized in the posterior silk gland. The gland is located in the posterior abdomen and the chimeric *L-chain/GFP* gene is expressed in the posterior silk gland of the transgenics.

4. Notes

1. The nucleotide sequence of the PCR products of the *L-chain* gene should be determined by the dideoxy termination method carried out with an automatic DNA sequencer. This is necessary to ensure that PCR did not introduce mutations into exons of the long arm, the short arm, or the ligated final product in the completed transfer vector.
2. PCR products for the long arm and the short arm are electrophoresed on the agarose gel and cut out from the gel.
3. BD Pharmingen supplies a 108-page instruction manual on the use and production of recombinant baculovirus using their linear Baculogold AcNPV DNA.
4. The average duration of the larval and pupal stages in the uninfected animals are 10 and 15 d, respectively, but duration of the pupal stage of infected animals is a few days longer than in the uninfected animals. This phenomenon suggests that the *ecdysteroid UDP-glucosyltransferase* (*EGT*) gene of AcNPV is expressed, and that the secreted EGT altered growth of the infected animals ***(6)***.
5. Female and male of the silkworm can be divided by the detection of Ishiwata imaginal disk for female sexual organ and Herold imaginal disk for male sexual organ.
6. After making a hole to the abdominal leg by using a needle, hemolymph is collected and the hemocytes are recovered by centrifugation in the microtube at 1000 rpm for 5 min.

References

1. Luckow, V. A. and Summers, M. D. (1988) Trends in the development of baculovirus expression vector. *BioTechnology* **6,** 47–55.
2. Miller, L. K. (1988) Baculoviruses as gene expression vectors. *Ann. Rev. Microbiol.* **42,** 177–199.
3. Yamao, Y., Katayama, N., Nakazawa, H., Yamakawa, M., Hayashi, Y., Hara, S., et al. (1999) Gene targeting in the silkworm by use of a baculovirus. *Genes Dev.* **13,** 511–516.
4. Kikuchi, Y., Mori, K., Suzuki, S., Yamaguchi, K., and Mizuno, S. (1992) Structure of the Bombyx mori fibroin light-chain-encoding gene: upstream sequence elements common to the light and heavy chain. *Gene* **110,** 151–158.
5. Mori, H., Yamao, M., Nakazawa, H., Sugahara, Y., Shirai, N., Matsubara, F., et al. (1995) Transovarian transmission of a foreign gene in the silkworm, Bombyx mori, by Autographa californica nuclear polyhedrosis virus. *BioTechnology* **13,** 1005–1007.
6. Shikata, M., Shibata, H., Sakurai, M., Sano, Y., Hashimoto, Y., and Matsumoto, T. (1998) The ecdysteroid UDP-glucosyltransferase gene of Autographa californica nucleopolyhedrovirus alters the moulting and metamorphosis of a non-target insect, the silkworm, Bombyx mori (Lepidoptera, Bombycidae). *J. Gen. Virol.* **79,** 1547–1551.

19

Green Fluorescent Protein in Transgenic Plants

Brassica Transformation

C. Neal Stewart, Jr., Matthew D. Halfhill, and Reginald J. Millwood

1. Introduction

Until the heterologous expression of *Aequorea victoria* green fluorescent protein (GFP) was demonstrated, scientists working with transgenic organisms had no good alternative to using destructive visible genetic markers. Genes coding luciferase *(1)* and β-glucuronidase *(2)* are the most popular destructive marker genes that have been successfully used in transgenic plants. Although these markers code for sensitive enzymes that have linear dose responses, they require expensive substrates, and are limited to laboratory uses. Most of all, they cannot be used to assay living tissue directly.

GFP offers the possibility to assay vital cellular functions, to determine the transgenic status of plants, and to monitor plant transgene expression in real time, in live cells or intact plants. This chapter focuses on the use of GFP as an enabling biotechnology in the production of transgenic plants, especially *Brassicas*. GFP offers the plant biotechnologist the tool to produce plants in the absence of, or in conjunction with, antibiotic or herbicide markers for selection. It also offers a mechanism to quickly identify transgenic plants in mixed populations. GFP will prove to be an important tool for the making and monitoring of transgenic crops and trees, in the future *(3,4)*.

Several GFPs have been shown to be useful in plants. The earliest useful variant was mGFP4, a near-wild-type version that had an altered plant-recognized cryptic intron *(5)*. Unfortunately, this GFP was neither bright nor very stable. Improved versions of mGFP4 (mGFP5 and mGFP5-ER) have wild-type chromophores, but have the following mutations: V163A, S175G, and I167T *(5,6)*. These mutations confer increased folding at warm temperatures, equal

From: *Methods in Molecular Biology, vol. 183: Green Fluorescent Protein: Applications and Protocols*
Edited by: B. W. Hicks © Humana Press Inc., Totowa, NJ

and dual excitation peaks at 395 and 475 nm, and an emission peak at 509 nm *(6)*. The endoplasmic reticulum version has a signal sequence and HDEL retention signal for targeting GFP to the endoplasmic reticulum. Human codon-optimized S65T mutants have also been useful in plants *(7,8)*. Versions of S65T GFP have a single excitation peak at 489 nm and a red-shifted excitation optimum to (a green) 511 nm *(8)*. Another good choice for plants is the commercially available (Clontech) enhanced GFP, which has the S65T as well as the F64L and Y145F mutations, and is human codon-optimized ***(9)***. Other researchers have produced mutants that have been useful in plants ***(10,11)***. Recently, GFPs from other organisms have been cloned ***(12)***. Plant-optimized GFP, and yellow fluorescent proteins may be expected to be better in plant applications than those currently available. In fact, a priori, *Renilla reniformis* GFP, which has recently been made commercially available by Stratagene, has spectral qualities that should make it brighter in heterologous systems ***(13)***. Fluorescent proteins that emit in the yellow and orange spectra have promise in transgenic plant work.

GFP has been used in plant transformation systems as a transformation marker in soybean ***(14)***, sugarcane ***(15)***, orange ***(16)***, tobacco ***(17)***, wheat ***(18)***, and apple ***(19)***, to name a few species. In certain instances, GFP has been used as the sole selectable marker in transgenic plants, demonstrating that a visual marker could be used instead of antibiotic or herbicide selection. Thus far, GFP as the sole selection marker has been proven useful mainly in monocots such as sugarcane ***(20)***, barley ***(21)***, rice ***(22)***, and oats ***(23)***. The dicot exception in this case is citrus ***(16)***, in which the transformation frequency was compared between GFP-only and GFP plus antibiotic selection. The researchers found that the transformation frequency was the same, but curiously, there were fewer GFP-positive shoots per experiment, using GFP selection ***(16)***. One of the benefits of using GFP as the selectable marker is that high-expressing events can be selected very early in the tissue culture and regeneration process.

In this chapter, methods are described that the authors' group has used to transform members of the mustard family (*Brassicaceae*), using GFP-only and GFP in conjunction with antibiotic selection. This lab has produced transgenic *Brassica*s using antibiotic selection ***(24)***, and is now using GFP to show proof-of-principle in *Brassica napus*, and also to extend the *Brassica* transformation procedure to a wild relative of the same genus: *Raphanus raphanistrum* (syn. *Brassica kaber*). Various experiments have been performed to demonstrate the efficiencies of GFP-only, or GFP-plus-antibiotic selection. Experiments described here employ a plasmid with GFP and an antibiotic selectable marker, but the goal is to use GFP as the sole selectable marker. Avoiding the use of antibiotic selection could address the criticism of biotechnology opponents who fear that the horizontal transfer of antibiotic resistance genes could cause medical and ecological emergencies.

2. Materials

1. Surface-sterilized seeds (20% bleach solution for 5 min) from *B. napus* cv Westar.
2. Marashige and Skoog (MS) basal medium ***(25)*** for seed (hypocotyl explant source) germination. All plant tissue culture plates are produced using 0.2% Gelrite gellan gum (Sigma, St. Louis, MO) as a gelling agent. All agents are autoclaved, except kanamycin, before media is poured into plates.
3. MS basal medium with 1 mg/L, 2,4-D (MSD1) for 24 h preconditioning hypocotyls, and postco-cultivation.
4. *Agrobacterium tumefaciens* strain GV 3850 containing pBin *mgfp5-er* (35S promoter controlling *mGFPer* gene with linked NOS promoter-controlled *nptII* for kanamycin selection [**Fig. 1**]).
5. *Agrobacterium* solution (10^8 cells/mL in liquid MS basal medium with acetosyringone 0.05 m*M*) for co-cultivation with hypocotyls.
6. MSD1 media containing 400 mg/L Timintin to select against *Agrobacterium,* and with or without 20 mg/L kanamycin to select for transformed cells. No kanamycin is used for GFP-only selection.
7. CSRA: MS basal media containing 4 mg/L 6-benzylaminopurine, 2 mg/L zeatin, 5 mg/L silver nitrate, and with or without the above antibiotics to promote organogenesis.
8. CSRB: MS basal media containing 4 mg/L 6-benzylaminopurine, 2 mg/L zeatin, with or without antibiotics.
9. CSE: MS basal medium containing 0.05 mg/L 6-benzylaminopurine plus antibiotics for shoot elongation.
10. MSR: MS basal media containing 0.1% indole burtyric acid plus antibiotics to promote root development.
11. 100-mm Petri dishes and GA7 Magenta boxes for tissue culture.
12. Standard dissecting microscope and Spectroline BIB-150 UV lamp.
13. Laminar flow-hood.

3. Methods

3.1. GFP Transformation and Selection in Brassica (24)

1. Seeds are germinated on MS basal media. Zygotic hypocotyls were dissected and chopped into 1-cm-long segments. The hypocotyls segments were placed in a Petri dish containing the *Agrobacterium* inoculum in liquid MS basal medium for 30 min. Periodically shake the segments gently during the 30 min inoculation time Transfer the explants to MSD1 for 1 d, then to MSD1 plus one or no antibiotics (no kanamycin was in the media when using GFP selection only).
2. After 3 d, transfer the tissue to CSRA to initiate shooting. There is a considerable time delay (a few weeks) between shoot initiation and shoot formation using this procedure.

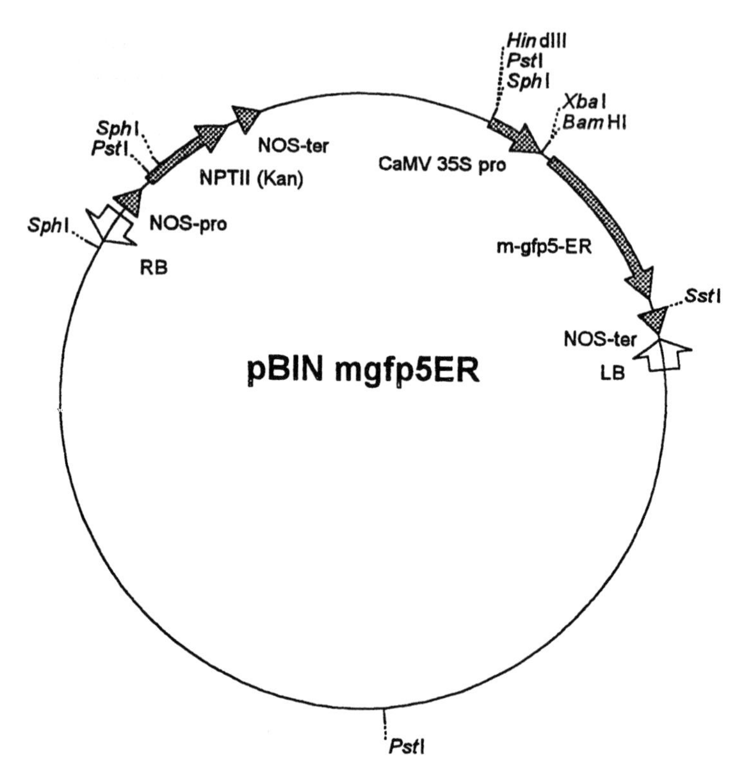

Fig. 1. The binary plasmid, pBin mgfp5ER, which was used for the plant transformation experiments (courtesy of Jim Haseloff). Kanamycin selectable (*nptII*) gene is under the control of the NOS promoter, and the endoplasmic reticulum targeted *GFP* gene is under the control of the 35S promoter from the cauliflower mosaic virus.

3, After another 7 d, (10 d after *Agrobacterium* transformation) transfer the tissue to CSRB. Between 2 and 4 wk GFP fluorescence will appeared in calli, then in shoots (*see* **Notes 1–3**).
4. At this point, weekly monitoring with a UV light is required to track transgenic events.
5. When the event callus (fluorescing uniformly green) is approx 0.5 cm in diameter, it is safe to isolate it from the greater tissue and transfer it onto fresh CSRB (*see* **Note 4**). Alternatively, shoots can be transferred to fresh CSRB.
6. Transgenic shoots are transferred to CSE as needed for elongation, then to MSR for rooting.
7. Visually assay for relative transgene expression by comparing GFP emission under UV illumination, thereby allowing selection of the highest-expressing events very early in the transformation process. **Figure 2** shows the product of this method for the transformation of the *Brassica* relative, wild radish (*Raphanus raphanistrum*) on CSRB.

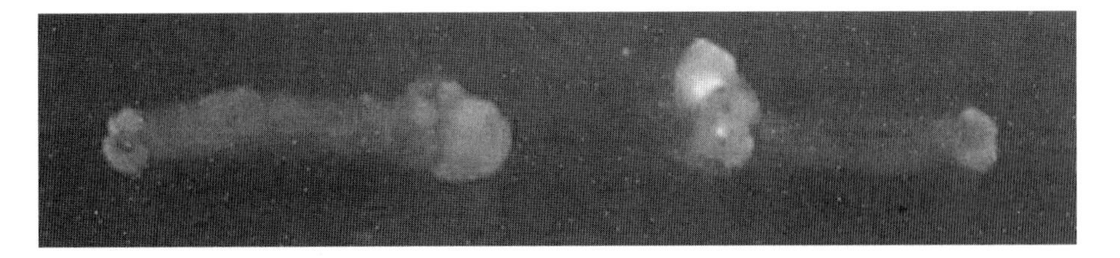

Fig. 2. *Raphanus raphanistrum* hypocotyls segments producing callus stably transformed with mGFP5er under the control of a constitutive promoter. Notice the variation of fluorescence between cut ends. GFP is visualized under UV (365 nm) illumination with no emission filter. (For optimal, color representation please see accompanying CD-ROM.)

4. Notes

1. Much of the success of GFP as an enabling technology in transgenic plants hinges on the success of seeing its production in plants. For lab work, most researchers use epifluorescence microscopes fitted with mercury lamps (~100 W) with blue filters (e.g., 470/40 nm) with 515 nm long-pass emission filters. Of course, without emission filters, one only sees blue reflectance (*see* **refs. *26–28***) for details. In using such arrangements, several researchers have reported background fluorescence that interferes with observing GFP ***(14,22,27)***. Altering filter choices, such as choosing emission filters of narrower bandwidth, or alternative emission filters should help ***(15,21)***. Empirical optimization by plant species and tissue types may need to be performed when using blue-light-excited GFPs. The choice of UV-excited GFPs, such as mGFP5, is often ignored as a viable choice by plant scientists. For example, there may be background fluorescence when excited by blue light, but not when excited by UV wavelengths.
2. If one desires to visualize whole plants or organs, then a microscope is not the best tool. For blue-excited GFPs, one can use the photonics of a microscope system, and indeed, Opti-Sciences (Tyngsboro, MA) produces a blue light source with the proper cutoff or bandpass filters for measuring GFP-transgenic plants (GF probe). For UV-excited GFPs, the authors' group and others typically use a portable UV lamp (UVP 100 AP, Upland, CA) with no emission filter, or the lighter Spectroline BIB-150 produced by Spectronics (Westbury, NY). These lamps have a 100 W mercury bulb and a 365-nm filter. The authors group and others have attempted to use less powerful UV lamps with little success. On the other side, we have combined 2–3 of the Spectroline UV lamps, to boost photon excitation irradiation, for more spectacular photographs. To effectively visualize GFP in transgenic plants, the lamp should be very bright and at the proper wavelength. Although the Spectroline or UVP lamps work well for UV excitation of GFP, they would be even more effective if they used a 395 nm filter instead of the 365-nm filter, since the former better matches GFP excitation.
3. UV protective eyewear should be used.

4. There are few tricks to keep in mind when using GFP as a selection for transformation of plants. Tracking transgenic events as early as possible, and keeping the events segregated is desirable. Isolating high-expressing events is important. However, if one excises green fluorescent tissue from the mother explant source, it may die. The authors have been unsuccessful if fluorescent *Brassica* callus is isolated, if the tissue piece is much smaller than 0.5 cm. The UV lamp makes it easy to screen several plates once per week. It also adds the additional benefit of "lighting-up" contaminants that are otherwise hard to see on Petri dishes.

Acknowledgments

We would like to thank Dow AgroSciences, the US Department of Agriculture Biotechnology Risk Assessment Program, and USDA Plant Pathology Special Grant for support.

References

1. Ow, D. W., Wood, K. V., de Luca, M., deWet, J. R., Helinski, D. R., and Howell, S. H. (1986) Transient and stable expression of the firefly luciferase gene in plant cells and transgenic plants. *Science* **234,** 856–859.
2. Jefferson, R. A. (1989) The GUS reporter gene system. *Nature* **342,** 837–838.
3. Leffel, S. M., Mabon, S. A., and Stewart, C. N., Jr. (1997) Applications of green fluorescent protein in plants. *BioTechniques* **23,** 912–918.
4. Harper, B. K., Mabon, S. A., Leffel, S. M., Halfhill, M. D., Richards, H. A., Moyer, K. A. et al. (1999) Green fluorescent protein as a marker for expression of a second gene in transgenic plants. *Nat. Biotechnol.* **17,** 1125–1129.
5. Haseloff, J., Siemering, K. R., Prasher, D. C., and Hodge, S. (1997) Removal of a cryptic intron and subcellular localization of green fluorescent protein are required to mark transgenic *Arabidopsis* plants brightly. *Proc. Natl. Acad. Sci. USA* **94,** 2122–2127.
6. Siemering, K. R., Golbik, R., Sever, R., and Haseloff, J. (1996) Mutations that suppress the thermosensitivity of green fluorescent protein. *Curr. Biol.* **6,** 1653–1663.
7. Heim, R., Cubitt, A. B., and Tsien, R. Y. (1995) Improved green fluorescence. *Nature* **373,** 663–664.
8. Chiu, W. L., Niwa, Y., Zeng, W., Hirano, T., Kobayashi, H., and Sheen, J. (1996) Engineered GFP as a vital reporter in plants. *Curr. Biol.* **6,** 325–330.
9. Yang, T.-T., Cheng, L., and Kain, S. R. (1996) Optimized codon usage and chromophore mutations provide enhanced sensitivity with the green fluorescent protein. *Nucl. Acid Res.* **24,** 4592–4593.
10. Pang, S.-Z., DeBoer, D. L., Wan, Y., Ye, G., Layton, J. G., Neher, M. K., et al. (1996) An improved green fluorescent protein gene as a vital marker in plants. *Plant Physiol.* **112,** 893–900.
11. Davis, S. J. and Vierstra, R. D. (1998) Soluble, highly fluorescent variants of green fluorescent protein (GFP) for use in higher plants. *Plant Mol. Biol.* **36,** 521–528.

12. Matz, M. V., Fradkov, A. F., Labas, Y. A., Savitsky, A. P., Zaraisky, A. G., Markelov, M. L., et al. (1999) Fluorescent proteins from nonbioluminescent Anthozoa species. *Nat. Biotechnol.* **17,** 969–973.
13. Ward, W. W. (1998) Biochemical and physical properties of green fluorescent protein, in.*Green Fluorescent Protein: Properties, Applications, and Protocols* (Chalfie, M. and Kain, S. R., eds.) Wiley and Sons, Chichester, England, pp. 45–75.
14. Ponappa, T., Brzozowski, A. E., and Finer, J. J. (2000) Transient expression and stable transformation of soybean using jellyfish green fluorescent protein (GFP). *Plant Cell Rep.* **19,** 6–12.
15. Elliot, A. R., Campbell, J. A., Dugdale, B., Brettell, R. I. S., and Grof, C. P. L. (1999) Green-fluorescent protein facilitates rapid in vivo detection of genetically transformed plant cells. *Plant Cell Rep.* **18,** 707–714.
16. Ghorbel, R., Juarez, J., Navarro, L., and Pena, L. (1999) Green fluorescent protein as a screenable marker to increase the efficiency of generating transgenic woody fruit plants. *Theor. Appl. Genet.* **99,** 350–358.
17. Molinier, J., Himber, C., and Hahne, G. (2000) . Use of green fluorescent protein for detection of transformed shoots and homozygous offspring. *Plant Cell Rep.* **19,** 219–223.
18. McCormac, A. C., Wu, H., Bao, M., Wang, Y., Xu, R., Elliot, M. C., et al. (1998) The use of visual marker genes as cell-specific reporters of *Agrobacterium*-mediated T-DNA delivery to wheat (*Triticum aestivum* L.) and barley (*Hordeum vulgare* L.). *Euphytica* **99,** 17–25.
19. Maximova, S. N., Dandekar, A. M., and Guiltinan, M. J. (1998) Investigation of *Agrobacterium*-mediated transformation of apple using green fluorescent protein: high transient expression and low stable transformation suggest that factors other than T-DNA transfer are rate-limiting. *Plant Mol. Biol.* **37,** 549–559.
20. Elliot, A. R., Campbell, J. A., Brettell, I. S., and Grof, P. L. (1998) *Agrobacterium*-mediated transformation of sugarcane using GFP as a screenable marker. *Aust. J. Plant Physiol.* **25,** 739–743.
21. Ahlandsberg, S., Sathish, P., Sun, C., and Jansson, C. (1999) Green fluorescent protein as a reporter system in the transformation of barley cultivars. *Physiol. Plant.* **107,** 194–200.
22. Vain, P., Worland, B., Kohli, A., Snape, J., and Christou, P. (2000) The green fluorescent protein (GFP) as a vital screenable marker in rice transformation. *Theor. Appl. Genet.* **96,** 164–169.
23. Kaeppler, H. F., Menon, G. K., Skadsen, R. W., Nuutila, A. M., and Carlson, A. R. (2000) Transgenic oat plants via visual selection of cells expressing green fluorescent protein. *Plant Cell Rep.* **19,** 661–666.
24. Stewart, C. N., Jr., Adang, M. J., All, J. N, Raymer, P. L., Ramachsndran, S., and Parrott, W. A. (1996) Insect control and dosage effects in transgenic canola, *Brassica napus* L. (Brassicaceae), containing a synthetic *Bacillus thuringiensis cry IAc* gene. *Plant Physiol.* **112,** 115–120
25. Murashige, T. and Skoog, F. (1962) A revised medium for rapid growth and bioassays with tobacco tissue cultures. *Physiol. Plantarum* **15,** 473—497.

26. Ellenberg, J., Lippincott-Schwartz, J., and Presley, J. F. (1998) Two-color green fluorescent protein time-lapse imaging. *BioTechniques* **25,** 838–846.
27. van der Geest, A. H. M., and Petolino, J. F. (1998) . Expression of a modified green fluorescent protein gene in transgenic maize plants and progeny. *Plant Cell Rep.* **17,** 760–764.
28. Rizzuto, R., Carrington, W., and Tuft, R. A. (1998) Digital imaging microscopy of living cells. *Trends Cell Biol.* **8,** 288–292.

V

Green Fluorescent Protein Biosensors

20

Green Fluorescent Protein Calcium Biosensors

Calcium Imaging with GFP Cameleons

Anikó Váradi and Guy A. Rutter

1. Introduction

Cytosolic and organellar free (Ca^{2+}) concentrations ([Ca^{2+}]) are among the most important and dynamic intracellular signals. Ca^{2+} signals are most often measured using Ca^{2+} sensitive fluorescent dyes ***(1–3)***, such as Fura-2 or Indo-1, or the bioluminescent protein, aequorin ***(4–6)***. Whereas synthetic fluorescent chelators are easily imaged, these are difficult to target precisely to specific subcellular locations (**Table 1**). By contrast, aequorin is easily targeted, but requires the incorporation of a cofactor, coelenterazine. Moreover, the photon intensity from aequorin is extremely low, so that single cell imaging requires specialized photon-counting systems ***(4)*** (**Table 1**).

In an attempt to overcome these problems, Miyawaki et al. ***(7)*** and Romoser et al. ***(8)*** developed the first dynamically responsive biochemical Ca^{2+} indicators based on green fluorescent protein (GFP), a spontaneously fluorescent protein from the jellyfish *Aequorea victoria*. Miyawaki et al. ***(7)*** fused enhanced cyan fluorescent protein (ECFP) to the N-terminus of calmodulin (CaM), and enhanced yellow fluorescent protein (EYFP) to the C-terminus of M13, the CaM-binding peptide from skeletal muscle myosin light chain kinase. ECFP-CaM and the M13-EYFP were fused via two glycines (**Fig. 1**). The binding of Ca^{2+} ions to the CaM domain causes it to associate with the M13 peptide. This leads to a conformational change in the molecule, bringing the two GFPs into closer molecular proximity, which in turn enhances the efficiency of fluorescence resonance energy transfer (FRET) between the two GFPs. FRET occurs when excitation energy from the higher-energy GFP (the donor, in this case ECFP) is passed to the lower-energy GFP (the acceptor, EYFP), causing

From: *Methods in Molecular Biology, vol. 183: Green Fluorescent Protein: Applications and Protocols*
Edited by: B. W. Hicks © Humana Press Inc., Totowa, NJ

Table 1
Advantages and Disadvantages of the Most Commonly Used Fluorescent Ca^{2+} Indicators

Ca^{2+} indicator	Advantages	Disadvantages
Synthetic fluorescent chelators (e.g., Fura-2 or Indo-1)	Easy to load in isolated cells and image Suitable for single-cell imaging	Hard to target Gradually leak out of cells Hard or impossible to load in thicker tissues
Recombinant aequorin	Easy to target Single-cell imaging requires specialized photon-counting system *(13)*	Gene transfer is required Requires the incorporation of cofactor coelenterazine Very difficult to image, because its luminescence produces <1 photon/molecule
Cameleons	Easy to target and image Suitable for single-cell imaging	Gene transfer is required The maximum change in emission ratio is less than for small-molecule dyes EYFP could be sensitive to changes in pH

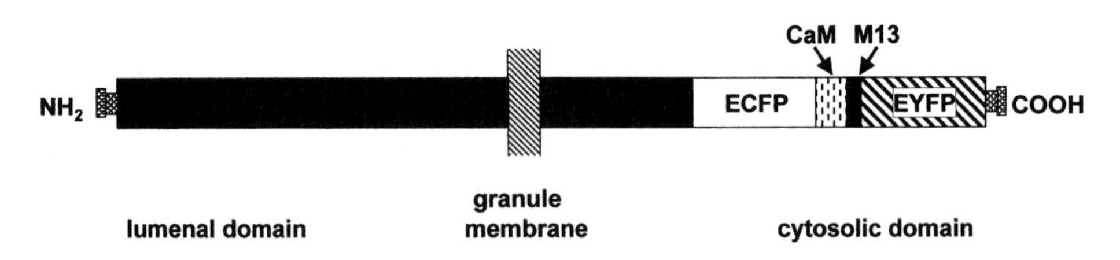

Fig. 1. Domain structure of phogrin–cameleon construct (phogrin-Ycam-2). The fluorescence indicator, yellow cameleon 2 (Ycam-2), comprising a fusion between the enhanced cyan fluorescent protein (ECFP), calmodulin (CaM), the calmodulin-binding peptide, M13, and enhanced yellow fluorescent protein (EYFP) *(7)*, was fused to phogrin, a protein localized to secretory granule membranes. Changes in Ca^{2+} concentration were monitored through alterations in FRET between ECFP and EYFP in Ycam-2. The Ca^{2+}-binding sites of the targeted Ycam-2 are predicted to be localized within 30 nm of the cytosolic surface of the secretory granule. Adapted with permission from **ref.** ***9***.

an increase in emission fluorescence of the acceptor GFP. This process is fully reversible when Ca^{2+} dissociates from CaM. This Ca^{2+} indicator has been termed 'cameleon' *(7)*, where 'cam' is derived from the common abbreviation for calmodulin (CaM), and because the molecule readily changes color and retracts and extends a long "tongue" (M13) into and out of the mouth of the CaM.

A great advantage of cameleons is that they are bright enough to be introduced into cells by DNA transfection rather than protein microinjection. The dynamic range of present cameleons is significantly lower than that of many ratiometric and nonratiometric fluorescent indicators (e.g., Fura-2, Indo-1), but it is sufficient to permit valuable measurements with a suitably equipped dual-emission microscope. A crucial advantage of cameleons, compared to Fura-2 and so on, is that they can readily be targeted to specific intracellular sites (e.g., nucleus, endoplasmic reticulum etc.) by fusing the corresponding cDNAs to appropriate organellar targeting signals or localized host proteins to observe local Ca^{2+} dynamics *(7)*. Moreover, their targetability allows Ca^{2+} measurements at previously inaccessible sites such as the immediate vicinity of secretory vesicles *(9)* (**Fig. 1**). Furthermore, the Ca^{2+} affinities of cameleons are adjustable by mutation of the CaM moiety. Thus, even the high Ca^{2+} of the endoplasmic reticulum can readily be measured, being 60–600 μ*M* in unstimulated cells *(7,10)*; (Váradi and Rutter, in press), decreasing to 1–50 μ*M* in cells treated with Ca^{2+}-mobilizing agonists.

Despite the potential versatility and power of the recombinant cameleons for measuring intracellular Ca^{2+}, up to now, their use in imaging experiments has been limited, mostly because of the small changes in fluorescence ratio that are achievable with these probes. Here is described the methodological approaches to monitoring Ca^{2+} in subdomains of living cells with targeted cameleons, as used in the authors' studies (*9*; **Fig. 2**) *(10a)*.

2. Materials

1. Poly-L-lysine: Prepare 0.2 mg/mL stock solution in water and store at –20°C, prepare the working solution at a concentration of 0.01 mg/mL in water and filter it before use, and store at +4°C.
2. Phosphate-buffered saline: 136 m*M* NaCl, 2.68 m*M* KCl, 6.46 ml Na_2HPO_4, 1.47 KH_2PO_4 (pH 7.4).
3. Complete growth medium for MIN6 pancreatic β-cells: Dulbecco's modified Eagle's medium (DMEM) (Gibco-BRL) containing 25 m*M* glucose supplemented with 15% (v/v) fetal bovine serum, 2 m*M* L-glutamine, 100 U/mL penicillin, and 100 μg/mL streptomycin.
4. LipofectAmine™ Reagent (Gibco-BRL): 2 mg/mL stock solution, store at +4°C, do not freeze, mix gently before use.

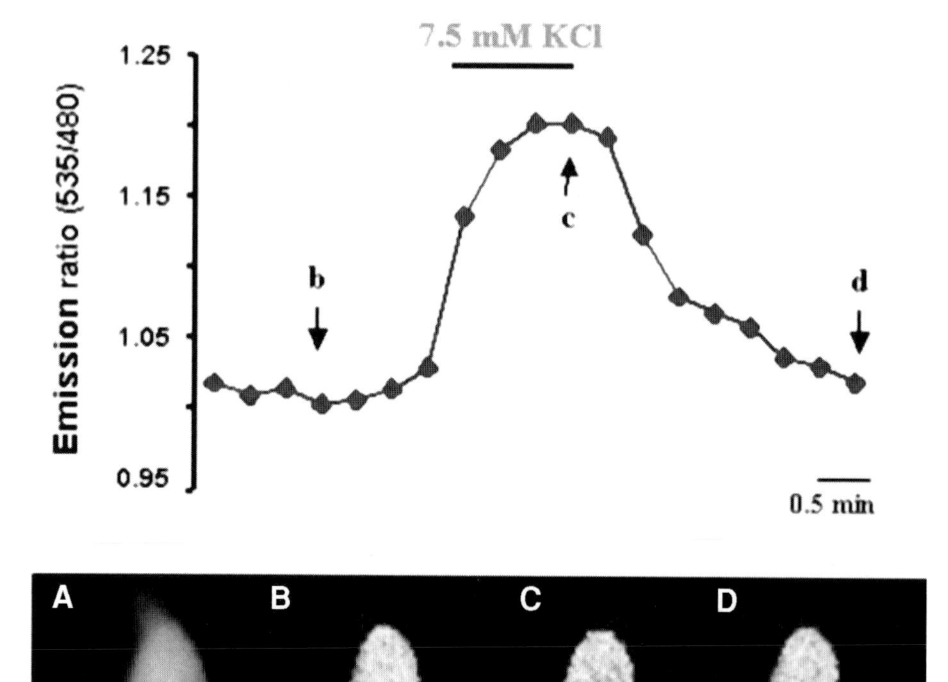

Fig. 2. Changes of $[Ca^{2+}]_i$ in a single MIN6 pancreatic β-cell expressing untargeted Ycam-2. MIN6 cells were transiently transfected with plasmid encoding Ycam-2, as described in **Subheading 3.2.1.** Cells were excited at 440 ± 21 nm, then emission of ECFP at 480 ± 30 nm and EYFP at 535 ± 25 nm were recorded, and the ratio of 535:480 were generated. The top panel shows the emission ratios from Ycam-2 in a single MIN6 cell monitored every 10 s by digital imaging. The corresponding ratio images, taken at the indicated time-points **(B–D)** together with the fluorescence image (440 excitation and 480 emission) of the cell **(A)** are shown at the bottom. (For optimal, color represention please see accompanying CD-ROM.)

5. OptiMEM I (Gibco-BRL): Store at +4°C in the dark, and handle as potentially infectious.
6. 0.2X TE buffer: 2 m*M* Tris-HCl, 0.2 m*M* EDTA (pH 8.0).
7. Krebs-Ringer-HEPES-bicarbonate (KRHB) buffer: 140 m*M* NaCl, 3.6 m*M* KCl, 0.5 m*M* NaH_2PO_4, 0.5 m*M* $MgSO_4$, 1.5 m*M* $CaCl_2$, 2.0 m*M* $NaHCO_3$, 3 m*M* glucose, and 10 m*M* HEPES. Equilibrate with O_2/CO_2 (95:5, v/v) before use and adjust the pH to 7.4.

8. Ionomycin: Prepare 2 mg/mL stock solution in dimethyl sulfoxide, and store at +4°C.
9. AdEasy system (AdEasy Application Manual Version 2.1 by Quantum Biotechnologies), for preparation of recombinant adenovirus-encoding phogrin-yellow cameleon (Ycam-2).
10. Fluorescence microscope with proper objective and filters: (e.g., Leica DM/IRBE) by using a X100 PL Apo 1.4 NA oil-immersion objective, 440 ± 21 nm excitation filter (e.g., Omega Optical, Glen Spectra, Middlesex, UK, or Chroma, Brattleboro, VT), a 455 DRLP dichroic mirror, and two emission filters (480DF30 for ECFP and 535DF25 for EYFP) alternated by a filter wheel (e.g., Ludl, Hawthorn, NY; Lamda 10-2, Sutter, San Rafael, CA, or Chroma).

3. Methods

3.1. Preparation of Pancreatic β-Cell Lines

1. Coat 13- or 24-mm-diameter cover slips with 0.01 mg/mL poly-L-lysine for 10–20 min at room temperature, then rinse the cover slips with phosphate-buffered saline once.
2. Seed the cells as a 25–35-µL droplet onto the cover slips at a density of 0.4–0.6 $\times 10^6$/mL. For other cell types, titrate the cell density to achieve 50–60% confluency after overnight culture (*see* **Note 1**).
3. Leave the cells to adhere to the cover slips for 45–60 min at 37°C in the presence of 5% CO_2.
4. Layer 2 mL complete growth medium very gently onto the cells and culture them overnight.

3.2. Transfection with Plasmids Encoding Yellow Cameleons

3.2.1. Transient Transfection

1. For each cover slip, add 1 µg plasmid encoding the cameleon construct (e.g., Ycam-2 or Ycam-4er), and 100 µL OptiMEM-I, into tube A, and 10 µL LipofectAmine and 100 µL OptiMEM-I into tube B (*see* **Note 2**).
2. Combine the contents of tubes A and B into tube C and mix them very gently. Do not vortex.
3. Incubate the mixture for 30 min at room temperature to allow DNA–lyposome complexes to form. Do not agitate the mixture at this stage.
4. Just before the 30-min incubation is over remove the growth medium from the cells and rinse them with 1–2 mL OptiMEM-I once.
5. For each cover slip, add 1.8 mL OptiMEM-I to tube C (**step 2**), then pipet the mixture up and down once and overlay 2 mL onto each cover slip.
6. Incubate the cells with the complexes for 3–4 h at 37°C in the presence of 5% CO_2, then replace the serum-free transfection medium with complete growth medium (*see* **Note 3**).
7. Culture the cells for 2–4 d before imaging and change medium every second day. For MIN6 β-cells, replace the high-glucose complete growth medium with 3 m*M* glucose containing DMEM, 12 h prior Ca^{2+} measurements.

3.2.2. Microinjection (see **Note 4**)

1. Prepare the cells as described under **Subheading 3.1.** Immediately prior to microinjection prepare the plasmid DNA encoding the cameleon construct at a final concentration of 0.3 mg/mL in 0.2X TE buffer, then centrifuge for 30 min at 13,000*g* at 4°C to sediment any particulate matter.
2. Pull micropipets (e.g., Flaming/Brown micropipet puller P-97, Sutter) from 1.2 mm external diameter glass borosilicate capillaries (GC120F-10; Clark Electromedical, Reading, UK) to give a microinjection diameter of 0.2–0.5 μ*M*. Pipet 2–3 μL DNA solution into the capillary needle. During intranuclear microinjection, maintain the injection pressure at ~500 hPa and retain injection time below 1.0 s. Inject (Eppendorf 1571/5246 semiautomatic microinjection system) ~300 cells to give ~30 productive injections. The cell nucleus must be viewed with a phase contrast objective (30–40× magnification) and may require maintenance on a thermally controlled stage (37°C). For long periods of injection (>45 min) CO_2-insensitive medium should be used (e.g., through the addition of 20 m*M* HEPES buffer).
3. Following microinjection, replace the medium with fresh complete growth medium, and culture the cells for 2 d before imaging. Change the growth medium every day to minimize the risk of contamination.

3.2.3. Adenovirus-Mediated Infection of Cells (see **Note 5**)

Recombinant adenovirus encoding phogrin-Ycam-2 was prepared using the AdEasy system ***(11,12)***.

1. Prepare the cells as described under **Subheading 3.1.** Infect cells with the recombinant phogrin-Ycam-2 at multiplicity of 30–100 viral particles/cell in growth medium then layer 100 μL onto each cover slip.
2. Incubate the cells for 1–4 h at 37°C in the presence of 5% CO_2 then remove the infection medium, and replace it with complete growth medium.
3. Culture the cells for 24–48 h prior imaging.

3.3. Single Cell Ca^{2+} Imaging

1. Rinse the cells with KRHB buffer and mount the cover slip on the microscope in a suitable preheated (to 37°C) superfusion chamber.
2. Perifuse the cells at constant rate of 1–2 mL/min with KRHB buffer and retain them in the same buffer until the addition of stimulating solutions (*see* **Note 6**).
3. Select a good region on the cover slip (*see* **Note 7**), then record the bright field image of the cells. Image the cells on a fluorescence microscope, by using a high numerical aperture (NA) objective, a 440 nm excitation filter, a 455-DRLP dichroic mirror, and two emission filters (480DF30 for ECFP and 535DF25 for EYFP), alternated by a filter wheel.
4. Acquire images on a cooled charge coupled device (CCD) camera every 3–10 s and allow data acquisition for 10–15 min (*see* **Notes 8** and **9**) controlled by a software (e.g., Metamorph/Metafluor, Universal Imaging, West Chester, PA).
5. Generate ratio images of 535/480.

3.4. Calculation of $[Ca^{2+}]$

1. For *in situ* calibration of $[Ca^{2+}]$, use the following equation:

$$[Ca^{2+}]=K'd[(R - R_{min})/(R_{max} - R)]^{1/n},$$

where $K'd$ is the apparent dissociation constant corresponding to the Ca^{2+} concentration at which R is midway between R_{max} and R_{min}, and n is the Hill coefficient ***(1)***.
2. To obtain the maximal ratio, R_{max}, add 10 µ*M* of Ca^{2+} ionophore ionomycin followed by a high concentration (20 m*M*) of extracellular Ca^{2+}. To establish the minimal ratio, R_{min}, clamp the external Ca^{2+} to zero with 20 m*M* EGTA, in the presence of 10 µ*M* ionomycin.

4. Notes

1. Transfection efficiency is sensitive to culture confluence, so it is important to maintain a standard seeding protocol from experiment to experiment. Do not add antibacterial agents to the media during transfection.
2. For transient transfection (*see* **Subheading 3.2.1.**), use 1 µg plasmid encoding the cameleon construct with lipofectamine, at a final concentration of 10 µg/mL in a final volume of 2 mL serum-free medium per 24-mm-diameter cover slip. This protocol normally provides 50–60% transfection efficiency 2–4 d after transfection.
3. The length of transfection can be shortened if the cell type used for experiments does not tolerate the absence of serum for 4 h. The transfection could be terminated when the DNA–lipid complexes become visible under microscope as small black precipitates on the cover slip, after ~2 h. This may reduce the transfection efficiency to ~30–40%.
4. When transfection is problematic (e.g., primary cells) or the transfection efficiency is low, intranuclear pressure microinjection of plasmid DNA could be a good alternative method (*see* **Subheading 3.2.2.**). The great advantage of this technique is that antisense oligonucleotides, antibodies, or other plasmids could be co-injected with the cameleon DNA construct. However, microinjection requires an expensive injection system, extensive practice, and dedicated experimentalist, before representative amounts of cells can be successfully injected. For more detailed protocol on instrumentation and microinjection technique see the review by Rutter et al. ***(13)***.
5. Adenovirus-mediated infection of cells could be another alternative to transfection and provides many advantages, because it is efficient, and can be used in a wide range of host-cell types. Using the protocol outlined in **Subheading 3.2.3.**, the majority, >90% of cells were infected with the adenoviral phogrin-Ycam-2 construct 24–48 h after infection. Phogrin-Ycam-2 chimera was correctly (>80%) targeted to insulin-containing vesicles ***(9,12)***.
6. The stimulating solutions could be added either manually with a syringe, or by perfusion. Since the maximum change in emission ratio of cameleons is less than for small-molecule dyes, any sudden change, such as adding or removing solutions, could generate significant background noise. Hence, for imaging cameleons, constant perifusion is strongly recommended.

7. The concentration of cameleons inside the cells could be heterogeneous whichever of the abovementioned methods (**Subheadings 3.2.1.**, **3.2.2.**, or **3.2.3.**) are used for gene transfer. Therefore, for imaging, select a region where the majority of cells express similar levels of cameleon. Do not choose cells with high cameleon concentration (brightest cells). Because each cameleon can bind four Ca^{2+}, in the case of high expression levels, they may significantly buffer $[Ca^{2+}]_i$ dynamics (e.g., efficient expression of the cameleon may give 1×10^6 molecules/cell, equivalent 1.0 μ*M* for 1 pL cell, or 4 μ*M* binding sites). However, if the cameleon concentration is very low, the cells will be too dim to give a good signal-to-noise ratio in dynamic single-cell imaging. Increasing the intensity of illumination cannot compensate because of the EYFP photochromism/photobleaching (*see* **Note 8**).
8. EYFPs can be photobleached by intense illumination at their main absorbance peak. If photobleaching occurs, reduce duration and/or intensity of exposure to the lowest possible levels that still provide adequate signal-to-noise characteristics in detection.
   a. Usually, altering a few parameters in the software can dramatically decrease the level of excitation needed to get a good signal. With some digital cameras, it is possible to combine the charges in adjacent pixels so that they form a single pixel. This process is known as "binning." It reduces readout time and increases sensitivity at the expense of resolution. Most software also allows improving the contrast of the images. Use contrast enhancement function to remap either the display or the pixel values, so that they use all the available dynamic range of the screen.
   b. Use oil-immersion objectives, which have far more light-gathering power than dry objectives, because of their larger numerical aperture (which translates into brighter fluorescence emission for the same level of excitation).
   c. Placing a neutral density filter in front of the exciting light could also effectively reduce photobleaching.
   d. If confocal fluorescent microscopy is used, give consideration to changing to conventional nonconfocal fluorescent microscopy. Because the confocality is improved, the signal is collected from a smaller and smaller volume of specimen, and, thus, from fewer and fewer molecules of excited fluorophore. Higher-intensity light sources provide the needed increase in output signal, but also increase the risk of photobleaching.
9. The first generation of cameleons is sensitive to changes in pH ***(7,14)***. Using these cameleons requires the careful checking of pH to calibrate the Ca^{2+} signal, to avoid pH-related artifacts. The pH sensitivity of cameleons could be examined by selectively recording the EYFP fluorescence using 480DF30 nm excitation, 535DF25 nm emission, and 505 DRPL dichroic mirror. In this condition, Ca^{2+}-dependent FRET does not occur and the probe reports pH, instead of Ca^{2+} changes. The ultimate solution to eliminate the pH-related artifact is to use the second generation of cameleons ***(14)***.

Acknowledgments

We thank to Roger Tsien (University of California at San Diego) for providing Ycam-2 cDNA, and Alan Leard and Mark Jepson of the Bristol MRC Imaging Facility for assistance with image analysis. This work was supported by the Biotechnology and Biological Sciences Research Council (BBSRC).

References

1. Grynkiewicz, G., Poenie, M., and Tsien R. Y. (1985) A new generation of Ca^{2+} indicators with greatly improved fluorescence properties. *J. Biol. Chem.* **260,** 3440–3450.
2. Hofer, A. M. and Schulz, I. (1996) Quantification of intraluminal free [Ca^{2+}] in the agonist-sensitive internal calcium store using compartmentalised fluorescent indicators: some considerations. *Cell Calcium* **20,** 235–242.
3. Golovina, V. A. and Blaustein, M. P. (1997) Spatially and functionally distinct Ca^{2+} stores in sarcoplasmic and endoplasmic reticulum. *Science* **275,** 1643–1648.
4. Rutter, G. A., Burnett, P., Rizzuto, R., Brini, M., Murgia, M., Pozzan, T., et al. (1996) Subcellular imaging of intramitochondrila Ca^{2+} with recombinant targeted aequorin: significance for the regulation of pyruvate dehydrogenase activity. *Proc. Natl. Acad. Sci.* USA **93,** 5489–5494.
5. Miller, A. L., Karplus, E., and Jaffe, L. F. (1994) Imaging $[Ca^{2+}]_i$ with aequorin using photon imaging detector, in *Practical Guide to the Study of Calcium in Living Cells.* (Nuccitelli, R., ed.) Academic, New York, pp. 305–338.
6. Knight, K. R., Read, N. D., Campbell, A. K., and Trewavas, A. J. (1993) Imaging calcium dynamics in living plants using semi-synthetic recombinant aequorins. *J. Cell. Biol.* **121,** 83–90.
7. Miyawaki, A., Llopis, J., Heim, R., McCaffery, M. J., Adams, J. A., Ikura, M., and Tsien, R. Y. (1997) Fluorescence indicators for Ca^{2+} based on green fluorescent proteins and calmodulin. *Nature* **388,** 882–887.
8. Romoser, V. A., Hinkle, P. M., and Persechini, A. (1997) Detection in living cells of Ca^{2+}-dependent changes in the fluorescence emission of an indicator composed of two green fluorescent protein variants linked by calmodulin-binding sequence. *J. Biol. Chem.* **272,** 13,270–13,274.
9. Emmanouilidou, E., Teschemacher, A. G., Pouli, A. E., Nicholls, L. I., Seward, E. P., and Rutter, G. A. (1999) Imaging Ca^{2+} concentration changes at the secretory vesicle surface with recombinant targeted cameleons. *Curr. Biol.* **9,** 915–918.
10. Foyouzi-Youssefi, R., Arnaudeau, S., Borner, C., Kelley, W. L., Tschopp, J., Lew, D. P., et al. (2000) Bcl-2 decreases the free Ca^{2+} concentration within the endoplasmic reticulum. *Proc. Natl. Acad. Sci. USA* **97,** 5723–5728.

10a. Varadi, A. and Rutter, G. A. (2002) Dynamic imaging of enodplasmic reticulum (Ca^{2+}) in pancreatic b-cells using recombinant targeted cameleons-roles of SERCA2 and Ryanodine receptors. *Diabetes* **51,** S190–S201.

11. He, T.-C., Zhuo, S., DaCosta, L. T., Yu, J., Kinzler, K. W., and Vogelstein, B. (1998) A simplified system for generating recombinant adenoviruses. *Proc. Natl. acad. Sci. USA* **95,** 2509–2514.
12. Tsuboi, T., Zhao, C., Terakawa, S., and Rutter, G. A. (2000) Simultaneous evanescent wave imaging of insulin vesicle membrane and cargo during a single exocytic event. *Curr. Biol.* **10,** 1307–1310.
13. Rutter, G. A., White, M. A. R., and Tavaré, J. M. (1998) Analysis and regulated gene expression by microinjection and digital luminescence imaging of single living cells, in *Imaging Living Cells* (Rizzuto, R. and Fasolato, C., eds.), Springer-Verlag, Heidelberg, pp. 301–328.
14. Miyawaki, A., Griesbeck, O., Heim, R., and Tsien, R. Y. (1999) Dynamic and quantitative Ca^{2+} measurements using improved cameleons. *Proc. Natl. Acad. Sci. USA* **96,** 2135–2140.

21

Green Fluorescent Protein Fluobody Immunosensors

Immunofluorescence with GFP-Antibody Fusion Proteins

Arjen Schots and Jan M. van der Wolf

1. Introduction

Immunofluorescence (IF) is widely used both in research and diagnosis *(1–4)*. Using IF, it is possible to localize antigens in tissues and individual cells. Antibodies (Abs), to which a fluorochrome is chemically added are extensively used for this purpose. However, conventional organic fluorochromes such as fluorescein isothiocyanate (FITC) or tetrarhodamine isothiocyanate, are sensitive to photobleaching, upon illumination. Also, if conjugation occurs with an antigen-binding site, a partial or complete loss of reactivity can occur *(2)*. To circumvent these disadvantages, it is useful to explore alternative fluorochromes and coupling procedures.

The green fluorescent protein (GFP) from the jellyfish, *Aequorea victoria*, offers such an alternative. As described by Tsien *(5)*, there are a vast number of GFP variants present today. These variants differ in their spectral characteristics, fluorescence intensity, stability, and other properties. Recently, six GFP homologs have been cloned (*6*, and *see* Chapter 1) from nonbioluminescent *Anthozoan* species, extending the range of alternatives. Although the quantum yield of the latter fluorescent proteins is lower when compared to GFP or GFP variants, they will, like the wild-type GFP, undoubtedly be improved upon.

An alternative to chemical coupling is available through genetic fusion of antibody and GFP-encoding genes. It will always result in a 1:1 ratio between Ab and fluorochrome, which should enhance accuracy in quantitative work, and GFP cannot be easily inactivated. Advances in molecular immunology

From: *Methods in Molecular Biology, vol. 183: Green Fluorescent Protein: Applications and Protocols*
Edited by: B. W. Hicks © Humana Press Inc., Totowa, NJ

(7–11) now allow cloning and expression of *Ab* genes in *Escherichia coli* and other heterologous production hosts. Thus, single-chain variable fragment Abs (scFv), which contain the variable regions from the heavy and light chain connected by a short linker sequence ***(7)***, or antigen-binding fragments, can be selected from phage Ab libraries ***(12)***. Alternatively, Ab genes can be cloned directly from hybridoma cell lines ***(8)***. This chapter describes the application of scFv Abs, selected from a large phage Ab library, genetically fused to a GFP variant in a controlled expression system. The usefulness of these fusion proteins is demonstrated through an application for the detection of an important plant pathogen, *Ralstonia solanacearum*, the causative agent of bacterial wilt in a wide variety of host plants (**Figs. 1** and **2**).

2. Materials

1. pSK-GFPmut1 (or other GFP-containing vector).
2. Restriction enzymes *Sfi*I and *Not*I.
3/ 0.8–1.0% 1% Agarose gel containing 1 μg/mL ethidium bromide, 10X TBE (1 L): 108 g Tris, 55 g boric acid, 9.3 g EDTA.
4. Concert™ Gel Extraction System (Life Technologies).
5. DNA ligase and 10X ligase buffer.
6. Luria Bertoni (LB)-ampicillin (AMP) broth: 10 g/L Bacto-tryptone, 5 g/L Bacto-yeast extract, 10 g/L NaCl (pH 7.0) with 100 μg/mL AMP. 2TY-AMP: 17 g/L Bacto-tryptone, 10 g/L Bacto-yeast extract, 5 g/L NaCl with 100 μg/mL AMP. LB-AMP plates: LB-AMP broth containing 15 g/L agar.
7. *E. coli* strain XL1-Blue-MRF' Kan (Stratagene).
8. Immobilized metal affinity chromatography resin (IMAC): Ni-NDA beads (Amersham-Pharmacia).
9. Prepare a stock solution of 200 μg/mL anhydrotetracyclin and store at –20°C.
10. Pre-extraction buffer: 50 m*M* Tris-HCl, pH 8.0, containing 30% sucrose and 1 m*M* EDTA.
11. Extraction buffer: 5 m*M* $MgSO_4$.
12. Loading buffer: 10 m*M* imidazole, 10 m*M* Tris-HCl, pH 8.0, 300 m*M* NaCl.
13. Elution buffer: 100 m*M* Tris-HCl, pH 8.0, containing 100 m*M* EDTA.
14. 12% Sodium dodecyl sulfate-polyacrylamide gel electrophoresis (SDS-PAGE) gel. Nitrocellulose membrane. Electrophoresis buffers.
15. Phosphate-buffered saline (PBS): (1 L) 0.2 g KCl, 8.0 g NaCl, 0.2 g KH_2PO_4, 1.15 g Na_2HPO_4, pH 7.4. PBS+T20: PBS containing 0.1% Tween-20. PBS+T20+milk: PBS+T20 containing 1–5% powdered milk. PBS+T20 containing 5% fetal calf serum (FCS).
16. Abs: GFP-specific polyclonal rabbit Abs (Clontech); alkaline phosphatase conjugated goat antirabbit polyclonal Abs.
17. 5-bromo-4-chloro-3-indolyl phosphate (BCIP, dissolved at 25 mg/500 μL in dimethyl formamide); nitroblue tetrazolium (dissolved at 25 mg/350 μL in dimethyl formamide + 150 μL); Western substrate buffer (10 mL): 0.1 *M*

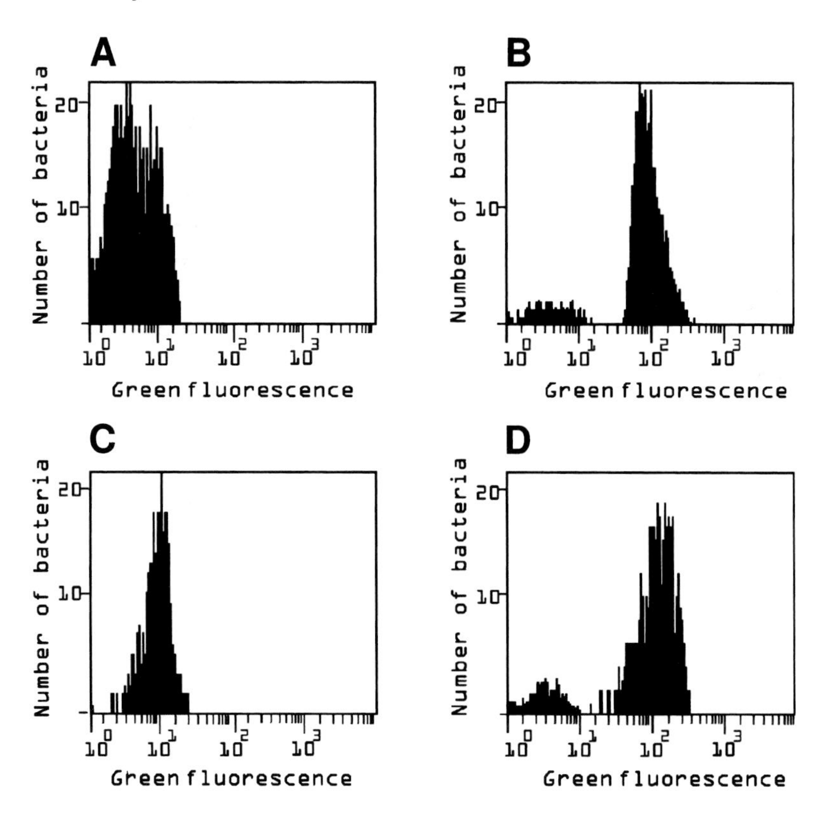

Fig. 1. Analysis on a Fluorescent Activated Cell Sorter (FACS) of *Ralstonia solanacearum* bacteria labeled with a specific antibody fused to GFPmut1. **(A)** The negative control. Non fused GFPmut1 incubated with *R. solanacearum.* **(B)** The scFv LPS 12-GFPmut1 fluobody incubated with *R. solanacearum.* **(C)** Control for cross reactivity, the scFv LPS12-GFPmut1 incubated with *R. pickettii* bacteria. **(D)** Positive control. LPS12 scFv-antibodies with Cmyc tag detected by a secondary anti-cMyc specific antibody conjugated to FITC. The scales with number of bacteria and fluorescence intensity are relative. It has to be noted that the fluorescence intensities as seen in **(B)** and **(D)** are similar, showing that the LPS12-GFPmut1 fusion protein has identical properties as LPS12-cMyc.

NaCl, 5 m*M* $MgCl_2$, 0.1 *M* Tris-HCl, pH 9.5, containing 34 µL BCIP and 68 µL nitroblue tetrazolium.

18. Slide warmer set (50°C). Microscope multiwell slides (e.g., Nutacon no. 10-342). 12 × 60-mm cover slips.
19. Fluorescence microscope equipped with a high-pressure mercury or xenon lamp, an appropriate filter set and minimally a X40 Neofluor objective.
20. Mounting buffer (Vectashield, Vector, Burlingham, UK, CA 94101).
21. Ethanol (96%).

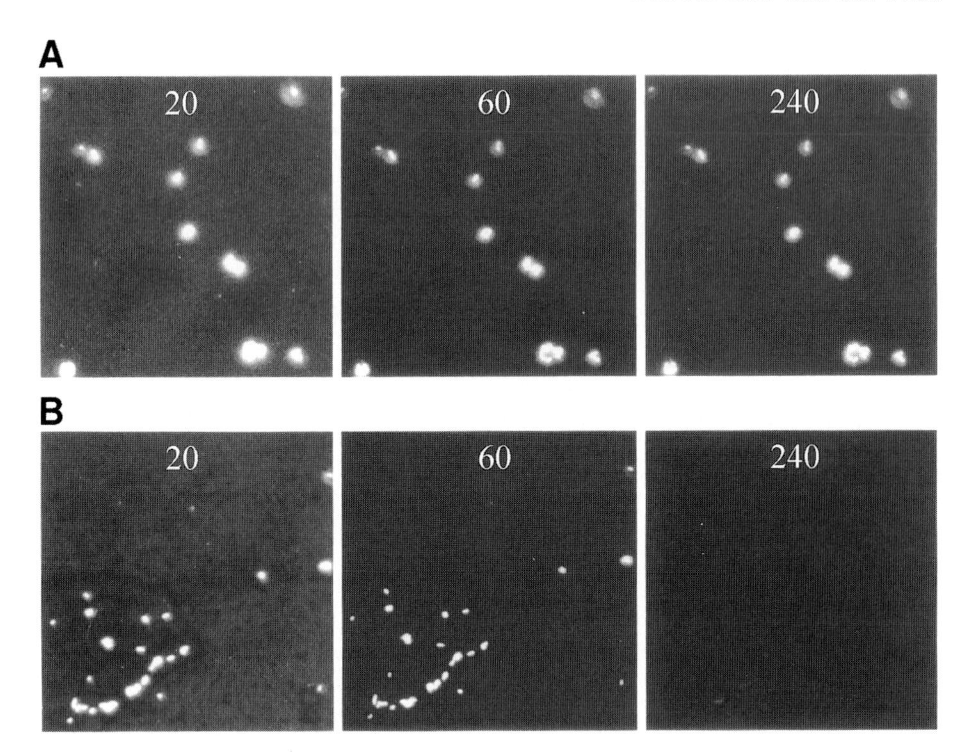

Fig. 2. Immunofluorescence cell staining of *Ralstonia solanacearum* bacteria using the fluobody LPS7-GFPmut1 (**A**) and an FITC conjugated polyclonal rabbit antiserum (**B**). The number indicated in each micrograph refers to the photographic exposure time in seconds. It has to be noted that FITC labeled antibodies 'suffer' from severe photobleaching while GFP is not at all affected.

3. Methods

3.1. Antibody Formats Available

Genes encoding for either scFv or antigen-binding fragments can be used for fusion with GFP. If antigen-binding fragments are used, it is a prerequisite that a bicistronic messenger is transcribed. A second ribosomal binding site should be present between variable regions of the light and heavy chains. In several Ab libraries, the *chimeric Ab* gene is inserted between a *Sfi*I and a *Not*I restriction site. In the vector, pSK-GFPmut1 (*see* **Fig. 3**), this is also the case. If Abs are inserted in a vector using other restriction sites, it is necessary to either change these in a PCR reaction using primers having *Sfi*I and *Not*I restriction sites, or to alter the vector pSK-GFPmut1 by introducing other restriction sites. If recombinant mouse Abs are used and one intends to introduce a *Xho*I site it is advised to thoroughly screen for the presence of this site in the heavy chain (8–10% do have it).

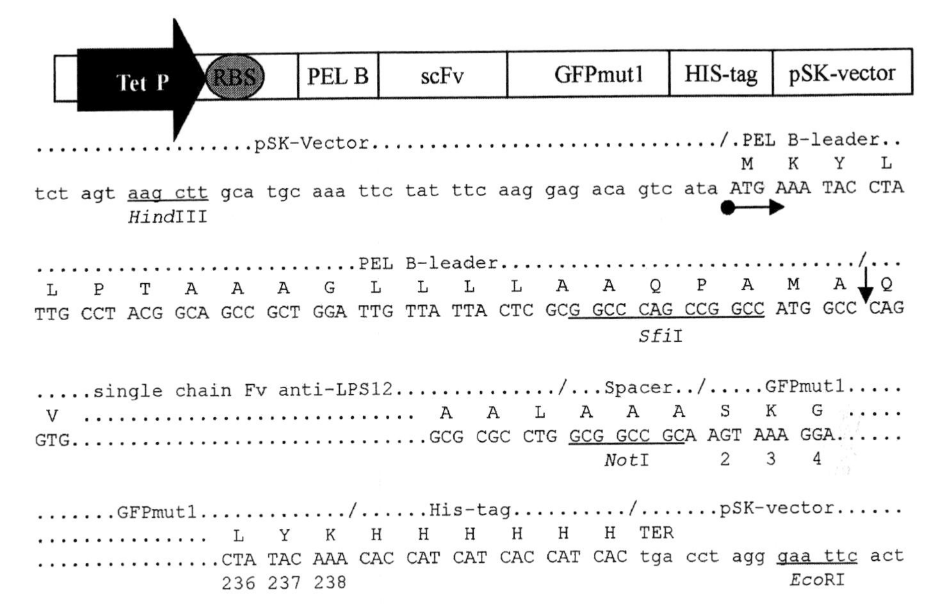

Fig. 3. Schematic representation of the vector pSK-GFPmut1. The position of the promoter (Tet P), the ribosomal binding site (RBS), the leader sequence (pel B), the antibody insertion site (scFv), the GFP (GFPmut1) and the HIS(6)-tag (HIS-tag) are indicated. The *Sfi*I and *Not*I sites are underlined. Translated codons are in capitals, the start codon is indicated by a horizontal arrow, the signal sequence cleavage site by a vertical arrow and the stop-codon by TER.

3.2. Construction of Ab–GFP Fusions

1. Digest the GFP vector with the restriction endonucleases *Sfi*I and *Not*I. Use the buffers and reaction conditions as recommended by the supplier.
2. Check for appropriate digestion on a 0.8–1% agarose gel run in TBE. We often use the vector with an inserted scFv as a stuffer fragment. This should be visible on the agarose gel as a fragment of approx 700 bp.
3. Digest the *Ab* gene of interest from its vector with *Sfi*I and *Not*I.
4. Gel-purify the gene fragment and vector for instance, by using Concert Gel Extraction System or similar kit.
5. Ligate the purified gene fragment in the vector according to the instructions of the manufacturer of the ligase used.
6. The resulting plasmid can then be transferred to *E. coli* XL1-Blue-MRF' Kan cells.
7. The transformed bacteria are plated on selective LB plates (containing ampicillin) and grown overnight at 25°C. Individual colonies are then toothpicked and grown in 0.75 mL LB-amp broth in 48-well plates (shaking at 250 rpm) at 25°C. When the OD_{600} reaches 0.5, the temperature is lowered to 16°C. After 1 h, anhydrotetracyclin is added to the medium (0.2 µg/mL final concentration), and

growth is resumed. After 48 h, samples are taken and the fluorescence is measured at 488 nm in a fluorometer. Alternatively, the amount of produced Ab–GFP fusion protein can be assessed on an immunoblot using anti-GFP Abs.

8. The reactivity of the Ab–GFP fusion protein can be assessed in various immunoassays e.g., using anti-GFP fusion proteins as a marker. Thereto, the bacteria are pelleted and resuspended in 0.5 mL PBS+T20. The bacteria are then lysed by repeated freezing and thawing (5×). The bacterial debris is removed by centrifugation, and the crude homogenate is then used.
9. If the Abs react satisfactorily, the culture can be scaled-up to a volume of 100 mL or larger. The same media and culture conditions are advised.

3.3. Purification of Fluobodies

1. The bacteria from the culture are pelleted by centrifugation and resuspended in one-twentieth vol (referring to original culture size) of pre-extraction buffer and incubated for 5 min at 0°C.
2. Cells are pelleted and 1/20 vol (referring to original culture size) of extraction buffer, is added and incubated for 45 min at 0°C.
3. The 6X His-tagged proteins are then purified from the periplasmic fraction, by immobilized metal affinity chromatography (IMAC), using Ni-NDA beads. The extraction buffer is changed into loading buffer by dialysis or Sephadex G25.
4. The sample is loaded on an IMAC column (size depending on the amount of protein, use a column as small as possible, and follow manufacturer's instructions).
5. The column is washed with 20 column vol of loading buffer.
6. Bound proteins are eluted with elution buffer.
7. The amount and concentration of eluted protein is assessed using a standard protein assay (e.g., Bradford *[13]*).
8. 2–3 μg protein are loaded on a 12% SDS-PAGE gel, and subsequently transferred onto nitrocellulose using standard protocols.
9. After blocking unoccupied sites for 30 min with PBS+T20 containing 5% skimmed milk powder, the membrane is incubated for 1 h with GFP-specific polyclonal rabbit Abs diluted 1/25,000 in PBS+T20 containing 2% skimmed milk powder. After washing (4×, PBS+T20), the blot is incubated for 1 h with alkaline phosphatase conjugated goat-antirabbit Abs diluted in PBS+T20 containing 2% skimmed milk powder (dilution varies per manufacturer and batch). To visualize the proteins, the blot is incubated with BCIP/NBT substrate.

3.4. Flow Cytometry

Purified Ab–GFP fusion proteins are added to a cell suspension in appropriate dilutions. The cell suspension can contain bacteria, animal cells (mammalian, insect), yeast cells, or even plant protoplasts. In the latter case, background fluorescence may proof somewhat problematic using the GFP in the vector described. However, new GFP variants are available which should result in lower background. The authors normally dilute the cells in PBS+T20 and 5% FCS. Cells and Ab–GFP fusion protein are incubated for 1 h while rotating.

Cells are washed twice with PBS+T20, resuspended in 1–2 mL PBS, and analyzed for fluorescent staining using a flow cytometer, or a fluorescent activated cell sorter (FACS) (**Fig. 1**).

3.5. IF Cell Staining

IF cell staining, like flow cytometry, can be applied to a wide variety of cell and tissue types. FITC has remained the most popular fluorochrome for IF until the present time. However, one of the main disadvantages of FITC has been its rapid fading under intense illumination ***(14)***. Furthermore, it is difficult to standardize the traditional conjugation procedure for FITC, in which purified Abs are conjugated in a carbonate buffer at high pH. Often, significant differences of fluorochrome to protein ratios are found between different conjugates using the same batch of Abs (unpublished results). The use of fluobodies can circumvent these disadvantages of the use of FITC-conjugated Abs.

For the preparation of the various cell and tissue types different protocols do exist. The protocol provided below is what we use for IF cell staining of bacteria with fluobodies.

1. Prepare 2 10-fold serial dilutions of the bacterial sample in water (autofluorescent particles may be present in samples and an excess of target cells in undiluted samples will result in a weak staining, up to 100 cells/microscope field give a good staining).
2. Place 10 µL each of the undiluted and both of the 10-fold, serially diluted samples onto separate wells of a multiwell microscope slide.
3. Air-dry samples on a slide warmer set at 50°C.
4. Fix cells by adding an excess of ethanol to the wells. Slides are soaked in alcohol for 5 min, rinsed 3× for 1 min with water and air-dried on the slide warmer. Avoid cross-contaminations between adjacent wells during washings.
5. Add 10 µL fluobodies in PBS and incubate for 30 min in a moist chamber at 37°C in the dark. For detection of *R. solanacearum*, a concentration of 0.5 µ*M* fluobody was used, but the optimal concentration will vary and will be Ab-dependent.
6. Rinse slides 3× for 1 min with distilled water.
7. Add 50 µL of mounting buffer per slide and place a cover slip over the well.
8. Observe preparations for typical fluorescent cells using a fluorescence microscope.

4. Notes

1. For general molecular biological techniques we refer to **ref. *14***). Regarding general serological and immunological techniques we refer to **ref. *15***).
2. In this vector, we have used the *TetA* tetracycline promoter/operator of the *TN10Tc*^{R} gene instead of the lac promoter. The reason for this is that the latter is leaky, and is therefore not well-controlled. In contrast, the *TetA* promoter is strongly controlled. Altering the concentration of anhydrotetracyclin can regulate the level of protein expressed. This is of particular interest if the *Ab* gene used is toxic for *E. coli* ***(16)***.

3. Although the vector, pSK-GFPmut1 is designed for secretion, most Ab–GFP fusion proteins are usually found in the periplasm. This has to be checked for every Ab fragment expressed anew. This is also the reason why cell lysis is used. Because the produced proteins are soluble, it may be desirable to include protease inhibitors in the lysis buffer. The protocol described is for small volumes. If large volumes (>1 L) are used, switching to mechanical lysis may be necessary, using a French press.
4. Most IF microscopes are equipped with a filter system for detection of FITC, which has a maximum excitation wavelength of 488 nm. It was found that a GFP mutant (GFPmut1), containing two amino acid substitutions (Phe64 to Leu and Ser65 to Thr), fluoresces 35× more intensely than the wild-type GFP when excited at this wavelength ***(17)***. For applications in IF, it is therefore advised to use GFPmut1.
5. Fluobodies have also been tested for use in immunofluorescence colony staining in which test samples are agar-mixed and plated in wells of a 24-well tissue culture plate, and incubated to grow bacteria ***(18)***. Incubation of agar preparations with fluobodies against *R. solanacearum*, however, resulted in a nonspecific staining of nontarget bacteria present in potato tuber extracts. Possibly, the hydrophobicity of the fluobodies resulted in nonspecific interactions with bacterial products.

References

1. Lingenfelter, B., Fuller, T. C., Hartung, L., et al. (1995) HLA-B27 screening by flow cytometry. *Cytometry* **22,** 146–149.
2. Reimann, K. A., Waite, B. C., Lee-Parritz, D. E., et al. (1994) Use of human leukocyte-specific monoclonal antibodies for clinically immunophenotyping lymphocytes of rhesus monkeys. *Cytometry* **17,** 102–108.
3. Salinas, J. and Schots, A. (1994) Microscopical studies of the infection of gerbera flowers by *Botrytis cinerea. Phytopathology* **84,** 351–356.
4. Griep, R. A., Van Twisk, C., Van der Wolf, J. M., et al. (1998) Selection of *Ralstonia solanacearum* race 3 specific monoclonal antibodies from combinatorial libraries. *Phytopathology* **88,** 795–803.
5. Tsien, R. Y. (1998) The green fluorescent protein. *Ann. Rev. Biochem.* **67,** 509–544.
6. Matz, M. V., Fradkov, A. F., Labas, Y. A., et al. (1999) Fluorescent proteins from nonbioluminescent Anthozoa species. *Nat. Biotechnol.* **17,** 969–973.
7. Huston, J. S., Levinson, D., Mudgett-Hunter, M., et al. (1988) Protein engineering of antibody binding sites: recovery of specific activity in an anti-digoxin single-chain Fv analogue produced in *Escherichia coli. Proc. Natl. Acad. Sci. USA* **85,** 5879–5883.
8. Orlandi, O., Gussow, D. H., Jones, P. T., et al. (1989) Cloning immunoglobulin variable domains for the expression by the polymerase chain reaction. *Proc. Natl. Acad. Sci. USA* **86,** 3833–3837.
9. Sastry, L., Alting-Mees, M., Huse, W. D., et al. (1989) Cloning of the immunological repertoire in *Escherichia coli* for generation of monoclonal catalytic anti-

bodies: construction of a heavy chain variable region-specific cDNA library. *Proc. Natl. Acad. Sci. USA* **86,** 5728–5732.
10. Griffiths, A. D., Williams, S. C., Hartley, O., et al. (1994) Isolation of high affinity antibodies directly from large synthetic repertoires. *EMBO J.* **13,** 3245–3260.
11. Vaughan, T. J., Williams, A. J., Pritchard, K., et al. (1996) Human antibodies with sub-nanomolar affinities isolated from a large non-immunized phage display library. *Nat. Biotechnol.* **14,** 309–314.
12. Hoogenboom, H. R., Marks, J. D., Griffiths, A. D., et al. (1992) Building antibodies from their genes. *Immunol. Rev.* **130,** 41–68.
13. Bradford, M. (1976) A rapid and sensitive method for the quantification of microgram quantities of protein utilizing the principle of protein dye binding. *Anal. Biochem.* **72,** 248–254.
14. Ausubel, F. M., Brent, R., Kingston, R. E., et al. (1998) *Current Protocols in Molecular Biology* Wiley, New York.
15. Harlow, E. and Lane, D., eds. (1999) *Antibodies: A Laboratory Manual* Cold Spring Harbor Laboratory, Cold Spring Harbor, NY.
16. Griep, R. A., Van Twisk, C., Kerschbaumer, R. J., et al. (1999) pSKAP/S: an expression vector for single-chain Fv alkaline phosphatase fusion proteins. *Prot. Express. Purific.* **16,** 63–69.
17. Cormack, B. P., Valdivia, R., and Falkow, S. (1996) FACS-optimized mutants of the green fluorescent protein (GFP). *Gene* **173,** 33–38.
18. Mahaffee, W. F., Bauske, E. M., van Vuurde, J. W. L., et al. (1997) Comparative analysis of antibiotic resistance, immunofluorescent colony staining, and a transgenic marker (bioluminescence) for monitoring the environmental fate of a rhizobacterium. *Appl. Environ. Microbiol.* **63,** 1617–1622.

22

Green Fluorescent Protein-Based Protein Kinase Biosensor Substrates

Scott Ulrich and Kevan Shokat

1. Introduction

Protein kinases (PKs) are a family of enzymes that catalyze the transfer of γ-phosphate from adenosine triphosphate (ATP) to tyrosine, serine, or threonine amino acid residues of substrate proteins. Phosphorylation alters the enzymatic activity, binding capability, or cellular localization of the substrate protein, as a means to relay environmental signals, such as the extracellular matrix, antigens, insulin, and growth factors *(1)*. Following the discovery of protein phosphorylation as a mechanism of signal transduction, the discovery of the v-*src* and v-*abl* oncogenes *(2,3)*, and the realization that PKs are an immense superfamily of proteins (2.1% of *Caenor elegans* genes are PKs); PKs have moved to center stage in the field of signal transduction. Because of their centrality in cell signaling, PKs have also become attractive therapeutic targets for such diverse diseases as diabetes and cancer *(4)*.

Biochemical study of PKs necessitates robust in vitro methods to assay kinase activity of recombinant and immunoprecipitated kinases. Often, exogenous peptide or protein substrates are added as phosphoacceptors in these assays. Random polypeptides such as poly Glu:Tyr (4:1) have been used in high-throughput kinase assays *(5–7)*. Peptide substrates can be optimized by screening a combinatorial peptide library for efficient phosphorylation sequences, yielding highly efficient substrates in a kinase-specific manner *(8–15)*. Kinase activity can then be assayed by quantitating transfer of ^{32}P to the peptide. However, peptides cannot be used in gel-based assays such as Western blotting. The protein substrates used for this purpose are often fortuitous, nonoptimal proteins such as histone H1, enolase, and immunoglobulin

From: *Methods in Molecular Biology, vol. 183: Green Fluorescent Protein: Applications and Protocols*
Edited by: B. W. Hicks © Humana Press Inc., Totowa, NJ

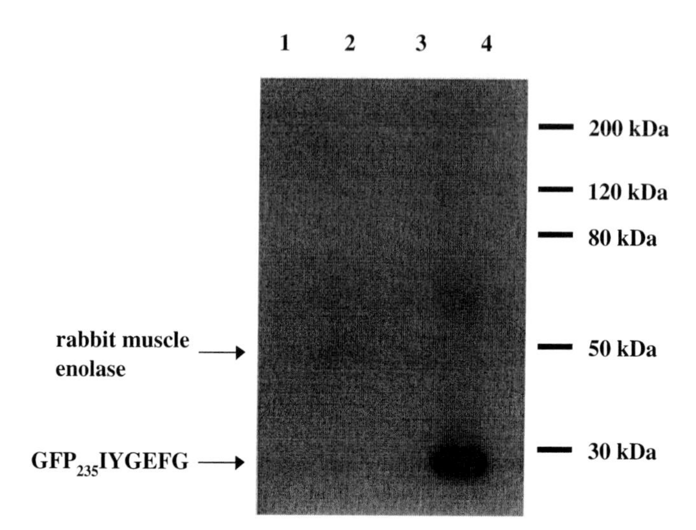

Fig. 1. Comparison of GFP, enolase, and GFP_{235}IYGEFG as XD4 substrates in a radioactive assay. Lane 1, 1 μg GFP; lane 2, 5 μg activated enolase; lane 3, 5 μg enolase; lane 4, 0.5 μg GFP_{235}IYGEFG. Reactions were run as described under Materials and Methods. Samples were run on a 12% SDS-PAGE gel. The gel was dried and exposed to an X-ray film at room temperature. Exposure time was 24 h.

heavy chain. Recognizing the need for efficient protein substrates for in vitro kinase reactions, our lab has developed a general method to transfer the advantages of a combinatorial peptide optimization strategy to the development of efficient green fluorescent protein (GFP), (this chapter refers to GFPMut1 *[16]* as GFP)-based PK substrates suitable for protein-based assays.

GFP is a compact protein that is easily expressed in prokaryotes and is very poorly phosphorylated by PKs (**Fig. 1**). The strategy used to transform GFP into an efficient kinase substrate involves starting with the optimal phosphorylation sequence of a given kinase gained through phosphorylation of a peptide library. The optimal phosphorylation sequences of many well-studied kinases are known *(8–15)*. This efficient phosphorylation sequence is then appended to the C-terminus of GFP as a flexible tail. This laboratory has demonstrated that these modified GFP proteins are efficient protein substrates that can be easily generated in a kinase-specific manner. The resulting GFP constructs are easy to express, easy to purify using standard 6× His affinity tags, and easy to use in protein-based kinase assays. The GFP constructs shown in **Table 1** retain the fluorescence and expression qualities of wild-type GFP (data not shown). The suitability of GFP as a carrier protein for an appended phosphorylation sequence is also verified by its lack of endogenous phosphorylation sites, which, if present, would complicate antiphosphotyrosine immunoblotting

Table 1
GFP Constructs

Kinase	GFP Sequence-Optimal Phosphorylation Sequence	Name of Construct
	^{215}RDHMVLLEFVTAAGITHGMDELYK^{238}	GFP
Src	RDHMVLLEFVTAAGITHGMDEIYGEFGGS	GFP$_{235}$IYGEFG
Src	RDHMVLLEFVTAAEIYGEFG	GFP$_{227}$EIYGEFG
Abl	RDHMVLLEFVTAAGITHGMDAIYAAPF	GFP$_{235}$AIYAAPF
PKA	RDHMVLLEFVTAAGITHGMDRRRRSII	GFP$_{234}$RRRRSII
PKA	RDHMVLLEFVTAARRRRSII	GFP$_{227}$RRRRSII
PKA	RDHMVLLEFVTAARRSII	GFP$_{227}$RRSII
	β sheet → Flexible	

(**Fig. 1**, lane 1). GFP is then purified from its prokaryotic expression host as an unphosphorylated, well-expressed, fluorescent protein. These constructs have proven to be much more efficient phosphoacceptors than proteins commonly used as kinase substrates. Serine/threonine kinases and tyrosine kinases have both had phosphorylatable GFP constructs made and the method appears to be general and useful to rapidly generate sensitive in vitro substrates for a variety of PKs.

In searching for a versatile, nonradioactive kinase assay system, the authors sought to combine the advantages of highly efficient peptide substrate sequences with the convenience of protein substrates useful in polyacrylamide gel electrophoresis (PAGE)-based kinase assays. GFP was chosen as a protein host for optimal substrate sequences derived from combinatorial peptide libraries. Its small size (28 kDa), ensures that it will separate easily from most, if not all, protein components isolated in immune complex kinase assays (majority >35 kDa). GFP is a highly expressed protein that is easy to purify and handle. By combining the efficiency of optimal phosphorylation sequences with these qualities, C-terminal tagged GFP constructs have been shown to be excellent substrates for a variety of PKs.

2. Materials

1. DNA plasmids: Plasmid pGFPMut1, which contains the *GFPMut1* gene in a pKEN vector, was a gift from B. P. Cormack (Stanford University). Plasmid vector pQE8 is from Qiagen (Chatsworth, CA). pGEX-KT-XD4 and the c-Abl expression vector have been described ***(16,17)***.
2. Qiaex II kit were purchased from Qiagen.
3. Restriction enzymes: *Bam*HI.

4. Calf intestinal alkaline phosphatase (CIP), and T4 DNA ligase (New England Biolabs, Beverly, MA).
5. Luria-Bertoni (LB) broth and LB agar (Bio101, La Jolla, CA).
6. *Escherichia coli* strain JM109.
7. Isopropyl β-D-thiogalactoside.
8. Protein kinase A (PKA) (Life Technology, Gaithersburg, MD).
9. Talon resin (Clontech, Palo Alto, CA).
10. [γ-^{32}P]ATP (Du Pont NEN, Boston, MA).
11. Antiphosphotyrosine monoclonal antibody 4G10 was a gift from Brian Druker (Oregon Health Science Center, Portland, OR).
12. SuperSignal chemiluminescent substrate for horseradish peroxidase (Pierce, Rockford, IL).
13. *Pfu* polymerase, deoxyribonucleoside triphosphate (dNTPs), and polymerase buffer (Stratagene, La Jolla, CA).
14. 1% Agarose gels.
15. 12% Sodium dodecyl sulfate (SDS)-PAGE gels.
16. Nitrocellulose transfer membrane.
17. Electrospray mass spectrometry was performed by the Mass Spectrometry Facility, Department of Chemistry, Princeton University, with a Hewlett-Packard 5989B spectrometer.
18. Oligonucleotide synthesis and automatic DNA sequencing were done in the Synthesis and Sequencing Facility at Princeton University. Polymerase chain reaction (PCR) reverse primers used in this work are listed below. The forward primer for all constructs was 5'-TCTAGGGATCCGGCATGAGTAAAGGA-3'.
    a. pGFP$_{235}$IYGEFG
       5'-TCTAGGATCCGCCGAATTCGCCGTATATTTCATCCATGCCATG-3'.
       The resulting construct replaces the last three codons of GFP with **IYGEFGGS.**
    b. pGFP$_{227}$EIYGEFG
       5'-TCTAGGATCCTTAGCCGAATTCGCCGTATATTTCAGCAGCTGT-TACAAACTCAA-3'.
       The resulting construct replaces the last 11 codons of GFP with **IYGEFG.**
    c. pGFP$_{235}$AIYAAPF
       5'-ACGTATTCGAATTAGAACGGCGCCGCATAGATCGCTTCATC-CATGCCATGTGTAATC- 3'.
       The resulting construct replaces the last three codons of GFP with **AIYAAPF.**
    d. pGFP$_{234}$RRRRSII
       RP, 5'-GATAGGATCCTTAGATGATAGATCTAGGCCGGCGATCCATG-CCATGTGTAATC-3'.
       The resulting construct replaces the last four codons of GFP with **RRRRSII**.
    e. pGFP$_{227}$RRRRSII
       5'-TCTAGGATCCTTAGATGATAGATCTACGCCGGCGAGCAGCTGT-TACAAACTCAA-3'.
       The resulting construct replaces the last 11 codons of GFP with **RRRRSII**.

f. $pGFP_{227}RRSII$

The last 11 codons of GFP are replaced with **RRSII,** the serine residue of which was at the same distance from the GFP core structure as that of tyrosine in $GFP_{227}EIYGEFG$.

19. Src buffer: 100 m*M* HEPES, pH 7.4, 10 m*M* $MgCl_2$, 10 μg/mL bovine serum albumin, 0.1 m*M* dithiothreitol, 1 m*M* sodium orthovanadate, 0.2 m*M* ATP.
20. c-Ab1 buffer: 50 m*M* Tris-HCl, pH 8.1, 10 m*M* $MgCl_2$, 200 m*M* ATP.
21. PKA (Sigma).
22. PKA buffer: 50 m*M* Tris-HCl, pH 7.4, 10 m*M* $MgCl_2$, 20 mM ATP.
23. Tris buffer: 20 m*M* Tris-HCl, pH 8.0.
24. Elution buffer: Tris buffer plus 200 m*M* imidazole.
25. 5% Dried milk in phosphate-buffered saline.

3. Methods

3.1. Procedures for Plasmid Construction (see Note 1)

1. Using the PCR, add a *Bam*HI site to the end of *GFPMut1* gene in pGFPMut1.
2. Add a nucleotide sequence coding for the peptide of choice, followed by a *Bam*HI site to the 5' end of the gene, corresponding to the C-terminus of the GFP.
3. Perform PCR using *Pfu* polymerase and dNTPs, according to manufacturer's instructions.
4. Digest the PCR product with *Bam*HI.
5. Separate the nucleic acid digest by agarose gel electrophoresis, and extract the appropriate fragment, using the Qiaex II kit.
6. Similarly, digest the vector, pQE8, which contains a 6X His coding sequence before a *Bam*HI site, treat with CIP.
7. Separate the pQE8 digest by agarose gel electrophoresis, and extract the appropriate fragment.
8. Ligate the 6X His vector and the GFP-kinase sensor insert using T4 DNA ligase according to manufacturer's instructions.
9. Use the ligation mixture to transform competent JM109 *E. coli*. Screen colonies using a hand-held ultraviolet lamp, and pick green fluorescent colonies.
10. Isolate plasmid DNA from these colonies and analyze by restriction digest. Confirm positive clones by sequencing.

3.2. Expression and Purification of GFP Kinase Sensors

1. Grow the transformed *E. coli* harboring one of the constructs overnight at 37°C in LB broth with 75 μg/mL ampicillin. Dilute the culture 1:100 into 50 mL of same medium, and grow at 37°C until OD_{596} reaches 0.4 (~3 h).
2. Add IPTG to a final concentration of 0.5 m*M*, and grow the culture for an additional 8–10 h.
3. Harvest the cells by centrifugation and lyse them by sonication.
4. Clear the lysate by centrifugation and recover the supernatant.
5. Add Talon resin to the supernatant and purify in batch fashion following the manufacturer's protocol.

6. Check the purity of the isolated proteins by SDS-PAGE (95% purity based on Coomassie staining). Although not essential in our work, the molecular weights of each protein were verified by electrospray mass spectrometry.

3.3. Expression and Purification of Kinases

1. XD4 is a v-Src lacking the SH3 domain and the first 80 residues of the SH2 domain (a Δ[77-225] truncation of v-Src), and is highly active and suitable for prokaryotic expression.
2. Grow bacterial strain DH5α harboring pGEX-KT-XD4.
3. Harvest and lyse the cells as outlined above.
4. Purify XD4 from the lysate using glutathione–agarose beads ***(16,17)***.
5. Plasmid construction, expression, and purification of c-Abl are carried out similarly, as described ***(16,17)***.

3.4. PK Assays

3.4.1. XD4 (Src) Assay

1. Two sets of reactions are run in Src buffer with 2 μL purified XD4, varying amounts of GFP_{235}IYGEFG, and with or without 2.5 μCi [γ-^{32}P]ATP, at 30°C for 15 min. The total volume of each reaction was 20 μL.
2. At the end of the incubation, use one set of reactions (with [γ-^{32}P]ATP) for autoradiography, and use the second set of reactions (without [γ-^{32}P]ATP) for Western blotting.
3. For autoradiography, add 5 μL 53 Laemmli loading dye to each tube and heat the mixture at 90°C for 5 min.
4. Load sample solutions onto a 12% SDS-PAGE gel and perform electrophoresis until the bromophenol blue dye migrates out of the gel.
5. Stain the gel with Coomassie brilliant blue R-250.
6. Destain, dry, and expose the gel to X-ray film at room temperature.
7. For Western blotting, separate GFP_{235}IYGEFG from other proteins in the reaction mixture that may interfere with antiphosphotyrosine immunoblotting.
8. Mix 10 μL of Talon resin in 30 μL 20 m*M* Tris buffer, pH 8.0, and add the slurry to each reaction.
9. Separate the resin by brief centrifugation (15,000*g*), and wash 3X with the Tris buffer.
10. Elute GFP_{235}IYGEFG with 20 μL elution buffer and separate from the resin by brief centrifugation (15,000*g*).
11. Add 5 μL 53 Laemmli loading dye to each supernatant and heat the mixture at 90°C for 5 min. Spin the tubes to remove insoluble debris (15,000*g*, 5 min).
12. Load the sample solutions onto a 12% SDS-PAGE gels and perform electrophoresis until the bromophenol blue dye migrates out of the gel.
13. Transfer the proteins to a nitrocellulose membrane and incubate the membrane overnight in 5% dried milk in phosphate-buffered saline.
14. Probe the membrane with antiphosphotyrosine antibody 4G10, wash the blot, then visualize with horseradish peroxidase-conjugated secondary antibody using chemiluminescence ***(19)***.

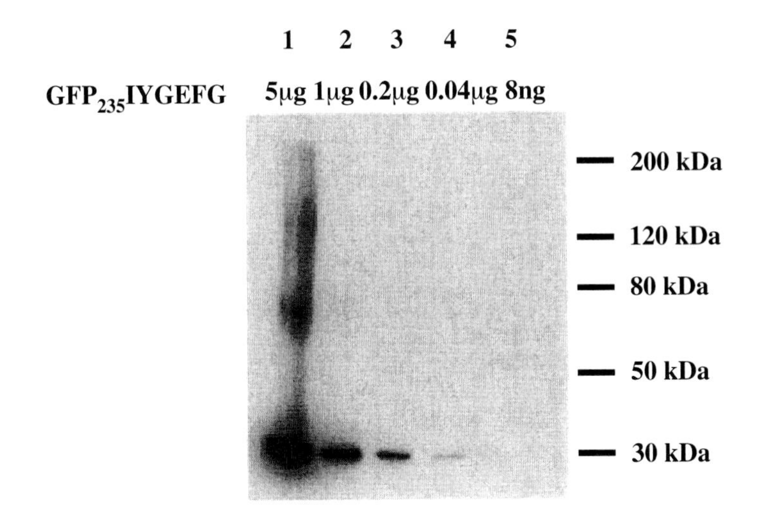

Fig. 2. Autoradiogram of XD4 phosphorylation reactions, using different amounts of GFP_{235}IYGEFG as indicated. Reactions were run in Src buffer with 2.5 μCi[γ-^{32}P]ATP for 15 min at 30°C. Laemmli loasing dye was added, and the mixture was heated to inactivate the kinase. Sample solutions were loaded onto a 12% SDS-PAGE gel, and electrophoresis was performed. The gel was dried and exposed to an X-ray film at room temperature. Exposure time was 14 h.

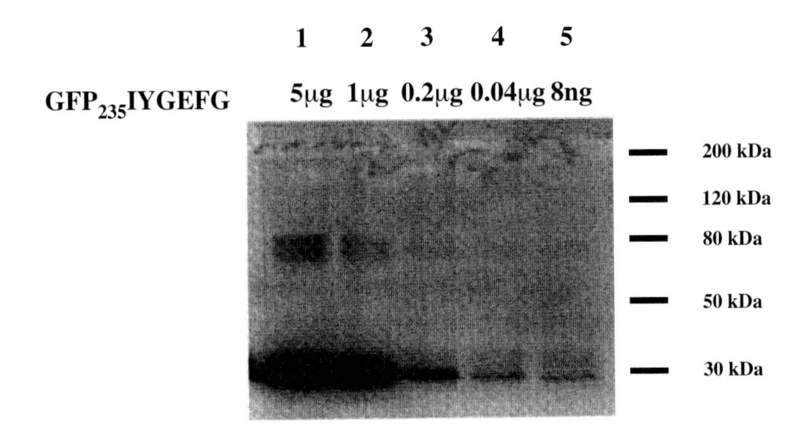

Fig. 3. Western blot of XD4 phosphorylation reactions with antiphosphotyrosine antibody, using different amounts of GFP_{235}IYGEFG. Reactions were run as described in **Fig. 2**, with the exception of [γ-^{32}P]ATP. GFP_{235}IYGEFG was separated from the reaction by addition of Talon resin (10 μL each reaction). The resin was separated and washed. GFP_{235}IYGEFG was eluted from the resin with 20 μL 200 m*M* imidazole in Tris buffer. Laemmli loading dye was added to the eluent. After electrophoresis, the proteins were transferred to a nitrocellulose membrane and immunoblotted with antiphosphotyrosine antibody.

The abovementioned strategy was employed to construct the putative Src substrate, GFP_{235}IYGEFG. This construct was then tested for its ability to act as a Src substrate. As can be seen in the autoradiogram in **Fig. 1**, GFP alone is not a Src substrate (lane 1). Addition of the optimal Src phosphorylation sequence ***(10)*** IYGEFG to the C-terminus of GFP results in an exceedingly efficient Src substrate (lane 4), compared to the common Src substrate, activated enolase (lanes 2 and 3).

Figure 2 shows another autoradiogram using varying amounts of GFP_{235}IYGEFG in a Src kinase assay. This experiment demonstrates that phosphorylation of the GFP construct can be detected using nanogram quantities of GFP_{235}IYGEFG.

Figure 3 is a similar experiment used to determine whether this sensitivity can be duplicated using a generally less sensitive method, antiphosphotyrosine Western blotting. Again, the sensitivity is carried over and only 8 ng phosphoprotein can be detected, demonstrating that the phosphorylated tail of these GFP constructs is efficiently recognized by the monoclonal antibody, 4G10.

3.4.2. c-Abl Assay

1. Carry out duplicate reactions as described above, except c-Ab1 buffer and substrate (GFP_{235}AIYAAPF) are used, and the reaction is performed for 1 h at room temperature.
2. Resolve the samples by SDS-PAGE and transfer to a nitrocellulose membrane.
3. Visualize using antiphosphotyrosine antibody 4G10 and chemiluminescence.
4. GFP alone is not an Abl substrate (data not shown). However, appending the Abl optimal phosphorylation site, AIYAAPF at position 235 produced an excellent Abl substrate (**Fig. 4**).

3.4.3. PKA Assay

1. Reactions are run, and autoradiography is performed as in the XD4 assay. Exceptions are that a PKA buffer is used and the substrate is altered for PKA detection (GFP_{234}RRRRSII or GFP_{227}RRRRSII). Add 7 U PKA to each reaction in place of XD4.
2. Be sure to test unmodified GFP as a kinase substrate. This was done for PKA and GFP is shown to be a poor PKA substrate (**Fig. 5**, lane 6). However, simple C-terminal attachment of the PKA optimal phosphorylation sequence RRRRSII, from $pGFP_{227}$RRRSII, as identified by Zhou et al. ***(8)***, also proved to be a poor PKA substrate by this assay, so that $pGFP_{227}$RRRSII was produced and shown to be efficient (data not shown, *see* **Note 3**).

4. Notes

1. Previously, the last six C-terminal residues of GFP have been deleted without altering expression or fluorescence ***(20)***. The authors pursued the reasonable strat-

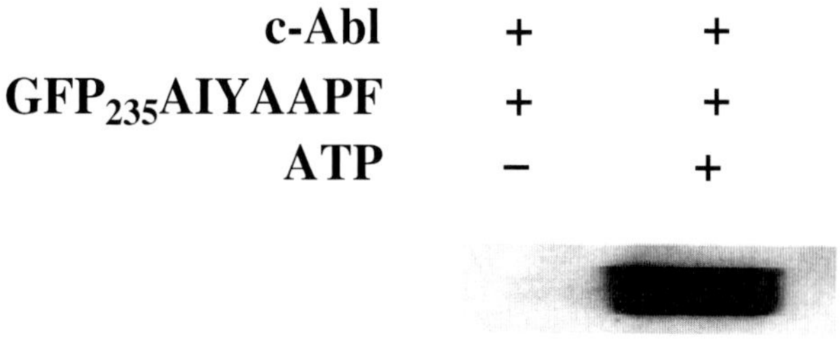

c-Abl	+	+
GFP_{235}AIYAAPF	+	+
ATP	–	+

Fig. 4. Phosphorylation of GFP_{235}AIYAAPF by c-Abl. Each reaction was carried out with 50 m*M* Tris (pH 8.1), 10 m*M* $MgCl_2$, 200 µ*M* ATP, and 0.5 µg GFP_{235}AIYAAPF. The samples were resolved on an SDS-PAGE gel, and transferred to nitrocellulose membrane. Immunoblotting was performed with the antiphosphotyrosine antibody.

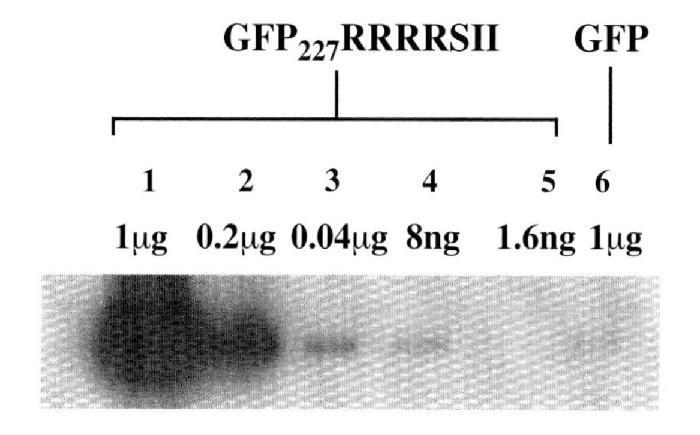

Fig. 5. Autoradiogram of PKA phosphorylation reactions with GFP_{235}RRRRSII. Each reaction was run in 50 m*M* Tris (pH 7.4), 10 m*M* $MgCl_2$, 20 µ*M* ATP, 2.5 µCi [γ-^{32}P]ATP, and different amounts of GFP_{235}RRRRSII or GFP, as labeled above each lane. PKA (7 U) was added, and the mixture was incubated for 15 min at 30°C. Laemmli loading dye was added and the reaction mixture was heated to inactivate the kinase. Electrophoresis was run on a 15% SDS-PAGE gel. The gel was dried and exposed to an X-ray film at room temperature for 14 h.

egy of attaching the phosphorylation sequence C-terminal to the last β strand, effectively replacing the abovementioned nonessential residues.

2. Creating an efficient GFP-based PK substrate for other kinases should follow a protocol analogous to those mentioned above. First, a peptide that has an optimal phosphorylation sequence must be identified. This can be determined experimentally ***(8,11)***, and the optimal phosphorylation sequences of many kinases are well known ***(8,9,11,14)***. Addition of this sequence to the C-terminus of GFP, using standard PCR techniques described in Materials and Methods, should rapidly generate an efficient kinase substrate. The success of this

procedure relies on the fact that GFP is a carrier protein for the peptide sequence, which remains unstructured.

3. Considering that the addition of this particular phosphorylation site also introduced a presumably efficient trypsin cleavage site into a known flexible region of the sequence, the protein product was subjected to mass spectrometry and was shown to be missing the phosphorylation sequence after the first arginine. To remedy this situation, the phosphorylation sequence was shifted to begin at position 227, closer to the last sequence containing secondary structure. GFP_{227}RRRRSII is resistant to proteolysis as judged by mass spectrometry and is an efficient PKA substrate (**Fig. 5**). Problems with proteolytic degradation as seen with the PKA substrate can be dealt with successfully by moving the peptide sequence a few positions toward the N-terminus, yielding a less flexible, proteolytically resistant protein.

References

1. Hunter, T. (2000) Signaling-2000 and beyond. *Cell* **100,** 113–127.
2. Hunter, T. and Cooper, J. A. (1985) Protein tyrosine kinases. *Ann. Rev. Biochem.* **54,** 897–930.
3. Bishop, J. (1985) Viral oncogenes. *Cell* **42,** 23–28.
4. Schlessinger, J. (1988) Signal transduction by allosteric receptor oligomerization. *Trends Biochem. Sci.* **13,** 443–447.
5. Cooper, J. A., Esch, F. S., Taylor, S. S., and Hunter, T. (1984) Phosphorylation sites in enolase and lactate dehydrogenase utilized by tyrosine protein kinases in vivo and in vitro. *J. Biol. Chem.* **259,** 7835–7841.
6. Braunwalder, A. F., Yarwood, D. R., Sills, M. A., and Lipson, K. E. (1996) Measurement of the protein tyrosine kinase activity of c-Src using time resolved fluorometry of europium chelates. *Analyt. Biochem.* **238,** 159–164.
7. Braunwalder, A. F., Yarwood, D. R., Hall, T., Missbach, M., Lipson, K. E., and Sills, M. A. (1996) A solid phase assay for the determination of protein tyrosine kinase activity of c-Src using scintillating microtitration plates. *Analyt. Biochem.* **234,** 23–26.
8. Wu, J., Ma, Q. N., and Lam, K. S. (1994) Identifying substrate motifs of protein kinases by a random library approach. *Biochemistry* **33,** 14,825–14,833.
9. Till, J. H., Annan, R. S., Carr, S. A., and Miller, W. T. (1994) Use of synthetic peptide libraries and phosphopeptide selective mass spectrometry to probe protein kinase substrate specificity. *J. Biol. Chem.* **269,** 7423–7428.
10. Zhou, S., Carraway, K. L., III, Eck, M. J., Harrison, S. C., Feldman, R. A., Mohammadi, M., et al. (1995) Catalytic specificity of protein tyrosine kinases is critical for selective signaling. *Nature* **373,** 536–539.
11. Zhou, S., Blechner, S., Hoagland, N., Hoekstra, M. F., Piwnica-Worms, H., and Cantley, L. C. (1994) Use of an oriented peptide library to determine the optimal substrates of protein kinases. *Curr. Biol.* **4,** 973–982.
12. Dente, L., Vetriani, C., Pelicci, G., Lanfrancone, L., Pelicci, P. G., and Cesareni, G. (1997) Modified phage peptide libraries as a tool to study specificity of phos-

phorylation and recognition of tyrosine containing peptides. *J. Mol. Biol.* **269,** 694–703.
13. Nishi, T., Budde, R. J., McMurray, J. S., Obeyesekere, N. U., Safdar, N., Levin, V. A., and Saya, H. (1996) Tight-binding inhibitory sequences against pp60 (c-Src) identified using a random 15 amino acid peptide library. *FEBS Lett.* **399,** 237–240.
14. Schmitz, R., Baumann, G., and Gram, H. (1996) Catalytic specificity of phosphotyrosine kinases Blk, Lyn, c-Src, and Syk as assessed by phage display. *J. Mol. Biol.* **260,** 664–677.
15. Chan, P. M., Keller, P. R., Connors, R. W., and Leopold, W. R., and Miller, W. T. (1996) Amino terminal sequence determinants for substrate recognition by platelet derived growth factor receptor tyrosine kinase. *FEBS Lett.* **394,** 121–125.
16. Cormack, B. P., Valdivia, R. H., and Falkow, S. (1996) FACS optimized mutants of the green fluorescent protein (GFP). *Gene* **173,** 33–38.
17. Shah, K., Liu, Y., Deirmengian, C., and Shokat, K. M. (1997) Engineering unnatural nucleotide specificity for Rous Sarcoma virus tyrosine kinase to uniquely label its direct substrates. *Proc. Natl. Acad. Sci. USA* **94,** 3565–3570.
18. Liu, Y., Shah, K., Yang, F., Witucki, L., and Shokat, K. M. (1998) Engineering Src family protein kinases with unnatural nucleotide specificity. *Chem. Biol.* **5,** 91–101.
19. Friedman, J. D. (1997) Undergraduate Thesis, Department of Molecular Biology, Princeton University, Princeton, NJ.
20. Dopf, J. and Horiagon, T. M. (1996) Deletion mapping of the *Aequorea victoria* green fluorescent protein. *Gene* **173,** 39–44.

23

Green Fluorescent Protein Urea Sensors

Uropathogenic Proteus mirabilis

Christopher Coker, Hui Zhao, and Harry L. T. Mobley

1. Introduction

Proteus mirabilis is a uropathogenic bacterium that is responsible for causing urinary tract infections in patients with structural abnormalities of the urinary tract, in the elderly, and catheterized individuals ***(1,2)***. A major virulence factor produced by this organism is the urea inducible urease ***(3,4)***, which is located in the cytosolic compartment of the bacterium ***(5,6)***. The molecular mechanism of urea induction has been described ***(5,7,8)***. The genes encoding urease have been cloned and sequenced, and comprise seven open reading frames (*ureD, ureA–G*), which encode polypeptides that make up the urease structural subunits and accessory proteins and one open reading frame (*ureR*), which encodes the AraC-like transcriptional gene activator (controls arabinose operon), UreR ***(5)***.

In the presence of urea, UreR promotes transcription of itself and the urease gene cluster ***(7,8,9)***. Indeed, UreR has been shown to interact directly with DNA encoding the transcriptional regulatory regions for both *ureR* and *ureD*, the first gene in the urease gene cluster ***(7,9)***. *P. mirabilis* produces minimal amounts of urease in the absence of urea in vitro, but after urea induction urease activity increases 5- to 25-fold ***(3,6,10,11)***. Urease induction by urea is concentration-dependent with peak urease induction achieved in the presence of 200 m*M* urea ***(11)***.

Initially, gene fusion technology, using β-galactosidase activity encoded by *lacZ* as a reporter gene, was utilized to study the mechanism of urea induction of the urease gene cluster. Plasmid constructs that encode *ureR* and *ureD* fused to *lacZ* were created. When expressed in *Escherichia coli* β-galactosidase activ-

From: *Methods in Molecular Biology, vol. 183: Green Fluorescent Protein: Applications and Protocols*
Edited by: B. W. Hicks © Humana Press Inc., Totowa, NJ

ity from these constructs was urea-inducible *(7,8)*. In another study, using a *ureA–lacZ* fusion, showed that inducibilty of the *urease* gene cluster by urea was highly specific; urea anologs and other inhibitors of urease were unable to induce β-galactosidase activity from this construct in concentrations ranging from 5–200 m*M* *(11)*. These data suggest that if UreR binds urea, the binding site is different from that found in urease which is able to bind urea analogs. In fact, when urea is used with the inhibitor acetohydroxamic acid to induce urease in wild-type *P. mirabilis* the level of induction is higher than that observed when urea is used as inducer alone *(3)*, suggesting that urease inhibitors do not compete with urea for activation of the urease.

Green flourescent protein (GFP) translationally fused to UreD was used as a urea-inducible reporter in the authors' laboratory *(12)*. A gene fusion encoded by plasmid pURE-RD-GFP (**Fig. 1**) was constructed between *ureD* and a variant of the gene encoding wild-type GFP (GFP[S65T/V68/LS72A]) *(13)*. The variant protein has increased fluorescent capability and does not form nonflourescent inclusion bodies to the same extent as the wild-type protein. *P. mirabilis* carrying pURE-RD-GFP were fluorescent in the presence of urea in a urea concentration dependent manner (as assessed by amount of fluorescence and GFP levels as detected by Western blot). As observed with urease expression, GFP expression is not completely shut off in the absence of inducer as observed by Western blot and fluorescence emission spectra. pURE-RD-GFP is a self-contained urea-sensing system. Both *ureR* (the urea "sensor" required for activation of *ureD*) and *ureD::gfp* (the reporter) are encoded on this plasmid. Additionally, the native promoter elements from *P. mirabilis* required for urea induction of both UreR and UreD:GFP are present in the intergenic region located between the *ureR* and *ureD* start codons. *E. coli* transformed with pURE-RD-GFP are fluorescent only in the presence of urea *(12)*. Fluorescence can be detected by direct examination of bacteria on a glass slide or the bacteria can be lysed and the resulting lysate measured for fluorescent activity using a spectrophotofluorimeter. Also, bacteria can be detected *in situ*, in tissue section samples from infected animals using fluorescence microscopy.

The protocol in this chapter describes the procedure for urea induction and detection based on the authors' in vitro and *in situ* experiments, using a mouse model of infection, with *P. mirabilis* harboring pURE-RD-GFP. Although the in vitro protocol describes urea induction by the addition of exogenous urea to culture medium the bacteria can be used as urea sensors in other media. Because pURE-RD-GFP uses a plasmid vector backbone with a gene encoding for ampicillin resistance, it can be used to transform other bacteria within the Enterobacteriaceae by traditional procedures. The resulting transformants can be used as urea sensors.

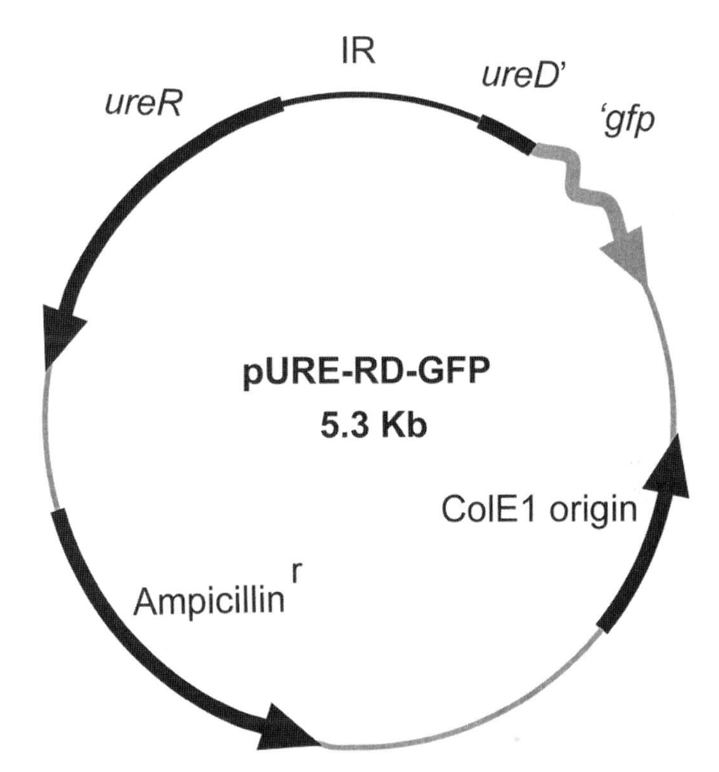

Fig. 1. Physical map of pURE-RD-GFP. The plasmid backbone is derived from pBluescript, and encodes full length UreR, the intergenic region (IR) between *ureR* and *ureD*, which contains the native regulatory regions responsible for urea induction of both UreR and UreD–GFP, and the UreD–GFP fusion protein. The plasmid is ~5.3 kb in size, encodes resistance to AMP, and is replicated via the ColE1 origin of replication.

2. Materials

2.1. Growth Media and Urea Induction

1. Nonswarming agar (10 g tryptone, 5 g yeast extract, 5 mL glycerol, 0.4 g NaCl/L, 20 g agar, pH 7.0. Autoclave 20 min at 15 lb/sq. in. ***(14)***.
2. Luria-Bertani medium: 10 g tryptone, 5 g yeast extract, 10 g NaCl/L deionized water, pH 7.0. Add 15 g agar for solid media. Autoclave 20 min at 15 lb/sq./in. ***(15)***.
3. Ampicillin (Amp).
4. Urea stock solution, 6*M* (make fresh, stable at 4°C for 1 wk) (*see* **Note 1**).
5. *P. mirabilis*, *E. coli* DH5α *P. mirabilis* and *E. coli* transformed with pERU-RD-GFP ***(12)***, a plasmid based on a pBluescript (Stratagene) vector with Amp resistance.

2.2. Solutions and Equipment for Spectrofluorimetry

1. French press.
2. Wash and resuspension buffer TNMD: Tris-HCl (pH 7.4), 100 m*M* NaCl, 1 m*M* $MgCl_2$, 10 m*M* dithiothreitol.
3. Spectrofluorimeter and cuvets, Spectronics AB-2 (Spectronics, Rochester, NY), or equivalent.

2.3. Solutions and Equipment for Fluorescence Microscopy of Bacteria

2.3.1. In Vitro Detection

1. Phosphate-buffered saline (PBS): 8 g NaCl, 0.2 g KCl, 1.44 g Na_2HPO_4, 0.24 g KH_2PO_4/L, (pH 7.4) ***(15)***.
2. Glass slides and cover slips.
3. Rubber cement.
4. Epifluorescence microscope equipped with camera (Zeiss Axiophot, or equivalent).
5. ASA400 film; Kodak.

2.3.2. In Situ Detection

1. Solutions and equipment for in vitro assay.
2. Dry ice.
3. OCT cryosectioning solution (Tissue-Tek, Miles).
4. Cryostat.

3. Methods

3.1. Cell Growth and Urea Induction

1. Culture *P. mirabilis* harboring the pURE-RD-GFP fusion construct, on non-swarming agar (*see* **Note 2**) containing 50 µg/mL Amp at 37°C. Amp is included for positive selection of pURE-RD-GFP.
2. Culture a single colony isolate in 5 mL L-broth with 50 µg/mL Amp at 37°C overnight in a shaking incubator.
3. Dilute overnight culture 1:100 in 100 mL L-broth containing 50 µg/mL Amp and incubate at 37°C in a shaking incubator to an OD_{600} of 0.1.
4. Add urea (final concentration 200 m*M*) and continue incubation at 37°C with shaking for additional 3 h (*see* **Notes 3–6**).

3.2. Detection of GFP

3.2.1. Spectrofluorimetry

1. Harvest the bacteria by centrifugation at 5000*g* for 5 min at 4°C.
2. Wash the bacterial pellet twice in 4 mL ice cold TNMD at 4°C. Centrifuge as in **step 1** to collect bacteria.
3. Resuspend the bacterial pellet in 4 mL ice cold TNMD.
4. Lyse cells by French press at 20,000 lb/in^2 (*see* **Note 7**).

5. Centrifuge lysate at 5,000*g* for 5 min at 4°C.
6. Collect supernatant and centrifuge at 27,000*g* for 15 min at 4°C (*see* **Note 8**).
7. Add supernatants to cuvets and measure emission (470-nm excitation wavelength, emission at 490–590 nm 2-nm slits) and corrected excitation (330–530-nm excitation wavelength, emission at 550 nm) spectra with spectrofluorimeter.

3.2.2. Fluorescence Microscopy of Bacteria

3.2.2.1. In Vitro Detection

1. After urea induction, harvest 1 mL cells in an Eppendorf tube by microcentrifugation at maximum speed for 3 min at room temperature.
2. Wash the cell pellet twice in 1 mL PBS at room temperature. Resuspend in 1 mL PBS.
3. Place a 10-µL loop of organisms onto a glass slide and let air dry. Do not heat-fix bacteria to slide, in order to prevent any denaturation of GFP.
4. Place a cover slip over the slide and seal with rubber cement (*see* **Note 9**).
5. View fluorescent bacteria on an epifluorescence microscope using filter sets for fluorescein isothiocyanate fluorescence (*see* **Note 10**). Fluorescent bacteria will appear green against a dark background. Using a negative control (bacteria not induced by urea) is important in order to ascertain the contrast between fluorescent and nonfluorescent bacteria.

3.2.2.2. *In Situ* Detection

This procedure is used after infection of an animal model host with the bacterial pathogen of interest harboring pURE-RD-GFP. Antibiotic selection for Amp must be used with the animal model, in order to select for bacteria that retain pURE-RD-GFP.

1. Collect tissues of interest from infected animal.
2. Embed tissue samples in OCT, and freeze on dry ice (*see* **Note 11**).
3. Cryosection samples into 5–10-µm sections.
4. Place sections on glass slide, and overlay with a glass cover slip (*see* **Note 12**).
5. Seal cover slip with rubber cement (*see* **Note 9**).
6. Analyze for fluorescent bacteria with a fluorescence microscope, as in **Subheading 3.2.2.1., step 5**.

4. Notes

1. Urea spontaneously breaks down during storage. Therefore, the stock solution must be made fresh and used within a period of ~1 wk. If the urea stock solution has an ammonia odor, make a fresh stock.
2. Nonswarming agar is used to prevent the swarming phenonemon exhibited by *P. mirabilis* ***(14)***, so that isolated colonies can be obtained.
3. Since *P. mirabilis* is ureaolytic the pH of the growth medium increases in the presence of urea, because of the ammonium ions generated by urea hydrolysis. Growth of the organism is halted at high pH. This can be overcome by using a

mutant of *P. mirabilis* that does not produce urease. Nonureaolytic organisms harboring pURE-RD-GFP do not present this problem.

4. There is low-level expression of GFP from pURE-RD-GFP when expressed in *P. mirabilis* in the absence of urea. This is not unexpected since low-level urease activity is also exhibited by *P. mirabilis* in the absence of urea ***(4)***. It is important to use a negative control (i.e., cultures grown in the absence of urea) to correctly interpret results. In contrast, GFP is not expressed from pURE-RD-GFP when expressed in *E. coli* DH5α as assessed by spectrophotofluorimetry and Western blot ***(12)***. Other Enterobacteriaceae probably exhibit this tight regulation of GFP expressed from pURE-RD-GFP.
5. The 3-h incubation, after addition of urea is included to allow an increase in bacterial density. However, we have found that fluorescent bacteria can be detected after only 30-min exposure to medium containing 200 m*M* urea.
6. Alternatively, in order to detect the presence of urea in other media or solutions, such as clinical samples, the bacteria can be harvested and resuspended in the media of interest. Incubation of the bacteria in the media should be continued under conditions that will not inhibit transcription and translation of the UreD-GFP fusion protein.
7. Although a French press was used in our studies to break open cells, other techniques such as multiple freeze–thawing procedures or sonication can be used.
8. Even though the GFP species used in these studies is more stable than wild-type GFP and is improperly folded to a lesser extent than wild-type GPF, two centrifugation spins are important. The first spin at 5000*g*, will remove cells that are not broken by the French press procedure. The second spin at 27,000*g*, will remove any insoluble inclusion bodies that contain improperly folded, overexpressed GFP protein. We have analyzed the insoluble fraction and found it does not fluoresce.
9. Other adhesives, such as nail polish, are not recommended since they are known to inhibit GFP fluorescence.
10. Generally, it is best to use freshly prepared samples to measure fluorescence. However, since GFP is a stable protein, we have found that samples stored at 4°C for at least 1 wk, are still fluorescent.
11. The samples may be perfused in order to get good embedding in OCT and sharper imaging of GFP rather than just placing the samples into OCT on dry ice.
12. Depending on the size of the tissue samples, one can add multiple sections per slide. A negative-control section should be included to compare and contrast fluorescence with experimental samples.

References

1. Warren J. W., Tenney, J. H., Hoopes J. M., Muncie H. L., and Anthony W. C. (1982) A prospective microbiologic study of bacteriuria in patients with chronic indwelling catheters. *J. Infect. Dis.* **146,** 719–723.
2. Mobley H. L. and Warren J. W. (1987) Urease-positive bacteriuria and obstruction of long-term urinary catheters. *J. Clin. Microbiol.* **25,** 2216–2217.

3. Rosenstein, I., Hamilton-Miller, J. M. T., and Brumfritt, W. (1980) The effect of acetohydroxamic acid on the induction of bacterial ureases. *Invest. Urol.* **18,** 112–114.
4. Jones, B. D., Lockatell C. V., Johnson D. E., Warren, J. W., and Mobley H. L. T. (1990) Construction of a urease-negative mutant of *Proteus mirabilis*: analysis of virulence in a mouse model of ascending urinary tract infection. *Infect. Immun.* **58,** 1120–1123.
5. Mobley, H. L. T., Island, M. D., and Hausinger, R. P. (1995) Molecular biology of microbiol ureases. *Microbiol. Rev.* **59,** 451–480.
6. Jones, B. D. and Mobley, H. L. T. (1988) *Proteus mirabilis* urease: genetic organization, regulation, and expression of structural genes. *J. Bacteriol.* **170,** 3342–3349.
7. D'Orazio, S. E. F., Thomas, V., and Collins, C. M. (1996) Activation of transcription at divergent urea-dependent promoters by the urease gene regulator UreR. *Mol. Microbiol.* **21,** 643–655.
8. Island, M. D. and Mobley, H. L. T. (1995) *Proteus mirabilis* urease: operon fusion and linker insertion analysis of *ure* gene organization, regulation and function *J. Bacteriol.* **177,** 5653–5660.
9. Thomas, V. J. and Collins, C. M. (1999) Identification of UreR binding sites in the *Enterobacteriacea* plasmid-encoded and *Proteus mirabilis* urease gene operons. *Mol. Micriobiol.* **31,** 1417–1428.
10. Mobley, H. L. T., Chippendale, G. R., Swihart, K. G., and Welch, R. A. (1991) Cytotoxicity of HpmA hemolysin and urease of *Proteus mirabilis* and *Proteus vulgaris* for cultured human renal proximal tubular epithelial cells. *Infect. Immun.* **59,** 2036–2042.
11. Nicholson, E. B., Concough, E. A., and Mobley, H. L. T. (1991) *Proteus mirabilis* urease: use of a UreA-LacZ fusion demonstrates that induction is highly specific for urea. *Infect. Immun.* **59,** 3360–3365.
12. Zhao, H., Thompson, R. B., Lockatell, V., Johnson, D. E., and Mobley, H. L. T. (1998) Use of green fluorescent protein to assess urease gene expression by uropathogenic *Proteus mirabilis* during experimental ascending urinary tract infection. *Infect. Immun.* **66,** 330–335.
13. Cormack, B. P., Valdivia, R. H., and Falkow, S. (1996) FACS-optimized mutants of the green fluorescent protein (GFP). *Gene.* **173,** 33–38.
14. Belas, R., Erskine, D., and Flaherty, D. (1991). Transposon mutagenesis in *Proteus mirabilis*. *J. Bacteriol.* **173,** 6289–6293.
15. Sambrook, J., Fritsch, E. F., and Maniatis, T. (1989) *Molecular Cloning, A Laboratory Manual*, 2nd ed. Cold Spring Harbor Laboratory, Cold Spring Harbor, NY.

VI

Viral Applications of Green Fluorescent Protein

24

Using Green Fluorescent Protein to Monitor Measles Virus Cell-to-Cell Spread by Time-Lapse Confocal Microscopy

W. Paul Duprex and Bert K. Rima

1. Introduction

The mechanisms by which viruses enter cells have been, and remain, an area of intensive study. Attaining a more comprehensive understanding of the mechanisms involved in cell attachment, penetration, and delivery of the genome into the cytoplasm or nucleus of the cell has potentially far-reaching implications for the prevention and therapy of virus infections. Similarly, elucidating the processes by which viruses spread from infected to uninfected cells within the host may also assist in the rational development of novel treatment strategies.

In the absence of an overt cytopathic effect (CPE), it can be difficult to determine if cells are infected by a virus. Techniques such as direct or indirect immunofluorescence, immunohistochemistry, *in situ* hybridization, and *in situ* reverse transcriptase-polymerase chain reaction (RT-PCR) have been used extensively to examine the types and numbers of cells infected in cultures, or within infected tissues. One disadvantage with these approaches is that they require the specimen to be fixed and permeabilized before the viral antigens or nucleic acids can be detected and this terminates the progress of the infection. Flow cytometry overcomes this difficulty, to a certain extent, by using fluorochrome–antibody conjugates that can directly or indirectly recognize virus proteins on the surface of living cells. However, this technique still has a limitation in that it only allows whole populations of cells to be studied and, therefore, gives little indication of any variation between the infected cells. Additionally, the very antibodies used to detect the presence of the virus may modulate the

From: *Methods in Molecular Biology, vol. 183: Green Fluorescent Protein: Applications and Protocols*
Edited by: B. W. Hicks © Humana Press Inc., Totowa, NJ

infection process. These techniques are therefore not suitable for studies of virus cell-to-cell spread in that, instead of observing an ongoing dynamic process they can only be used to examine a series of discrete steps. Thus, the analysis of the resulting data is rather like piecing together a film from a set of stills.

The possibility of stably inserting a gene encoding green fluorescent protein (GFP) into virus genomes has fundamentally altered the way in which the process of infection can be studied. One major advantage is that infected cells can be readily detected in living cell monolayers in the absence of CPE without using the standard indirect approaches that necessitate fixation. Additionally, these cells can be marked and relocated at regular intervals during the course of virus infection, thereby permitting the study of virus infection and spread in real time. A number of approaches have been used to introduce GFP and other fluorescent reporter protein-expressing genes into virus genomes and the specific strategy chosen depends on the manner in which the virus replicates and expresses its genetic material. Cell-to-cell spread of *Tobacco Mosaic Virus* has been examined by fusing GFP to the 30-kDa movement protein. This protein facilitates transport of virus infection between adjacent cells by modifying plasmodesmata and the GFP-tagged form has been used to investigate the mechanisms involved ***(1)***. GFP has also been fused to other viral proteins, e.g., a major tegument (VP22) gene of *Human Herpes Virus 1*. VP22 is an important component of the virus particle, having a number of functions. For example, the protein reorganizes and stabilizes the microtubule network. Construction of a GFP fusion protein has permitted the study of intracellular trafficking of VP22 ***(2)***. In other instances, the GFP gene has been inserted into the genome as an additional transcription unit, as is the case with *Rinderpest Virus* and *Simian Virus 5* ***(3,4)***.

Measles virus (*MV*) is a member of the *Paramyxoviridae*. These viruses have negative-stranded RNA genomes and express their genes from a single promoter at the 3' end of the genome. Upon infection, a gradient of viral messenger RNA transcripts is generated and this permits a degree of regulation in the expression of viral proteins. For example, the virus polymerase is required only in catalytic amounts and the gene is present at the 5' end of the genome, from where the least number of transcripts are derived (**Fig. 1A**). A reverse genetics system, comprising a full-length infectious clone and accompanying helper plasmids, is available for *MV* ***(5)***. The gene for the human codon-optimized, red-shifted variant of GFP, enhanced GFP (EGFP), has been inserted into the full-length clone of *MV* within an additional transcription unit (**Fig. 1B**). This insertion was made at the 3' end of the *MV* genome, prior to the *nucleocapsid* (*N*) gene and a recombinant virus, MVeGFP, has been recovered ***(6)***. The rationale behind this approach is that it should maximize the number

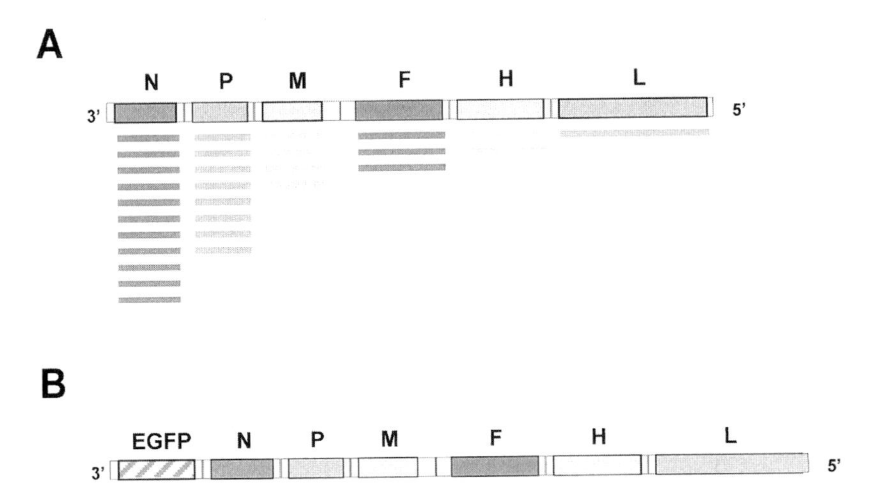

Fig. 1. **(A)** Schematic representation of the measles virus genome, showing the virus genes (N, nucleocapsid; P, phosphoprotein; M, matrix; F, fusion; H, hemagglutinin; L, polymerase). Capped, polyadenylated messages are transcribed from the 3' end of the genome in decreasing amounts, and are illustrated beneath the genome. **(B)** Schematic representation of the genomic organization of the recombinant virus, MVeGFP. The gene encoding EGFP is inserted at the 3' end of the virus genome. (Also on CD-ROM.)

of EGFP transcripts generated during infection which should increase the amount of EGFP generated in infected cells, thereby increasing sensitivity of virus detection. This is desirable in cell-to-cell spread studies. Additionally, presence of the EGFP gene at this position in the genome will not perturb the ratio of transcripts generated from the virus genes. MV infects and spreads from cell to cell by fusing to the plasma membranes. One potential disadvantage of using MVeGFP to study virus spread is that green fluorescence of a recently fused cell may not necessarily equate with presence of the virus genome. Rather the spread of EGFP fluorescence simply demonstrates that a cytoplasmic bridge exists between the cells. Therefore, certain questions cannot be addressed, such as, at which precise moment, and by what specific mechanism the virus genome enters a cell in these time-lapse experiments.

The recombinant virus has been used to demonstrate that living MV-infected cells can be detected during early stages of infection in the absence of CPE *(7)*. Interestingly, we have shown that some fluorescent cells are negative in immunocytochemical techniques for the most abundant viral antigen confirming that they are in the very early stages of infection. MVeGFP has also been used to examine cell-to-cell spread, both in vitro *(7)* and, more recently, in vivo *(8)*.

2. Materials

2.1. Equipment and Software

1. Routine tissue-culture equipment, including CO_2 incubators, laminar-flow hoods, and tissue-culture-grade plastics.
2. Leica DMIRB/E inverted research microscope, equipped with Plan (PL) Fluotar, X10/0.3 lens, fluorescence filter cube, transmitted light source, ultraviolet xenon lamp, and fitted with three-plate mechanical stage.
3. Leica heated temperature stage, maintained at 37°C, using a Linkam CO102 temperature controller.
4. Leica TCS/NT confocal scanning laser microscope, equipped with a krypton-argon laser as the source for the ion beam. EGFP is visualized by virtue of its fluorescence by excitation at 488 nm, with a 506–538-nm band-pass emission filter. TCS/NT software version 1.6.5.8.7.
5. Personal AVI Editor (FlickerFree, Denmark).

2.2. Virus, Cells, and Cell Culture

1. Recombinant MVeGFP was recovered from a full-length, modified clone of MV using a cell line (293-3-46) that expresses the nucleocapsid and phospho-proteins of MV as well as T7 RNA polymerase ***(5)***.
2. African green monkey kidney cells (Vero) were obtained from the American Type Culture Collection.
3. Human astrocytoma cells (GCCM), established from an anaplastic astrocytoma of normal adult brain, were obtained from the European Collection of Cell Cultures.
4. Dulbecco's modified Eagle's medium (DMEM) (Gibco).
5. Trypsin solution (Sigma).
6. Versene solution (Gibco).
7. Glucose British Drug House.
8. Newborn calf serum (NCS) (Gibco).
9. Fetal calf serum (FCS) (Gibco).

3. Methods

3.1. Cell Passage

3.1.1. Vero Cells

Vero cells were used for the production and titration of MVeGFP stock virus. Cells were grown in 75-cm^2 tissue culture flasks and cells were passaged when confluency was attained.

1. Rinse cell monolayer twice using preheated trypsin (5 mL).
2. Add trypsin (3 mL) and incubate cells at 37°C on a platform shaker until cells begin to detach from the plastic.
3. Add DMEM supplemented with 8% NCS (3 mL) and carefully pipet to a single-cell suspension.

4. Seed 1.5×10^7 cells into DMEM supplemented with 8% NCS (20 mL) and add the cell suspension to either a 75-cm^2 tissue culture flask, for routine cell passage, or to a 96-well tray (200 µL/well) for titration of virus.
5. Adjust atmosphere to 5% (v/v) CO_2, and incubate cells at 37°C until confluency is attained.

3.1.2. Astrocytoma Cells

Astrocytoma (GCCM) cells were used to study cell-to-cell spread of MVeGFP using real-time confocal microscopy. Subconfluent monolayers were produced to mimic the astrocyte network in the central nervous system and to facilitate the examination of the process-mediated cell-to-cell spread of MV.

1. Rinse cell monolayer using versene solution (5 mL) preheated to 37°C.
2. Rinse cell monolayer twice using trypsin solution (5 mL) preheated to 37°C.
3. Add trypsin (3 mL) and incubate cells for 5 min at 37°C on a platform shaker.
4. Add DMEM supplemented with 10% FCS (3 mL) to the trypsinized cells and carefully pipet to a single-cell suspension.
5. Seed 2×10^6 cells into a 25-cm^2 tissue culture flask containing DMEM supplemented with 10% FCS (10 mL).
6. Adjust atmosphere to 5% (v/v) CO_2, and incubate cells at 37°C until 60% confluency is attained.

3.2. Viral Infections

3.2.1. Titration of Virus

The titer of MVeGFP was determined using the 50% end point dilution assay *(9)*, and is expressed in 50% tissue culture infective doses ($TCID_{50}$).

1. Make logarithmic dilutions (10^{-1}–10^{-8}) of the stock MVeGFP in DMEM containing 2% NCS.
2. Carefully pipet off the growth medium from a 96-well plate containing confluent Vero cells.
3. Beginning with the highest dilution (10^{-8}), pipet 200 µL into each of the four wells in row A of the 96-well tray.
4. Continue until all the dilutions are plated in quadruplicate in separate rows (B–H).
5. Adjust the atmosphere to 5% (v/v) CO_2 and incubate cells at 37°C for 7 d.
6. Determine which wells contain syncytia using a low-power objective scoring simply for either presence or absence. Typical MV syncytia are shown in **Fig. 2**.
7. Calculate the viral titer using the method of Reed and Muench *(9)*.

3.2.2. MVeGFP Infection of Astrocytoma Cells and Location of Infectious Centers

1. Culture GCCM cells to 60% confluency in a 25-cm^2 tissue culture flask (*see* **Note 1**).
2. Rinse cell monolayer twice in DMEM supplemented with 2% NCS (10 mL).

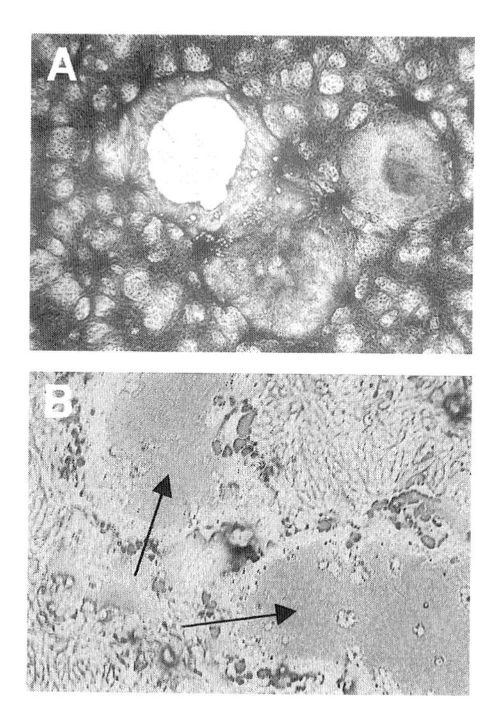

Fig. 2. **(A)** MV syncytia in infected Vero cells. **(B)** Higher magnification of two syncytia (*arrows*), demonstrating virus-mediated cell fusion. (Also on CD-ROM.)

3. Based on the $TCID_{50}$ of the virus stock, make a dilution of MVeGFP in DMEM, containing 2% NCS, (1 mL) to achieve a multiplicity of infection of 0.01.
4. Incubate the flask on a platform shaker for 60 min at 37°C. During this time, the virus infects the cells.
5. Remove the virus inoculum and add DMEM supplemented with 2% NCS (9 mL).
6. Incubate infected cells for 50 h at 37°C.
7. Observe the monolayers regularly for the appearance of single EGFP-positive cells by inverted fluorescence microscopy.
8. Mark the position of a number of discrete foci of infection on the lower surface of the tissue culture flask (*see* **Notes 2** and **3**).

3.3. Vital Fluorescent Microscopy

Because of the high levels of EGFP expression obtained upon infection with the recombinant MV, green fluorescence is readily detected in individual cells. This makes it possible to visualize the cell-to-cell spread of the virus indirectly by observing the accumulation of EGFP in real-time using confocal scanning laser microscopy.

3.3.1. Initialization of Confocal Microscope

1. Switch on the krypton-argon laser and UV lamp for 15 min, prior to use.
2. Launch the TCS/NT software, selecting the fluorescein-isothiocyanate filter channel, X10 lens, and 1024 × 1024 resolution mode.

3.3.2. Repeated Observation of Multiple Foci Within Infected Monolayers

1. Set thermostat on heated stage to 37°C, switch on and allow temperature to stabilize for at least 5 min.
2. Place the infected 25-cm^2 tissue culture flask on the heated stage and mark its position on the mechanical stage precisely.
3. Orientate a first focus of infection in the center of the field of view.
4. Defocus the image slightly and set this lower Z position as the lower threshold on the microscope.
5. Refocus the image and set the upper Z position as the upper threshold on the microscope. These two steps are necessary to allow the confocal electronics to precisely move the objective nosepiece of the Leica DMIRB/E microscope in Z-wide mode. This permits the collection of a Z series through infected fluorescent cells.
6. Switch the inverted microscope to the data collection mode and commence confocal scanning of the MVeGFP-infected cells.
7. Adjust the potential of the photomultiplier tube (PMT value) to either increase or decrease the intensity of the green fluorescence observed.
8. Choose the limits of the optical section by setting the beginning and ending of the Z-series. Typical Z-series vary between 10 and 30 μm for cell monolayers.
9. Collect optical sections every 2–5 μm, within the limits of the selected Z-series. Each section should be averaged at least 4× to reduce non-specific fluorescence.
10. View sections as composite images in the projected focus mode and save selected images as tagged-image format files.
11. Continue observations of other foci of infection within the monolayer.
12. Return flask to 37°C and incubate until the next observation time-point (*see* **Notes 2–4**).

Two representative time-courses of infection of GCCM cells are shown in **Fig. 3**. In the first, a focus of infection is shown at 66 h post-infection (**Fig. 3A**). After 8 h, rapid process-mediated cell-to-cell spread was observed and many more surrounding cells were infected. Interconnecting processes were clearly visible (arrowed). The second example was visualized over an extended period of 40 h (**Fig. 3B**). The initial focus of infection is shown at 66 hpi. Four hours later, at 70 hpi, a number of additional cells were infected (*arrowed*) and over the ensuing 20 h many more cells became infected. By 90 hpi, the resulting syncytium had lysed and many infected cells had detached from the plastic

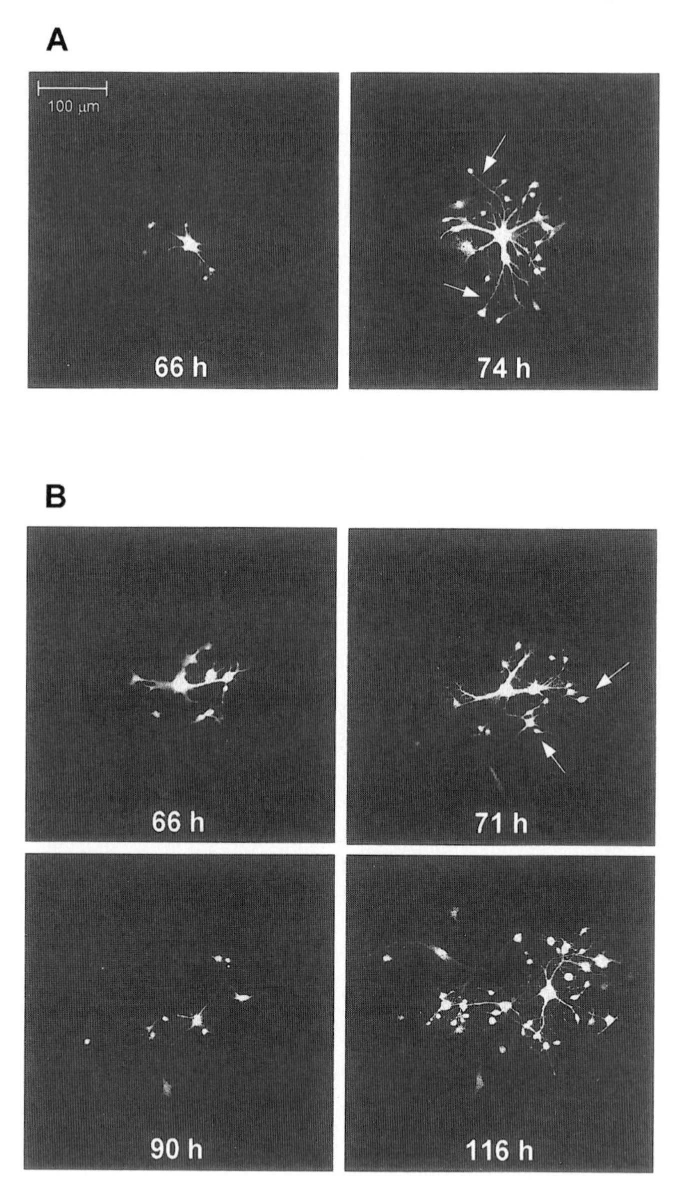

Fig. 3. Variation in the process-mediated cell-to-cell spread of MVeGFP. Two separate foci of infection in the same, 60%-confluent, GCCM cell monolayer, are shown (**A** and **B**). Infected cells were examined at indicated times (h) post-infection. (Also on CD-ROM.)

support, as can be seen by the drop in fluorescence. However, virus infection was not observed to terminate and, by 116 hpi, a large number of recently infected cells were observed. Additional time-courses, showing the variation in cell-to-cell spread, have been published *(7)*.

3.3.3. Time-Lapse Microscopy of Single Infectious Centers

Based on the above observations, this technique has been extended to permit the visualization of cell-to-cell spread at shorter intervals by making use of the time-lapse facility of the confocal microscope. This permits the generation of real-time movies of virus cell-to-cell spread (*see* **Note 5**).

1. Initialize the microscope and heated stage as before (*see* **Subheading 3.3.1.**).
2. Place the infected 25-cm^2 tissue culture flask on the heated stage and select a single focus of infection.
3. Set the confocal microscope to Z-wide mode and choose the limits of the optical sections as before (*see* **Subheading 3.3.2.**).
4. Record a time-lapse program defining the number of optical sections (usually 4–6), the number of times each optical section is averaged (usually 4) and the delay interval between observations (between 15 and 60 min).
5. Start time-lapse program that leads to the automated collection of optical sections at the preset time intervals.
6. Occasionally check images to ensure that the microscope has not drifted out of focus (*see* **Note 6**).
7. At the end of the time-lapse period, save the resulting images as a complete data set. This is comprised of a set of individual images representing all the optical sections collected throughout the time-course (*see* **Note 7**).

3.4. Data Analysis and Production of Digital Videos

The data files, which are generated during time-lapse observations, are very large, on the order of 150 MB. It is necessary to produce individual composite images for each of the many separate time-points, before cell-to-cell spread of the virus can be examined in any detail (*see* **Note 8**).

1. Select all the optical sections obtained at the first time-point, view in projected mode and save the image as a composite tagged-image file.
2. Repeat this process for all the images collected at each of the time-points. The resulting number of composite tagged-image files will depend on the time interval chosen during the process of data collection and the overall length of period of observation.
3. Import all the images, in order, into a new project in the Personal AVI Editor Program.
4. Copy the individual images, in order, into the video construction window.
5. Generate an uncompressed audio–video interleaved (AVI) file, setting the size of the image to 1200 × 1200 pixels, 24-bit color, and the frame rate to 5 frames per second.

6. The resulting AVI file can be viewed on the accompanying CD-ROM using any standard media player such as the Microsoft Media Player (Version 6.4.07.1112; *see* **Note 9**).

A representative video of MVeGFP cell-to-cell spread in astrocytoma cells is included in the compact disk that accompanies this volume. This shows the real-time cell-to-cell spread of MV, from an initial focus of six infected cells selected at 24 hpi. Images were collected every 15 min over 24 h. The virus was observed to rapidly infect neighboring cells by fusion of cells. Cell-to-cell spread was also observed to occur via the interconnecting astrocytic processes, confirming the data presented in **Fig. 2**. At the end of the period of observation, approx 60 cells were virus infected.

4. Notes

1. Because uniform subconfluent monolayers are rarely generated upon cell passage, it is important to examine the numbers and density of uninfected cells surrounding the foci of infection that have been chosen for repeated observation. This should be verified by light microscopy. Cells within the field of view should be close to 60% confluent and the single EGFP-positive cell should clearly be in intimate contact with neighboring cells.
2. In experiments in which multiple observations are made within the same monolayer, it is vital to precisely reposition the flask on the heated stage. This requirement is greatly facilitated by viewing the image that was collected at the previous time-point.
3. In the latter stages of time-courses, when cells exhibit obvious CPE and begin to detach from the tissue culture flask, it can be difficult to ascertain exactly which cells were previously observed. This is especially the case when large numbers of cells detach from the flask. If this occurs, the time-course should be discontinued.
4. It is vital to have a 37°C incubator close to the confocal microscope when multiple observations are made in the same monolayer over four hourly time-points. This limits the time taken to transport flasks from the stage to the incubator and lessens the likelihood of mechanically detaching infected cells from the tissue culture flask. Nevertheless, care should always be taken when moving the infected monolayers, even over short distances.
5. Before undertaking a complete time-course, it is essential to determine the approximate rate at which MVeGFP spreads from cell to cell. This is important, because it will determine the overall length of the time-course and the interval between observations. This is particularly relevant when observations are made in different cell types.
6. It is essential to have a microscope that does not drift from the selected focus: This is generally the case with the Leica DMIRB/E model. Nevertheless, drift can occur in longer time-lapse studies, and it is important to check the confocal microscope regularly.

7. A significant amount of condensation can develop within the tissue culture flask, in time-courses that last over 12 h. This has no noticeable effect on resolution of the resulting images. If this is perceived to be a problem, it can be easily resolved by the addition of additional medium to the flask at the outset of the time-course.
8. It is important to ascertain that sufficient disk storage space is available for data collection prior to the commencement of a time-lapse study.
9. It is necessary to use a high-specification personal computer, in order to efficiently edit and generate the AVI files. The basic minimum requirement is a system with a 500 MHz processor, with 128 MB of RAM and a 10.2 GB hard disk.

Acknowledgments

The expert assistance of Stephen McQuaid in many areas of confocal microscopy is gratefully appreciated. We thank Martin Ludlow for the images that comprise the digital video file, and Martin Billeter for valuable advice and continual support. These studies were supported by the Biotechnology and Biological Sciences Research Council Bioimaging Initiative (grant 81/BI11169) and the Wellcome Trust (grant 047245).

References

1. Reichel, C. and Beachy, R. N. (1998) Tobacco mosaic virus infection induces severe morphological changes of the endoplasmic reticulum. *Proc. Natl. Acad. Sci. USA* **95,** 11,169–11,174.
2. Elliott, G. and O'Hare, P. (1999) Live-cell analysis of a green fluorescent protein-tagged herpes simplex virus infection. *J. Virol.* **73,** 4110–4119.
3. He, B., Paterson, R. G., Ward, C. D., and Lamb, R. A. (1997) Recovery of infectious SV5 from cloned DNA and expression of a foreign gene. *Virology* **237,** 249–260.
4. Walsh, E. P., Baron, M. D., Anderson, J., and Barrett, T. (2000) Development of a genetically marked recombinant rinderpest vaccine expressing green fluorescent protein. *J. Gen. Virol.* **81,** 709–718.
5. Radecke, F., Spielhofer, P., Schneider, H., Kaelin, K., Huber, M., Dotsch, C., et al. (1995) Rescue of measles viruses from cloned DNA. *EMBO J.* **14,** 5773–5784.
6. Hangartner, L. (1997) M.Sc. Thesis, University of Zurich, Switzerland.
7. Duprex, W. P., McQuaid, S., Hangartner, L., Billeter, M. A., and Rima, B. K. (1999) Observation of measles virus cell-to-cell spread in astrocytoma cells by using a green fluorescent protein-expressing recombinant virus. *J. Virol.* **73,** 9568–9575.
8. Duprex, W. P., McQuaid, S., Roscic-Mrkic, B., Cattaneo, R., McCallister, C., and Rima, B. K. (2000) *In vitro* and *in vivo* infection of neural cells by a recombinant measles virus expressing enhanced green-fluorescent protein. *J. Virol.* **74,** 7972–7979.
9. Reed, L. J. and Muench, H. (1938) A simple method for estimating fifty percent endpoints. *Amer. J. Hygiene* **27,** 493–497.

25

Tracking and Selection of Retrovirally Transduced Murine Bone Marrow Cells Using Green Fluorescent Protein

Jessamyn Bagley and John Iacomini

1. Introduction

Murine replication-defective retroviral vectors are among the most well-studied, commonly used gene transfer vehicles *(1,2)*. The authors have used retroviruses as gene delivery tools, to genetically modify bone marrow (BM) hematopoietic progenitors, in order to induce immunological tolerance to various antigens following BM transplantation *(3–5)*. However, one of the major problems with retroviral infection of BM is poor efficiency, resulting in the low ratio of transduced to nontransduced bone marrow cells (BMCs) used for transplantation. This limitation of retroviral gene therapy may be overcome by including, within retroviral constructs, marker genes that would allow one to obtain relatively pure populations of transduced cells by cell sorting, based on marker-gene expression prior to BM transplantation. An additional benefit is that expression of a marker gene allows one to easily monitor virus transduction in vitro, and to track the fate of transduced BMCs and their progeny in vivo.

Green fluorescent protein (GFP) is a naturally fluorescent protein, derived from the bioluminescent jellyfish *Aequorea victoria*, which emits light when excited by ultraviolet light without the need for additional substrates or cofactors *(6)*. Many mutants of GFP are now commercially available and can be incorporated into retroviral constructs to facilitate selection and monitoring of transduced cells using conventional flow cytometers without the need for cell-surface staining or addition of substrate. Overcoming the need for cell-surface staining with antibodies to select for infected cells prior to transplantation is

From: *Methods in Molecular Biology, vol. 183: Green Fluorescent Protein: Applications and Protocols*
Edited by: B. W. Hicks © Humana Press Inc., Totowa, NJ

critical because the presence of antibodies bound to the surface of cells can lead to their elimination by macrophages, cells that can mediate antibody-dependent cell-mediated cytotoxicity, and the complement system. In addition, many genes that one may wish to introduce into cells may not be expressed on the cell surface, or specific antibodies may not be available. In such cases, GFP can be used as a surrogate marker of gene expression.

The authors have previously shown that GFP can be used to directly monitor transduction of murine BM, and to select for transduced donor BMCs prior to BM transplantation *(7)*. GFP can be expressed in multiple BM-derived hematopoietic cell lineages as well as Sca-1$^+$, lineage-marker-negative, early hematopoietic progenitors. In addition, GFP can be used to track the fate of transduced BM-derived cells and their progeny in vivo, following transplantation (**Fig. 1**). Using GFP expression as a marker, it is possible to enrich or purify, by cell sorting, populations of transduced cells prior to transplantation. Mice reconstituted with enriched populations of GFP-expressing cells show an increase in the percentage of cells that express GFP long-term in the periphery, compared to mice reconstituted with unenriched populations of transduced BM (**Fig. 2**). These studies demonstrate that GFP can be used as a marker to select for transduced cells in vitro, track the fate of transduced cells long-term in vivo, and enrich for transduced cells prior to transplantation. Here, the authors provide a detailed description of the methodology used to perform such studies in mice.

2. Materials

All reagents are analytical, tissue-culture, or molecular-biology grade. All reagents, buffers, and media are either purchased as sterile for tissue culture, or filtered through 0.2-μm disposable sterile filters.

2.1. Development of Retroviral Producer Cell Lines

1. 100X 20-mm tissue culture dishes (Becton Dickinson, Franklin Lakes, NJ).
2. AM12 cells: Amphotropic retroviral packaging cell line is maintained as described in **ref.** *8*.
3. 15P media: Dulbecco's modified Eagle's medium (Mediatech, Herndon, VA) containing 15% fetal calf serum, 1 m*M* sodium pyruvate, 0.1 m*M* MEM nonessential amino acids, 2 m*M* L-glutamine, 200 U/mL penicillin-streptomycin. All supplements purchased from Gibco-BRL, Gaithersburg, MD.
4. 2.5 *M* $CaCl_2$: Aliquots can be made in advance and stored at –20°C.
5. 2X BBS pH 6.95: 50 m*M* N,N-bis (2-hydroxyethyl)-2-aminoethanesulfonic acid, 280 m*M* NaCl, 1.5 m*M* Na_2HPO_4.
6. Hanks's balanced salt solution (HBSS) without Ca^{2+}, Mg^{2+} or phenol red (Gibco-BRL).
7. Trypsin: 0.5% trypsin, 0.53 m*M* EDTA in HBSS (Gibco-BRL).

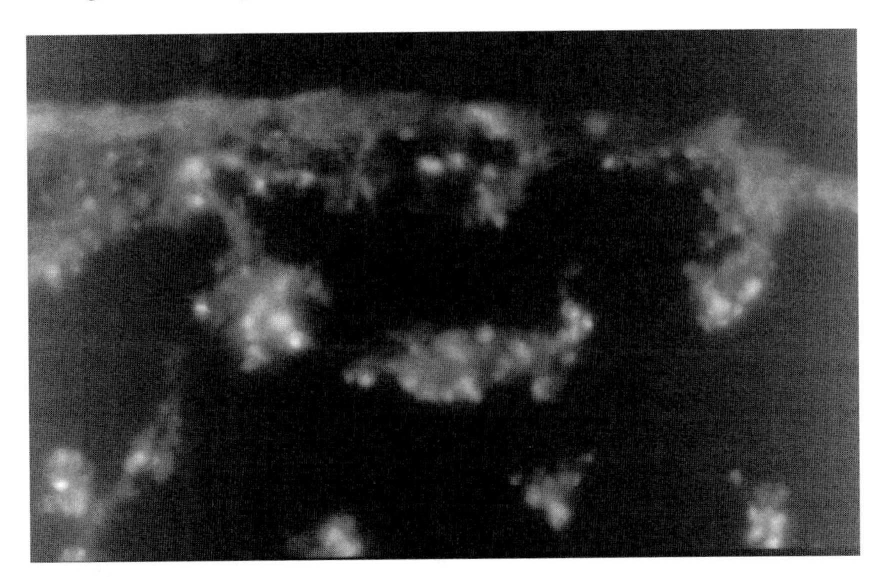

Fig. 1. Detecting progeny of transduced BMCs in vivo, using GFP. Shown is expression of GFP in splenic colonies formed 12 d after BM transplantation. Spleens were harvested on d 12 and frozen for tissue sectioning. Shown is a direct visualization of GFP-positive colonies in tissue sections using fluorescent microscopy. Data kindly provided by Jennifer L. Bracy. (For optimal, color representation please see accompanying CD-ROM.)

8. G418 (Sigma, St. Louis, MO): 100 mg/mL active concentration stock solution stored at –20°C.
9. Dimethyl sulfoxide (DMSO) (Sigma).
10. NIH3T3 cells (American Type Culture Collection, Manassas, VA): fibroblastoid cell line derived from NIH Swiss mice.
11. Polybrene (Sigma): 16 mg/mL stock solution in HBSS can be stored 1 mo at 4°C.
12. Fluorescence-activated cell sorting (FACS) staining buffer: HBSS containing 25 m*M* HEPES (Mediatech), 1% heat-inactivated normal rabbit serum (Gibco-BRL), and 0.1% sodium azide (Sigma). Store at 4°C.

2.2. Collection of Murine BM Cells

1. 5-flurouracil (5-FU) (Pharmacia, Kalamazoo, MI): Sterile 50 mg/mL stock stored at room temperature.
2. BM harvest media: 15P media supplemented with 25 m*M* HEPES and 10 μg/mL gentamycin.
3. Spleen mesh (Sefar, Briarcliff Manor, NY): mesh count, 102 threads/in, cut into 5 × 5 cm squares and autoclaved.
4. ACK red blood cell lysis buffer (BioWhittaker, Walkersville, MD).
5. 0.4% Trypan blue (Gibco-BRL).

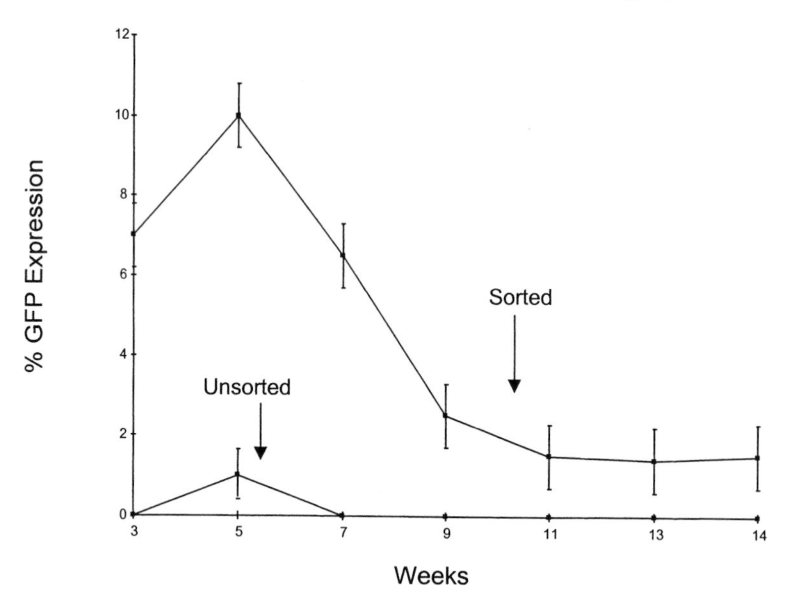

Fig. 2. Percentages of GFP+ cells in peripheral blood of mice reconstituted with unenriched or enriched populations of BMCs transduced with retrovirus containing the gene encoding GFP. 3 wk postreconstitution, and every 2 wk thereafter, peripheral blood samples were obtained by tail bleeds and GFP expression was analyzed by flow cytometry. Reprinted with permission from **ref.** *7*. (Also on CD-ROM.)

2.3. Retroviral Transduction of BM Cells

Transduction media: 15P supplemented with 100 ng/mL human interleukin-6 (IL-6), 100 ng/mL recombinant mouse stem cell factor (SCF), 50 ng/mL recombinant mouse thrombopoietin (TPO), (50 ng/mL) recombinant mouse Flt-3 ligand, and 8 μg/mL polybrene. All cytokines were purchased from R&D Systems, Minneapolis, MN.

2.4. Monitoring Long-term GFP Expression In Vivo

1. 0.5 *M* EDTA, pH 8.0.
2. Lineage-specific antibodies: anti-B220 (RA3-6B2) ***(9)***, biotinylated CD11c (HL3) ***(10)***, anti-Sca-1 (CT-6A.2) ***(11)***, CD117/c-Kit (2B8) and Ter119 ***(12)***, anti-CD11b/Mac-1 (M1/70) ***(13,14)***, leukocyte-specific CD5/Ly1 (53.7.3), granulocyte-specific Ly-6G/Gr-1 (8C5) ***(15)***.

3. Methods

3.1. Generating Retroviral Producer Cell Lines

To generate virus stocks, the retroviral construct carrying the gene encoding GFP, is transfected into packaging cell lines. Retroviral packaging cell lines

(16,17), such as AM12 carry the viral genes encoding structural products (*gag*, *pol* and *env*) provided in *trans*. Separating the structural genes and retroviral constructs makes it possible to package retroviral RNA into virions without generating replication-competent viruses. In addition to GFP, many retroviral constructs contain drug selection markers, such as the neomycin resistance gene, which can be used to select for transfected cells.

3.1.1. Transfecting Packaging Cells

1. 24 h before transfection, seed 100-mm tissue culture dishes with 5×10^6 AM12 cells in 10 mL 15P media. Grow the cells in an incubator at 37°C with 5% CO_2 to approx 70% confluence (*see* **Note 1**).
2. Dilute 10 µg supercoiled retroviral plasmid DNA to 450 µL with sterile distilled and deionized water (*see* **Note 2**).
3. Add 50 µL 2.5 *M* $CaCl_2$ and mix well by pipeting gently up and down.
4. Slowly add 500 µL 2X BBS to the DNA mixture. Mix the tube by flicking several times. Incubate 20 min at room temperature, to allow the DNA precipitate to form (*see* **Note 3**).
5. While the precipitate is forming, aspirate the media from the AM12 cells. Gently rinse plates with 10 mL HBSS, then add 8 mL fresh 15P media.
6. Add the DNA precipitate slowly to the AM12 cells drop by drop, then gently rock the plate to mix (*see* **Note 4**).
7. 12–16 h after transfection, aspirate media, and rinse the plates 3× with 10 mL HBSS. Add 8 mL fresh 15P media and continue to grow the cells for an additional 24 h.

3.1.2. Selecting Viral Producer Clones

1. To harvest cells, aspirate media from transfected cells and rinse the plates once with HBSS. Add 2 mL trypsin to each plate and return the cells to the incubator for 1 min. Add 3 mL 15P media to each plate and pipet media up and down vigorously to facilitate detachment of the cells from the plate.
2. Seed three 100-mm tissue culture dishes with 1 mL of the cell suspension. Add 7 mL 15P media to each plate, and continue to culture overnight.
3. To select for neomycin-resistant transfectants, on the following morning aspirate the media and replace it with 8 mL 15P media containing 0.8 mg/mL G418 (*see* **Note 5**).
4. Continue to grow cells in selection media for 10 d, at which point individual colonies of G418-resistant cells will appear in the plates.
5. When the colonies reach approx 50–100 cells, they can be visualized and picked, using an inverted phase-contrast microscope. To pick colonies, place the microscope in a tissue culture hood. Using the microscope, identify well-isolated colonies. Using a micropipet fitted with a sterile plugged pipet tip, gently scrape each colony from the plate and aspirate into the tip. Transfer single colonies into individual wells of a 24-well tissue culture plate containing 1.5 mL 15P media containing 0.8 mg/mL G418 per well (*see* **Note 6**).

6. When the clones reach 95% confluence, rinse each well with HBSS and replace the media with 15P media and culture an additional 24 h.
7. Transfer virus supernatants to a fresh 24-well plate and store at –80°C.
8. To freeze clones while the viral titer of each supernatant is being determined, add 1.5 mL 15P media containing 10% dimethyl sulfoxide to each well and store at –80°C (*see* **Note 7**).

3.1.3. Determination of Viral Titer

After individual virus-producer clones are established, it is necessary to determine the viral titer of supernatants harvested from these clones. Titer is expressed as the number of infectious virus particles per milliliter of cell culture supernatant. After infection with viruses carrying the gene encoding for GFP, successfully transfected cells can be detected using flow cytometry. The goal is to select the producer clones that make the greatest amount of virus. These clones can then be expanded to produce a large volume of viral supernatant for use in subsequent experiments. The titer of viral supernatants produced using retroviral vectors that also carry drug selectable marker genes can be determined by the ability to confer drug resistance on infected cells.

1. For each supernatant to be titered, seed two 6-well tissue culture dishes (35-mm wells) with 10^5 NIH3T3 cells in 2 mL 15P media. One plate will be used to titer based on GFP expression, the other based on neomycin resistance.
2. The next day, thaw the viral supernatants (*see* **Subheading 3.1.2., step 7**), and prepare five dilutions (4 mL each) of viral supernatants in 15P media containing 4 µg/mL polybrene (*see* **Note 8**).
3. Aspirate media from each well and replace it with 2 mL diluted viral supernatant. To one well, add only 15P media containing 4 µg/mL polybrene (mock-infected or negative control).
4. After 12 h, replace media in one plate with fresh 15P media. Culture the cells for an additional 48 h. Replace media in the second plate with 15P media containing 1.2 mg/mL G418. Culture the cells for an additional 10 d.
5. After 48 h, aspirate media from all wells in the plate for GFP analysis, and rinse each well with HBSS. Add 0.5 mL trypsin to each well and incubate at 37°C for 1 min. Add 2 mL 15P to each well and pipet cells vigorously to facilitate detachment from the plate. Transfer cells from each well to individual 15-mL centrifuge tubes.
6. Count the cells using a hemocytometer.
7. To wash cells, pellet by centrifugation at 500*g* for 5 min, and resuspend in FACS staining buffer. Wash twice more and resuspend 10^6 cells in 1 mL FACS staining buffer. Analyze the percentage of cells expressing GFP by flow cytometry. Titer can be calculated as: ([percent GFP positive cells] × [cell number])/(actual volume of viral supernatant added). This value can then be normalized to per mL (*see* **Note 9**).

8. After 10 d, aspirate selection media from the remaining plate and wash wells with HBSS. Count the number of G418 resistant colonies per plate using an inverted microscope. Titer can be calculated as: (number of G418-resistant colonies)/(actual volume of viral supernatant added). This value can then be normalized to virons per mL.

3.2. Harvesting BM Cells for Transduction

Typically, 3–5 million BMCs can be harvested per 5-FU-treated mouse. 5-FU treatment kills dividing cells, thereby enriching for relatively quiescent hematopoietic progenitors ***(18)***.

1. Using a 1-mL syringe fitted with a 25-gage needle, inject 6–8-wk-old donor mice via the tail vein, with 150 mg/kg 5-FU.
2. 7 d later, sacrifice the mice. Using sterile forceps and tweezers, dissect out the hind limbs by cutting below the hip and above the ankle. Remove the excess muscle tissue and the knee. Place the cleaned bones in tissue culture dishes containing BM harvest media.
3. Using a 35-mL syringe fitted with 25-gage needle, flush the bones with copious amounts of harvest media, into 15-mL centrifuge tubes. Pipet cells vigorously to dissociate the BM.
4. Filter the BM through sterile nylon monofilament mesh and pool all marrow into 50-mL conical tubes. Pellet cells by centrifugation at 500*g* for 5 min at room temperature.
5. Resuspend cells in 5 mL ACK red cell lysis buffer and mix vigorously by pipeting. Incubate 3 min at room temperature (*see* **Note 10**).
6. Add 45 mL harvest media, mix, then pellet cells by centrifugation at 500*g* for 5 min at room temperature to wash cells. Wash cells twice more in 50 mL harvest media, and resuspend in a final volume of 25 mL harvest medium.
7. Count the cells.

3.3. Retroviral Transduction of BMCs

In order to be permissive for infection with retroviruses, target cells must be in cell cycle. To force BMCs into cell cycle, cytokines known to promote cycling of relatively early hematopoietic progenitors are included in the BM transductions. Including IL-6, SCF, TPO, and recombinant mouse Flt-3 ligand in the transduction media allows for improved transduction of early hematopoietic progenitors ***(19)***. Co-culture of 5-FU-treated BM on retroviral producer clones helps maintain early progenitors while providing a source of virus. Transduction can also be performed with cell-free viral supernatants, but both cell viability and the percentage of transduced cells are reduced.

1. Seed 5×10^6 virus producer cells per 100-mm tissue culture dish in 15P media. Allow cells to grow to 80% confluence.

2. Aspirate the media from each plate and add 10^7 BMCs in 10 mL transduction media containing 100 ng/mL human IL-6, 100 ng/mL recombinant mouse SCF, 50 ng/mL recombinant mouse TPO, 50 ng/mL recombinant mouse Flt-3 ligand, and 8 µg/mL polybrene. Grow the cells at 37°C with 5% CO_2 for 96 h (*see* **Note 11**).
3. Change transduction media after 24 h and again after 72 h. To change media, transfer 7 mL culture supernatant from each plate to a sterile 50-mL conical tube. Do not disturb the majority of cells while removing media. Centrifuge at 500*g* for 5 min to recover any cells that may have been removed. Aspirate the supernatant from the tube and resuspend cells in fresh transduction media. Return the appropriate volume to each tissue culture dish (*see* **Note 12**).
4. After 96 h, harvest the cells from the plate by vigorous pipeting.
5. Transfer cells to fresh 100-mm tissue culture dishes, and incubate for 2 h at 37°C with 5% CO_2, to allow contaminating producer cells to adhere.
6. Harvest transduced nonadherent cells, by gentle pipeting.
7. Count the cells.

3.4. Direct Determination of Transduction Efficiencies by Flow Cytometry

1. Wash 5×10^5 transduced BMCs twice in FACS staining buffer.
2. Determine the percentage of GFP-positive cells, by flow cytometry. The percentage of GFP-positive cells corresponds to transduction efficiency.
3. Expression of GFP in various BMC lineages can be determined by staining cells with monoclonal antibodies (*see* **Subheading 2.4.**) specific for known lineage markers as described in **ref. 7**. All lineage marker specific antibodies must be revealed with fluorochromes, such as phycoerythrin, which will not interfere with the GFP signal. GFP lineage marker double-positive cells can be detected by flow cytometry.

3.5. Selection and Enrichment of Transduced BM by Flow Cytometry

Since most labs do not own their own cell sorter, most academic researchers must rely on core facilities at their institution to sort GFP-positive cells. The ability to enrich GFP-positive cells prior to BM transplantation is highly dependent on the skill of the cell sorter operator and the technology of the instrument in use. Prior to sorting GFP-positive cells, it is very important to discuss with the sorter operator what percentage of GFP-positive cells there will be after transduction, how bright the fluorescent signal will be, what level of purity one wishes to attain, and what types of samples will be needed, e.g., a negative control, to set up the cell sorter. In general, high-purity sorting is achieved at the cost of cell yield.

1. Resuspend transduced BMCs at 10^7/mL in HBSS containing 5% fetal calf serum. Including fetal calf serum to maintain high cell viability, is extremely important.

Provide sterile collection tubes containing 15P medium supplemented with 25 m*M* HEPES and 10 μg/mL gentamycin. The operator of the cell sorter will specify which types of collection tubes are needed, and how much media can be in each tube. Providing a mock-transduced negative control is advisable, to allow the instrument to be set up properly (*see* **Note 13**).
2. Sort GFP^+ cells (*see* **Note 14**).
3. Following sorting, collect cells by centrifugation and resuspend in HBSS.
4. Count the cells.
5. Reanalyze sorted cells by flow cytometry to determine percent GFP^+ cells and fold enrichment after sorting.

3.6. Conditioning and Reconstitution of Recipient Mice

1. Precondition 6–8-wk-old mice with 10.25 gy whole-body irradiation, 12–24 h prior to reconstitution.
2. To reconstitute mice, inject a minimum of 10^6 BMCs via the tail vein in 1 mL HBSS.

3.7. Monitoring Expression of GFP In Vivo

To monitor expression of GFP, one can simply analyze the presence of GFP-positive cells in the blood by flow cytometry. This allows one to follow the fate of transduced cells and their progeny over long periods of time without killing the reconstituted mice.

3.7.1. Analysis of GFP-Expressing Blood Cells

1. Collect 100 μL blood from reconstituted and control mice in tubes containing 5 μL 0.5 *M* EDTA, pH 8.0. Approved methods for blood collection vary between institutions. It is best to check with veterinarians at the institution to find which methods are preferable.
2. Add 1 mL ACK lysis buffer and incubate 4 min. Add 3 mL FACS staining buffer and pellet cells by centrifugation and wash twice more in FACS staining buffer.
3. Determine the percentage of GFP^+ cells by flow cytometry. Set analysis gate baseline using blood cells from a control mouse.

3.7.2. Analysis of GFP Expression in Distinct Blood Lineages and Hematopoietic Tissues

1. To analyze GFP-expressing cells of various hematopoietic lineages, add a saturating concentration of directly labeled lineage specific antibodies to blood cells prepared as described in **steps 1** and **2** of **Subheading 3.7.1.** (*see* **Note 15**).
2. Incubate at 4°C in the dark for 30 min.
3. Wash cells twice in FACS staining buffer.
4. Analyze cells by flow cytometry.
5. To analyze GFP expression in lymphoid tissue, sacrifice mice and prepare single cell suspensions from the spleen, thymus, BM, lymph nodes and peritoneal lavage. Repeat **steps 1–4**.

4. Notes

1. Plating cells 24 h in advance allows them to reach log-phase growth, and increases the likelihood of DNA uptake, because best transfection efficiencies are obtained in cells growing exponentially. The optimal number of cells will vary depending how quickly the cells grow, and should be adjusted according to how well the cells grow in one's lab.
2. The volume of DNA added to the water should not be greater that 45 µL. If multiple transfections are planned do not scale up the volumes given. Instead, prepare additional tubes using the volumes indicated.
3. Do not disturb the mixture while the DNA precipitate is forming. It may not be possible to see the precipitate form in the tube. RPMI is highly charged, and should not be used for $CaPO_4$ transfection.
4. After the DNA is added, a fine precipitate will be apparent in the culture dishes. If precipitate is not visible at this point, or is too dense, with aggregated clumps, too much DNA has been added.
5. Not all mammalian cell lines are equally sensitive to G418. The minimum lethal concentration of G418 can range from 100 µg/mL to 2.0 mg/mL. The correct concentration of G418 can be selected by determining the lowest dose capable of killing untransfected cells in 10 d. If the retroviral construct does not contain a drug-selectable marker, individual viral producer cell clones can be obtained by cloning the cells by limiting dilution. Plate 1000 cells/100 mm plate. Continue to grow cells for 10 d. At this point, discrete colonies will have formed, and can be selected as described in **Subheading 3.1.2., step 5**, using an inverted fluorescent microscope to visualize GFP. Alternatively, single cell sorting of GFP-positive cells can be performed using flow cytometry.
6. In general, it is advisable to pick at least 20 colonies in order to ensure the selection of a clone with high viral production.
7. Clones frozen in this way can be stored for at least 2 mo.
8. Polybrene (hexadimethrine bromide) is a positively charged molecule that neutralizes surface charge of the target cell and virus, allowing the virus to bind more efficiently to its receptors. Viral titers are generally in the range of 10^5 infectious particles/mL. The desired percentage of infection is 5–20%, because too many infection events may result in multiple copies per cell. Thus, an appropriate dilution might be 2, 4, 8, 16, and 32 µL viral supernatant/10^5 NIH3T3 cells.
9. Although a technical description of flow cytometry is outside the scope of this review, several comprehensive sources exist. The GFP mutant, S65T, is measured using an excitation wavelength of 488 nm and is detected in the FL1 channel of most flow cytometers (as for fluorescein isothiocyanate). When determining viral titer by FACS, make use of several dilutions of virus to ensure being in the linear portion of the titration curve. Titration by flow cytometry yields slightly higher titers than evaluation by drug selection, because of nonspecific killing by selection agents.
10. Excessive lysis with ACK buffer is deleterious to BM, so do not exceed 5 min. Failure to completely wash cells with media after this procedure will result in decreased yields. Do not treat more than 10^8 cells with ACK/tube.

11. Maintaining a high BMC density during in vitro transduction is important to maintain BM viability. There should be at least 10^7 BMCs/plate. BMCs harvested from one 100-mm plate are generally sufficient to reconstitute at least 10 conditioned host mice at 10^6 cells/mouse.
12. An increase in multiplicity of infection can be achieved by the addition of viral supernatants during transduction. Collect supernatant as described in **Subheading 3.1.2.**, and dilute 1:1 in 2X transduction media before adding to BMCs. The addition of viral supernatant will result in higher transduction efficiencies.
13. HBSS containing 5% fetal calf serum must be filtered through a 0.2-mm, sterile, disposable filter to prevent clogging of the flow cytometer.
14. When setting the cell-sorting gates, be sure to exclude cells that are autofluorescent, using a negative-control sample. These cells will not reconstitute a lethally irradiated mouse, and interfere with the ability to select viable GFP^+ cells.
15. In order to correctly set up the flow cytometer, it is important to include in the experiment samples from control mice stained individually with each lineage-specific antibody. It is also important to include an unstained sample of cells from a mouse reconstituted with GFP transduced BM. These controls will allow one to set up the compensation for the instrument.

Acknowledgments

We wish to thank Jennifer L. Bracy for sharing unpublished results and the Iacomini laboratory for helpful discussions. In addition, we thank Dr. Xandra Breakefield for providing GFP retroviral vectors and helpful advice. This work was supported by National Institutes of Health grants RO1 AI43619 and RO1 AI44268 to J.I.

References

1. Link, C. J. J. and Beecham, E. J. (1994) Human gene therapy trials for cancer. *J. Biotech. Healthcare* **2,** 183–195.
2. Miller, A. D. (1992) Genetic manipulation of hematopoetic stem cells, in *Bone Marrow Transplantation* (Forman, S. J., Blume, K. G., and Thomas, E. D., eds.), Blackwell, Cambridge, pp. 72–78.
3. Bracy, J. L., Sachs, D. H., and Iacomini, J. (1998) Inhibition of xenoreactive natural antibody production by retroviral gene therapy. *Science* **281,** 1845–1847.
4. Mayfield, R. S., Hayashi, H., Sawada, T., Bergen, K., LeGuern, C., Sykes, M., et al. (1997) The mechanism of specific prolongation of class I mismatched skin grafts induced by retroviral gene therapy. *Eur. J. Immunol.* **27,** 1177–1181.
5. Bagley, J. and Iacomini, J. (1998) Gene therapy in organ transplantation. *Cancer Res. Ther. Control* **7,** 33–36.
6. Chalfie, M., Tu, Y., Euskirchen, G., Ward, W. W., and Prasher, D. C. (1994) Green fluorescent protein as a marker for gene expression. *Science* **263,** 802–805.
7. Bagley, J., Aboody-Guterman, K., Breakefield, X., and Iacomini, J. (1998) Long-term expression of the gene encoding green fluorescent protein in murine hematopoietic cells using retroviral gene transfer. *Transplantation* **65,** 1233–1240.

8. Markovitz, D., Goff, S., and Bank, A. (1988) Construction and use of a safe and efficient amphotropic packaging cell line. *Virology* **167,** 400–406.
9. Coffman, B. (1982) Surface antigen expression and immunoglobulin rearrangement during mouse pre-B cell development. *Immunol. Rev.* **69,** 5–23.
10. Metlay, J. P., Witmer-Pack, M. D., Agger, R., Crowley, M. T., Lawless, D., and Steinman, R. M. (1990) The distinct leukocyte integrins of mouse spleen dendritic cells as identified with new hamster monoclonal antibodies. *J. Exp. Med.* **171,** 1753–1771.
11. Spangrude, G. J., Heimfeld, S., and Weissman, I. L. (1988) Purification and characterization of mouse hematopoietic stem cells. *Science* **241,** 58–62.
12. Ilkuta, K., Kina, T., MacNeil, I., Uchida, N., Peault, B., Chein, Y. H., et al. (1990) A developmental switch in thymic lymphocyte maturation potential occurs at the level of hematopoietic stem cells. *Cell* **62,** 863–874.
13. Ho, M. and Springer, T. A. (1982) Mac-1 antigen: quantitative expression in macrophage populations and tissues, and immunofluorescent localization in spleen. *J. Immunol.* **128,** 2281–2286.
14. Springer, T. A., Galfre, G., Secher, D. S., and Milstein, C. (1979) Mac-1: a macrophage differentiation antigen identified by monoclonal antibody. *Eur. J. Immunol.* **9,** 301.
15. Ikuta, K. and Weissman, I. (1992) Evidence that hematopoietic stem cells express mouse c-kit but do not depend on steel factor for their generation. *Proc. Natl. Acad. Sci. USA* **89,** 1502–1506.
16. Miller, A. D. (1990) Retrovirus packaging cells. *Hum. Gene Ther.* **1,** 5–17.
17. Danos, O. (1991) Construction of retroviral packaging cell lines, in *Methods in Molecular Biology* (Collins, M., ed.), Humana, Clifton, NJ, pp. 17–26.
18. Bodine, D. M., McDonagh, K. T., Seidel, N. E., and Nienhuis, A. W. (1991) Survival and retrovirus infection of murine hematopoietic stem cells in vitro: effects of 5-FU and method of infection. *Exp. Hematol.* **19,** 206–212.
19. Bracy, J. L. and Iacomini, J. (2000) Induction of B cell tolerance by retroviral gene therapy. *Blood* **9,** 3008–3015.

26

Green Fluorescent Protein as a Reporter of Adenovirus-Mediated Gene Transfer and Expression in the Hypothalamic–Neurohypophyseal System

Elisardo Corral Vasquez and Alan Kim Johnson

1. Introduction

The subfornical organ (SFO), the hypothalamic supraoptic nuclei (SON) and the paraventricular nuclei (PVN), and the neurohypophysis (NH) comprise an important central nervous system integrative network, which is involved in the maintenance of cardiovascular and body fluid homeostasis. Magnocellular SON and PVN neurons receive neural projections from the SFO and synthesize the hormone vasopressin, which is stored in the NH nerve terminals and released into the circulation in response to changes in blood volume, extracellular osmolarity, and blood pressure *(1)*. Dysfunction of these structures can result in pathophysiologic states, such as arterial hypertension and diabetes insipidus.

In recent years, the development of techniques of central nervous system gene transfer into rats and mice has provided a unique opportunity to study the physiology and pathophysiology of specific neural cells and functional systems *(2)*. A key vector for delivering functional and/or reporter genes to the brain has been replication-deficient adenovirus (Ad) because it is relatively safe, can be generated in high titers, can transfect nondividing cells, and can undergo retrograde transport from nerve terminals to somata *(3)*. Although there are several gene reporters to indicate successful transduction and transgene expression, bioluminescent green fluorescent protein (GFP) from the jellyfish, *Aequorea victoria,* has a major advantage in that it will fluoresce so that cells can be identified without the need for lysis or fixation. Therefore, it is an ideal gene marker for studying living cells *(4)*. The ability to visualize suc-

From: *Methods in Molecular Biology, vol. 183: Green Fluorescent Protein: Applications and Protocols*
Edited by: B. W. Hicks © Humana Press Inc., Totowa, NJ

cessful gene transfer in living cells is important because in many circumstances, the need for further imaging, or the application of additional methods, such as electrophysiology, require viable cells.

We have developed new approaches for transferring genes to SFO, SON, PVN, and NH cells, and have defined the Ad concentrations and conditions required to optimize gene expression in target cells. In our studies, we used a recombinant Ad serotype 5 (Ad5), in which the regions containing the *E1A* and *E1B* (early) genes required for virus replication were replaced by GFP cDNA which was subcloned from pGFPS65T and modified to maximize expression in eukaryotic cells. A Rous sarcoma virus (RSV) promoter was used to drive gene expression because of its ability to maintain transgene expression in the hypothalamic-NH system ***(5)***.

The following describes the approach used for GFP gene transfer to the SON (**Fig. 1A**). In order to preserve the normal anatomical and physiological environment of the SON cells, brain slices were taken from animals that had been injected with Ad vectors into the SON 4 d earlier. As illustrated in **Fig. 1B**, GFP-expressing cells are predominantly found within the SON, and are not found in adjacent areas. We have observed that high concentrations of Ad ($>8 \times 10^6$ pfu/site) can lead to immune and inflammatory responses in the region of the injection site.

An advantage of Ad vectors is that they are not taken up by fibers that pass through the injection site. After rats were injected in the SON with Ad5RSVgfp, the authors visualized GFP expression in SFO-cultured cells and verified the presence of the gene in the cells of this structure, by polymerase chain reaction, which indicates that viral vectors were taken up by nerve terminals in the SON and transported retrogradely to the soma of SFO neurons ***(5)***. The pattern of backlabeled cells in the SFO is of particular significance because the vectors taken up by nerve terminals in the SON were retrogradely transported specifically to cell soma located in the annulus of the structure. This discrete pattern indicates that the labeled cells in the annulus do not transduce cells located in the core. This pattern of retrograde labeling in the SFO (i.e., backlabeling in the annulus, but not in the core) is consistent with what has been observed with other retrograde tracers ***(6)***.

Since magnocellular neurons of the SON and PVN send their axonal projections to the NH, one can both directly transduce NH pituicytes and selectively target magnocellular neuron gene expression via retrograde transport by injecting vector directly into the posterior pituitary (**Fig. 2A**). **Fig. 2B** illustrates the GFP expression in the NH from a rat transfected in vivo with Ad5RSVgfp (8×10^6 pfu) 4 d earlier. GFP-positive SON and PVN neurons can be visualized in cultures, following in vivo injection of Ad5RSVgfp into the NH.

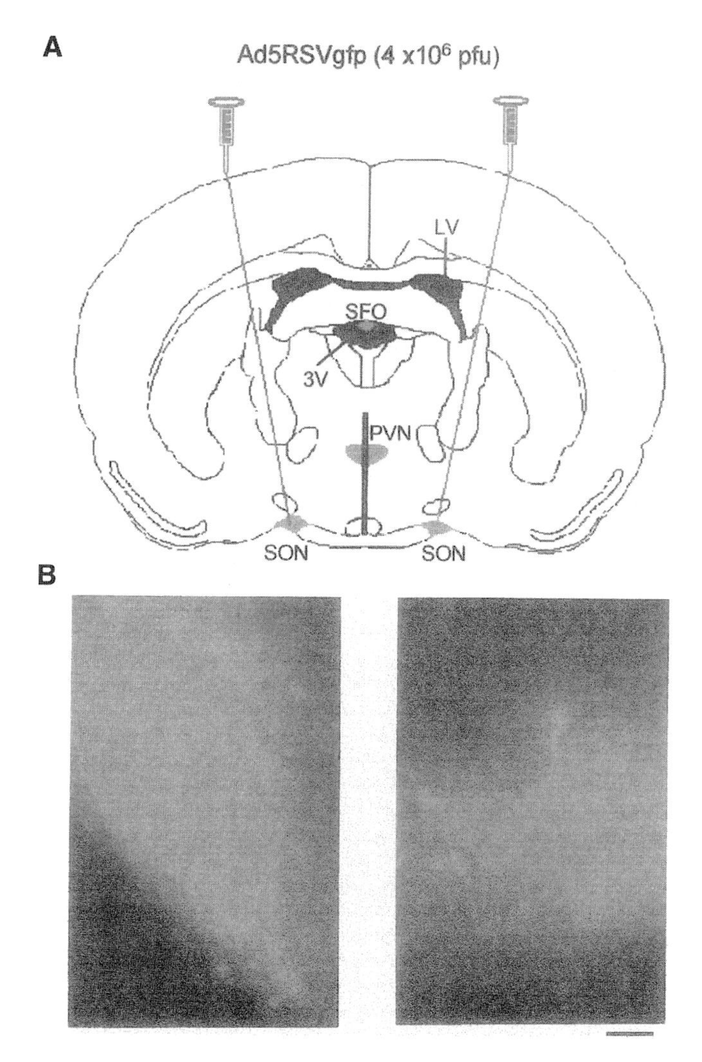

Fig. 1. **(A)** Coronal diagram of a rat brain (based on the atlas of Paxinos and Watson *[7]*) showing the position of the cannulas used to inject Ad solution into the supraoptic nucleus (SON). **(B)** Photomicrograph (scale bar: 50 μm) of both SON showing expression of GFP in a living coronal slice (400 μm thick) of a rat with bilateral Ad5RSVgfp injections made into the SON 4 d earlier. (For optimal, color representation please see accompanying CD-ROM.)

Attempts at transfection of the NH with Ad5RSVgfp of concentrations higher than 8×10^6 pfu should be avoided because of cytotoxicity to the pituicytes and an inflammatory response at the injection site.

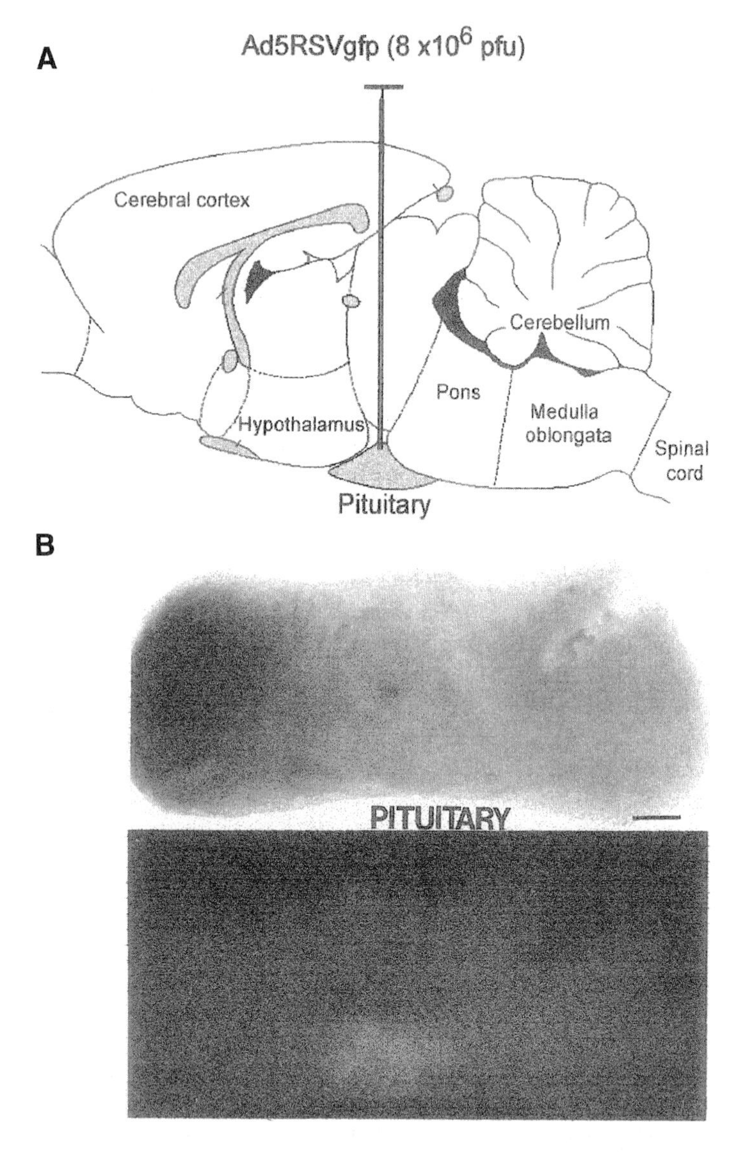

Fig. 2. **(A)** Sagittal diagram of a rat brain (based on the atlas of Paxinos and Watson *[7]*) showing the position of the cannula to inject the Ad solution into the pituitary. **(B)** Photomicrograph (scale bar: 250 μm) of a living pituitary showing the expression of GFP in the middle of the gland (NH) of a rat injected with Ad5RSVgfp delivered 4 d earlier. (For optimal, color representation please see accompanying CD-ROM.)

We have also defined the optimal conditions for in vitro gene transfer to cells that are dissociated, isolated and cultured from rat SFO, SON, PVN, and NH. Cell cultures exposed to graded concentrations of Ad5RSVgfp, 3 h after

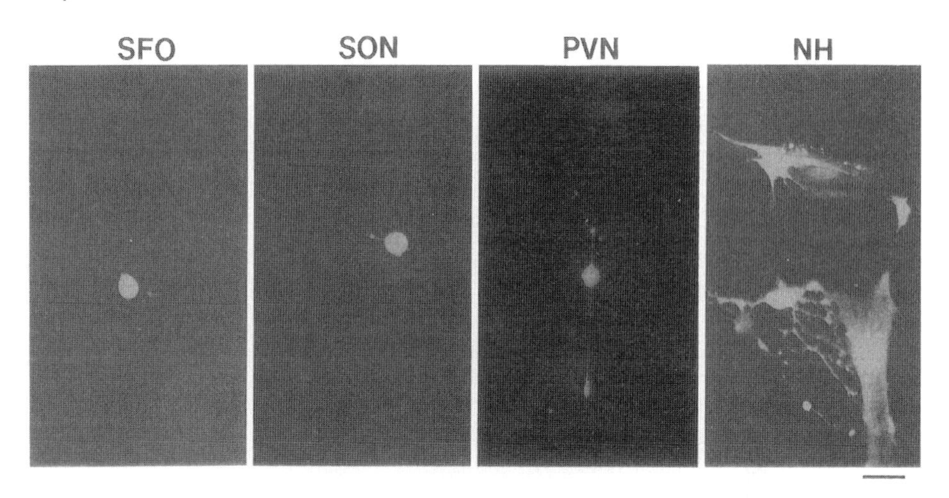

Fig. 3. Typical photomicrographs (scale bar: 50 μm) of viable subfornical organ (SFO), supraoptic nucleus (SON), paraventricular nucleus (PVN) neurons, and neurohypophysis (NH) pituicytes in culture expressing the GFP reporter gene. The cells are from rats and were dissociated and transduced in vitro with Ad5RSVgfp (2×10^7 pfu/well) 6 d earlier. (For optimal, color representation please see accompanying CD-ROM.)

dissociation show detectable GFP-positive cells 24 h later. **Figure 3** shows typical living SFO, SON, and PVN neurons and NH pituicytes expressing GFP 6 d after Ad transduction. The expression of GFP in dissociated, cultured cells from those areas is related to the Ad concentration. Ad concentrations up to 2×10^7 pfu/well do not affect survival of the cultured cells. They retain a normal morphology and remain adherent to the cover glasses. The percentage of GFP-positive cells reached an asymptote 2–4 d after transduction. We have also seen that this low concentration of Ad5RSVgfp does not adversely affect the properties of cell ion channels as observed when using whole-cell, patch-clamp and electrophysiological recordings *(5)*. At higherAd concentrations, less time for exposure to the fluorescent light source is required to visualize expression of the transgene. Although higher levels of gene expression are observed with higher titers of Ad5RSVgfp, the number of surviving cells remaining attached to the cover glasses significantly decreases over time, because of the cytotoxic effect of high concentrations of the vector. The percentage of GFP-positive cells from the SFO, SON, and PVN is higher for nonneuronal cells probably because they are more susceptible to Ad transfection than neurons and/or because they continue to divide in culture.

The preceding approaches using Ad vectors to deliver functional genes (plus GFP as a gene marker) to the SFO–hypothalamus–NH neuroendocrine axis are likely to provide new insights into the normal function of this system. Because of the biological importance of the neural circuitry controlling the synthesis and release of vasopressin, gene transfer may eventually provide a therapeutic approach for the treatment of diseases related to disordered regulation of body fluid balance and blood pressure (e.g., hypertension, congestive heart failure, diabetes insipidus).

2. Materials

2.1. In Vivo Transduction

1. Adult (45–60 d old) rats (Harlan, Indianapolis, IN).
2. Pentobarbital (Nembutal sodium, Abbott, North Chicago, IL).
3. 10 μL Hamilton microsyringe (Reno, Nevada).
4. PE-10 polyethylene tubing.
5. Stainless steel 30-gage tubing.
6. Kopf stereotaxic instrument (Model 900, David Kopf, Tujunga, CA).
7. Ice-cold, oxygenated, buffered salt solution: 125 m*M* NaCl, 25 m*M* $NaHCO_3$, 25 m*M* glucose, 2.5 m*M* KCl, 1.25 m*M* NaH_2PO_4, 2 m*M* $CaCl_2$, and 1 m*M* $MgCl_2$.
8. Cyanoacrylate adhesive, for sectioning.
9. Ad5RSVgfp. This should be divided in 50 μL aliquots and stored at –70°C.

2.2. Cell Dissociation and Culture for In Vitro Transduction

1. Postnatal (7–14 d old) rats (Harlan).
2. Methoxyflurane (Metofane, Mallingkrodt, IL).
3. 70% Ethanol.
4. Hank's balanced salt solution (HBSS) (H-8264, Sigma, St. Louis, MO) containing 20 m*M* HEPES buffer (H-0887, Sigma).
5. Earle's balanced salt solution (EBSS, Sigma) containing 20 m*M* HEPES.
6. Dispase I (Boehringer Mannheim, Indianapolis, IN): 1.5 U/mL stock solution.
7. Culture medium: Dulbecco's modified Eagle's medium nutrient mixture, F-12 HAM (DMEM) (D-6421, Sigma) containing 10% fetal calf serum (F-0643, Sigma) and 1% L-glutamine-penicillin-streptomycin solution (Sigma stock: 200 m*M* glutamine, 10,000 U penicillin, and 10 mg streptomycin, respectively).
8. Poly-L-lysine (P-5899, Sigma): 0.1 mg/mL stock solution.

3. Methods

3.1. In Vivo Gene Transfer to Supraoptic Nucleus or Neurohypophysis (see Notes 1 and 2)

1. Anesthetize the animal with an intraperitoneal injection of pentobarbital (35 mg/kg) (*see* **Note 3**).
2. Place the animal in a Kopf stereotaxic apparatus.
3. Level the skull between bregma and lambda, after a longitudinal incision of scalp and cleaning of the exposed dorsal cranium.

4. The stereotaxic coordinates *(6)* for the bone perforation targeting SON are: 1.3 mm caudal to bregma, 5.5 mm lateral on each side of the midline for the point of injector entry; the injector is angled 20 degrees from perpendicular, toward the midline, and inserted 9.6 mm below the skull, in order to avoid the perforation of the lateral ventricles en route to the SON (*see* **Notes 3** and **4**, **Fig. 1A**)
5. The stereotaxic coordinates *(7)* for the bone perforation and targeting the NH are: 5.5 mm caudal to bregma on the midline and 10.0 mm below the skull (*see* **Fig. 2A**).
6. Make the hole in the skull using a dentist-like drill equipped with a high-speed cutter, but be careful not to disrupt the dura.
7. Attach stainless steel 30-gage tubing to a 10-µL Hamilton microsyringe by a PE-10 polyethylene tubing (approx 30 cm long). The whole extension should be backfilled first with sterile saline followed by a small air bubble (approx 5.0 mm) to indicate successful delivery of the injectate, and finally by the Ad solution.
8. Ad solution (Ad5RSVgfp) is injected bilaterally into the SON (200 nL, $2–4 \times 10^6$ pfu; **Fig. 1A**) or in the center of the pituitary (400 nL, $4–8 \times 10^6$ pfu; **Fig. 2A**).
9. Suture the incision, and return the rats to their home cages (covered with a filter) after recovering from anesthesia.
10. 4 d postinjection, deeply anesthetize the animal, decapitate it, and remove the pituitary gland or whole brain to ice-cold, oxygenated, buffered salt solution for living tissues.
11. Using a stainless blade, block the brain and attach it to the chuck of a vibratome using cyanoacrylate adhesive and cut coronal slices (300–400-µm thick) close to the SON level. Transfer brain slices or the whole pituitary to a chamber and maintain them with oxygenated, buffered salt solution during the observation at the microscope.
12. Visualize the GFP expression in the injection sites at the SON (*see* **Fig. 1B**) or NH (*see* **Fig. 2B**), using an inverted microscope equipped with an epifluorescence unit and a fluorescein isothiocyanate filter.

3.2. In Vitro Gene Transfer to Cultured Cells

1. For each set of cultures, anesthetize six rat pups with methoxyflurane, decapitate them, and place each head in ice-cold 70% ethanol (*see* **Note 5**).
2. Quickly remove the whole brain or the pituitary gland, and place the tissue in ice-cold, aerated (95% O_2 and 5% CO_2) HBSS containing 20 m*M* HEPES.
3. Under a dissecting microscope, slice the brain coronally approx at the level of the optic chiasm *(5)*, and isolate the desired area (SFO, SON, or PVN), or dissect the NH (the circular structure in the middle of the pituitary).
4. Transfer the tissue to ice-cold EBSS containing 20 m*M* HEPES, and place on ice for approx 1 h.
5. Initiate digestion by adding Dispase I (1.5 U/mL) and transfer the tube to a water bath maintained at 35°C under aeration (95% O_2 and 5% CO_2) for 1 h.
6. Terminate enzymatic activity by washing tissue fragments 3× with the DMEM, supplemented with 10% fetal calf serum and 1% L-glutamine-penicillin-streptomycin solution (200 m*M*, 10,000 U, and 10 mg, respectively).

7. Prepare cell suspensions by gentle tituration of tissue fragments through a sterile, fire-polished Pasteur pipet until the cells are visibly dissociated (15–20×).
8. Plate the cells onto 18-mm cover glasses precoated with poly-L-lysine (0.1 mg/mL).
9. Place the cover glasses in 12-well culture dishes and maintain them in an incubator with humidified atmosphere of 5% CO_2 and 95% air at 37°C for at least 2–3 h, to allow attachment of the cells. At this time, add 2 mL culture medium to each culture dish and replace two-thirds of the medium every 2 d.
10. 3 h after tissue dissociation and plating of the cells, remove the culture medium, and replace it with 1 mL of either Ad vector solution or vehicle (saline–3% sucrose). After 30 min (average time required for vectors to enter the cells), the viral solution should be removed, to avoid subsequent cell transduction, and replaced with culture medium. This procedure is based on our observation that Ad vectors can survive up to 12 d in wells of culture medium in the absence of plated cells.
11. Visualization of cultured cells expressing GFP and cell counting should be made using an inverted microscope equipped with an epifluorescence unit and a fluorescein isothiocyanate filter (*see* samples in **Fig. 3**).
12. To unequivocally establish that a specific type of cell (e.g., neurons that synthesize vasopressin or pituicytes) was transduced, an immunohistochemical analysis is required.

4. Notes

1. All procedures for Ad-mediated gene transfer to the anesthetized animal brain or dissociated and cultured cells should be made in an isolated, sterile area.
2. In vivo direct Ad-mediated gene transfer to the SFO and PVN can be made, but we have noted that the viral vectors often reach the neighboring cerebral ventricles and GFP expression will occur in extraneous cells.
3. Postnatal (7–14 d old) rats can also be used for direct Ad-mediated gene transfer to the SON. Inhalation anesthesia is delivered to rat pups from a thin and short plastic tube which can easily be moved near or away from the nose. The stereotaxic apparatus can be the same one used for adult rats, but using the following coordinates: bilateral holes 1.3 mm caudal to bregma and 4.5 mm lateral to midline, allowing the insertion of the stainless-steel, 30-gage cannula 8.3 mm below the dura at a 20-degree angle.
4. These protocols of in vivo and in vitro gene transfer can also be used for mice *(8)*. The stereotaxic coordinates for adult mice are 2.3 mm caudal to bregma on the midline and 6.4 mm below the skull for gene transfer to NH, and 0.5 mm caudal to bregma, 3.5 mm lateral on each side of the midline, 6.4 mm below the skull, and the injector angled 20 degrees from perpendicular, toward the midline, for gene transfer to SON.
5. For cell culture and in vitro gene transfer to SFO, SON, or PVN cells, adult rats can also be used instead of rat pups, but we find that the culture will be more viable if brain tissue from 3-d-old postnatal rats is added to each set of structures dissected from six adult rats for dissociation and culture.

Acknowledgments

The authors wish to thank Terry G. Beltz for his superb technical assistance. Mr. Beltz has been a major contributor to the development of the approach described for gene transfer to rat pups. The work was supported by grants from the NIH (HL14388, HL57472, and DK54759) and the office of Naval Research (N00014-97-1-0145).

References

1. Johnson, A. K. and Gross, P.M. (1993) Sensory circumventricular organs and brain homeostatic pathways. *FASEB J.* **7,** 678–686.
2. Weihl, C., MacDonald, R. L., Stoodley, M., Lüders, J., and Lin, G. (1999) Gene therapy for cerebrovascular disease. *Neurosurgery* **44,** 239–253.
3. Davidson, B. L. and Bohn, M. C. (1997) Recombinant adenovirus: a gene transfer vector for study and treatment of CNS diseases. *Exp. Neurol.* **144,** 125–130.
4. Gerdes, H.-H. and Kaether, C. (1996) Green fluorescent protein: Applications in cell biology. *FEBS Lett.* **389,** 44–47.
5. Vasquez, E. C., Johnson, R. F., Beltz, T. G., Haskell, R. E., Davidson, B. L., and Johnson, A. K. (1998) Replication-deficient adenovirus vector transfer of *gfp* reporter gene into supraoptic nucleus and subfornical organ neurons. *Exp. Neurol.* **154,** 353–365.
6. Johnson, R. F., Beltz, T. G., Jurzak, M., Wachtel, R. E., and Johnson, A. K. (1999) Characterization of ionic currents of cells of the subfornical organ that project to the supraoptic nuclei. *Brain Res.* **817,** 226–231.
7. Paxinos, G. and Watson, C., eds. (1997) *The Rat Brain in Stereotaxic Coordinates*, 3rd ed., Academic, New York.
8. Vasquez, E. C., Beltz, T. G., Meyrelles, S. S., and Johnson, A. K. (1999) Adenovirus-mediated gene delivery to hypothalamic magnocellular neurons in mice. *Hypertension* **34 (part II),** 756–761.

27

Enhancement of Green Fluorescent Protein Expression in Adeno-Associated Virus with the Woodchuck Hepatitis Virus Post-Transcriptional Regulatory Element

Jonathan E. Loeb, Matthew D. Weitzman, and Thomas J. Hope

1. Introduction

Since the advent of recombinant DNA technology, maximization of exogenous gene expression has been an important issue for molecular biologists. Efforts at enhancing transgene expression have mostly been directed at improving the efficiency of delivery and increasing levels of transcription and translation. Less progress has been made in the application of post-transcriptional methods for improving gene expression. Here is described the use of an element derived from the woodchuck hepatitis virus (WHV) that possesses the ability to enhance the expression of heterologous genes post-transcriptionally. Green fluorescent protein (GFP) is frequently employed as a fusion protein to enable the detection, visualization, and quantification of molecules under study. Problems of expression are often encountered when using this strategy, making the woodchuck hepatitis virus post-transcriptional regulatory element (WPRE) a useful addition to vectors designed to express fusion proteins. This chapter discusses general issues of cloning and placement of the WPRE when designing such vectors.

Gene expression can be modulated at many levels, including transcription, post-transcriptional modification of RNA such as 5' and 3' end-processing, RNA export and stability, and translation. Additionally, the presence of introns, or the process of intron splicing itself, can increase gene expression ***(1–3)***. New insight continues to be acquired regarding how viruses ensure high levels of expression of their genes at the post-transcriptional level. Several viruses,

From: *Methods in Molecular Biology, vol. 183: Green Fluorescent Protein: Applications and Protocols*
Edited by: B. W. Hicks © Humana Press Inc., Totowa, NJ

including HIV, type D retroviruses, murine leukemia virus, cytomegalovirus (CMV), influenza virus, the herpesviruses, and the hepatitis B viruses, have been shown to possess *cis*- and *trans*-acting elements whose purpose is to post-transcriptionally increase the expression of viral genes by interacting with host-cell machinery ***(4–10)***. In some cases, such elements are supporting the expression of intronless viral genes. The expression of heterologous cDNAs, which are intronless messages, for experimental purposes, could likewise be enhanced by such elements. The post-transcriptional regulatory element of hepatitis B virus (HPRE) can functionally replace an intron in stimulating β-globin expression ***(11)***. Our studies of the WHV demonstrated that it contains a post-transcriptional regulatory element (WPRE), which is partially homologous to the HPRE ***(5)***. These PREs function in an orientation-dependent manner, suggesting that they are RNA elements. Consistent with this interpretation, the PREs contain conserved RNA stem-loop structures, which are required for maximal function, and whose structure has been confirmed by compensatory mutational analysis ***(12)***. The WPRE functions more efficiently than the HPRE, because of the presence of an additional *cis*-acting sequence in the WPRE, termed "γ" ***(5)***. The mechanism of this enhancement, although incompletely understood, is known to be post-transcriptional in nature and may involve the facilitation of RNA processing and/or export ***(13,14)***.

Our studies reveal that the post-transcriptional effect of the WPRE can significantly stimulate the expression of GFP in transient and stable transfections (**Fig. 1**; ***15***). Increases in GFP expression of greater than twofold are obtainable in 293 cells transiently transfected with GFP-expressing plasmids, 6–9-fold increases in expression have been obtained in the context of stably transfected 293 cells or primary human fibroblasts, as well as in other primary tissue ***(15)***. We surmise that this differential effect may result from the comparatively large number of plasmid copies being expressed in the context of transient, as opposed to stable, transfection, in which cells can contain as few as one integrated copy of the heterologous gene. In the transient context, high copy number may saturate those pathways of RNA metabolism influenced by the WPRE, thus minimizing its effect. The WPRE has also been shown to enhance expression in RNA and DNA viral vectors, including murine leukemia virus, HIV (**Fig. 2**), and adeno-associated virus (AAV), using various promoters and transgenes ***(15,17)***. The WPRE effect also appears to be species-independent, as evidenced by work in murine and avian cells. Thus, the WPRE is a broadly useful tool for increasing gene expression. Here, is described in detail the use of the WPRE in enhancing GFP expression in the context of AAV. Protocols are provided for preparing plasmids that express GFP, with and without the WPRE in the message, and for adapting them for use in an AAV shuttle vector. We further discuss production of recombinant AAV stocks, and verification of enhanced expression.

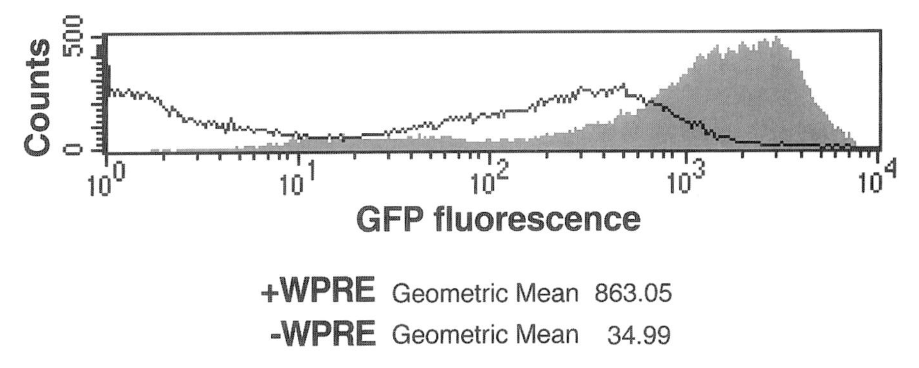

+WPRE Geometric Mean 863.05
-WPRE Geometric Mean 34.99

Fig. 1. WPRE enhances GFP expression in 293 cells stably transfected with GFP-expressing constructs. 293 cells were transfected with plasmids that included (+WPRE), or did not include (–WPRE), the WPRE in messages expressing GFP driven by the CMV promoter. Following selection in media containing 200 µg/mL G418 for 4 wk, 30,000 cells were analyzed for GFP expression by flow cytometry. Geometric mean fluorescence is shown for each population.

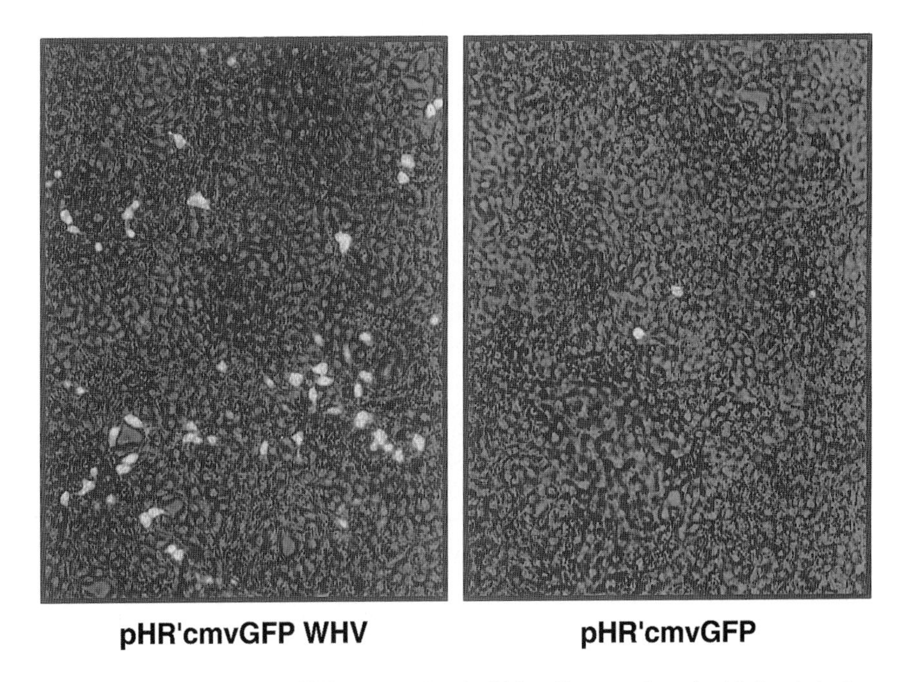

Fig. 2. WPRE enhances GFP expression in 293 cells transduced with lentiviral vectors. 293 cells were transduced with HIV-1-based vectors which included (+WPRE), or did not include (–WPRE), the WPRE in messages expressing GFP driven by the HIV-1 long terminal repeat (LTR). Cells were visualized by fluorescence microscopy. Only high-expressing cells are visualized here.

2. Materials

1. WPRE sequence: The WPRE has been functionally defined as encompassing nucleotides of position 1093–1684 of WHV 8 *(5)*. The viral DNA template of WHV 8 is available from the American Type Culture Collection (Genbank accession number J04514).
2. Plasmids: This sequence is available as a subclone in the plasmid pSK+WPRE-B11 *(5)*. The plasmid contains the WPRE inserted at the *Cla*I site (*see* **Note 1**). Also needed for this protocol are EGFP-C1 (Clontech), p*sub*201 ***(18)***, rep/cap packaging plasmid and adenovirus helper plasmid for AAV production ***(16)***.
3. Primers: If direct subcloning of the WPRE into the expression vector by restriction digest is not possible, the WPRE can be amplified from pSK+WPRE-B11 or from the WHV8 genome, using the following primers (*see* **Note 2**). 5'-AATCAACCTCTGGATTACAAAAT-3' and 5'-AGGCGGGGAGGCGGCC-CAAAGGGA-3'. Other synthetic oligonucleotides as listed in **Subheading 3.**
4. Enzymes: Restriction enzymes desribed here are obtained from New England Biolabs, and the buffers used are those recommended by the manufacturer. Required are: *Bgl*II, *Eco*RI, *Ssp*I, *Xba*I, *BspH*I, *Ase*I, *Cla*I; buffers 2, 3, 4, *Eco*RI, *Ssp*I; bovine serum albumin, provided with enzymes. Also needed are calf intestinal phosphatase (CIP) and its buffer (Gibco/Life Technologies) and T4 DNA ligase and its buffer (Gibco/Life Technologies).
5. Cell culture and viral preparation: Dulbecco's modified Eagle medium (DMEM) (BioWhittaker), fetal bovine serum (Gibco/Life Technologies) and penicillin-streptamycin-L-glutamine solution (BioWhittaker). For transfection: 2.5 *M* $CaCl_2$, 2X HEPES-buffered saline (20 m*M* HEPES, pH 7.8, 150 m*M* NaCl), 10 m*M* Tris-Cl, pH 7.6, DNaseI (Ambion) and RNase A 10 mg/mL (Ambion), 4 *M* ammonium sulfate, CsCl, 10% glycerol, HybQuest kit (Mirus) for dot blotting, 200 µg/mL Geneticin G418 (Gibco/Life Technologies), phosphate-buffered saline (PBS) (BioWhittaker), and PBS 5 m*M* EDTA, 5 m*M* propidium iodide.

3. Methods

A generally applicable strategy for using the WPRE in an expression vector consists of the digestion of pSK+WPRE-B11 at the appropriate restriction sites to allow the insertion of the WPRE into the expression plasmid. Alternatively, subcloning can be achieved by polymerase chain reaction amplification of the WPRE, using primers with convenient restriction sites incorporated at their 5' ends, followed by digestion and insertion into the target plasmid at the appropriate site (*see* **Note 3**).

3.1. Production of AAV Plasmids Containing WPRE for EGFP Expression

Plasmids used in these transfection and transduction experiments are based on the EGFP-C1 vector (Clontech). Since the vector was originally intended for the construction of eGFP fusion proteins, a stop codon must be inserted after the GFP reading frame, to allow the use of the multiple cloning site for the addition of untranslated elements. For maximum flexibility, the stop codon may be triple-frame.

1. Digest EGFP-C1 (Clontech) with *Bgl*II and *Eco*RI in *Eco*RI buffer (New England Biolabs). Isolate the opened plasmid by gel electrophoresis.
2. Insert the triple-frame stop as a pair of annealed synthetic oligonucleotides of sequences GATCTTAGCTAACTG (sense) and AATTCAGTTAGCTAA (antisense). The oligos may be annealed by incubating an equimolar mixture of the pair in 0.5 *M* NaCl for 2 min at 95°C followed by 5 min at 65°C, 10 min at room temperature, and 5 min on ice. The annealed fragment will possess sticky ends compatible with with *Bgl*II and *Eco*RI, and can be ligated directly into the EGFP-C1 vector, using T4 DNA ligase (Gibco/Life Technologies) according to the manufacturer's recommendations to produce the plasmid EGFP.stop. Transform *Escherichia coli*, and select on plates containing kanamycin; this and all subsequent selection of bacteria should be performed with kanamycin.
3. Optimum packaging of AAV occurs with an insert size of approx 4.5 kb. To adapt the construct for use in AAV, cut the plasmid, EGFP.stop, with *Ssp*I in *Ssp*I buffer, to remove the f1 ori and ampicillin resistance promoter, and religate with T4 DNA ligase, reducing the size of the recombinant AAV genome eventually generated, and producing plasmid eGFP.stop.ΔSsp.
4. To allow transfer of the expression cassette to the AAV shuttle vector, p*sub*201 ***(18)***, insert synthetic linkers consisting of oligonucleotides annealed as before, containing *Xba*I restriction sites and sticky ends compatible with the unique *Bsp*HI (oligo sequences: CATGACGTCTAGACG and CATGCGTCTAGA-CGT) and *Ase*I (oligo sequences: TAATCGTCTAGACG and TACGTCTAGA-CGAT) sites in EGFP.stop.ΔSsp, yielding EGFP.stopBXAX. These insertions must be performed one at a time in series; digest eGFP.stop.ΔSsp with *Bsp*HI in buffer 4, with *Ase*I in buffer 3. The oligos are designed so that one of the *Bsp*HI or *Ase*I sites flanking the *Xba*I will be destroyed. Therefore, at this stage, sequencing through the subcloning junctions is useful, to ascertain the orientation of the artificial *Xba*I site inserts. Additionally, if a control construct is desired in which the WPRE is introduced upstream of the CMV promoter so that it is included in the AAV insert, but not in the CMV-driven message (*see* **step 7**), then it is important to use a plasmid in which *Ase*I is intact 3' to the *Xba*I site.
5. Open plasmid EGFP.stopBXAX with *Acc*I in buffer 4, and treat the plasmid with CIP (Gibco/Life Technologies) according to the manufacturer's recommendations. Cut plasmid pSK+WPRE-B11 with *Cla*I in buffer 4+BSA, isolate the resulting 591 base fragment comprising the WPRE by gel electrophoresis, and ligate into the EGFP.stop.BXAX vector at the *Acc*I site, to produce the plasmid EGFP.stop.WPRE. Once again, orientation of the insert must be ascertained by sequencing or digestion.

6. Open p*sub*201 by digestion with *Xba*I in buffer 2 + bovine serum albumin. Treat with CIP as directed by the manufacturer. Excise the expression cassettes from EGFP.stop.BXAX and eGFP.stop.WPRE by digestion with *Xba*I and purify by gel electrophoresis. Ligate the cassettes, with and without WPRE, into the *Xba*I site of the AAV shuttle vector p*sub*201, yielding pAAV.EGFP.W and pAAV.EGFP.stop.
7. To generate a control plasmid, pAAV.EGFP.W(as), in which the WPRE is introduced upstream of the CMV promoter in the antisense orientation, so that it is included in the AAV insert but not in the CMV-driven message, bases 1093–1684 of WHV8 can be amplified by polymerase chain reaction to yield a product with *Ase*I restriction sites at each end, using synthetic oligos of sequences CGCGATTAATAATCAACCTCTGGATTACAAAAT and CGCGATTAAT-AGGCGGGGAGGCGGCCCAAAGGGA. The PCR product can then be cut with *Ase*I in buffer 3 and ligated into the *Ase*I site of the p*sub*201.EGFP.stop vector. Orientation of the insert should be confirmed by sequencing or digestion. All plasmids in the p*sub*201 vector should be grown in *recA' E. coli* HB101 to prevent recombination caused by the presence of the inverted terminal repeats.

3.2. Generation of Purified Recombinant AAV Stocks

1. Seed 10 × 10-cm plates with 293 cells in DMEM supplemented with 10% fetal bovine serum and 1% penicillin-streptamycin-L-glutamine solution so that they will be at 90–95% confluence at the time of transfection. Allow the cells to recover at least 24 h after passing and prior to transfection.
2. Change the media on the plates for 10 mL fresh media and place back into the incubator for at least 1 h to allow for pH stabilization.
3. Prepare the transfection solution. The conditions for each 10-cm plate should be a total of 25 μg plasmid DNA in a 1:1:3 ratio (p*sub*201.eGFP.stop/WPRE: rep/cap packaging plasmid: adenovirus helper plasmid). The transfection cocktail conditions for each plate are then as follows: 5 μg p*sub*201.eGFP.stop/WPRE, 5 μg rep/cap packaging DNA, 15 μg Adenovirus helper plasmid, 60 μL 2.5 *M* $CaCl_2$, 440 μL distilled H_2O, 500 μL 2X HEPES, for a total volume of 1 mL ***(16)***.
4. Add the plasmid DNA to the $CaCl_2$ and H_2O, and mix well. In a separate polystyrene tube, place the 2X HEPES. Add the $CaCl_2$ mixture drop-wise to the HEPES, while vortexing. Allow this to sit undisturbed for 5–15 min and monitor the precipitate formation.
5. Remove 1–2 10-cm plates at a time from the incubator to keep the pH as stable as possible and add 1 mL of the cocktail to each plate, drop-wise, distributing evenly over all of the plate and agitating gently to mix into the media. Place the plates back into the incubator as soon as possible.
6. About 15–20 h posttransfection, remove the media, wash once with PBS, then add fresh DMEM. Allow the cells to produce virus for 48–72 h.
7. The cells should easily detach from the plate at the time of harvest by pipeting. Collect in a centrifuge tube and spin the cell suspension in the range of 1500*g* for

15–20 min at 4°C. Spinning the cells at too high a *g*-force can cause the cells to lyse and release the virus into the media.

8. Remove and save supernatant if further processing is required, and resuspend the cell pellet in 15 mL 10 m*M* Tris-HCl, pH 7.6. Store lysate at –80°C, until purification of the virus begins.
9. Freeze–thaw the pellet from 10 10-cm plates 3–4×, alternating between an ethanol/dry ice bath and a 37°C water bath.
10. Spin at 1500*g*, for 15 min at 4°C.
11. Save the lysate and wash the cellular debris pellet in 10 mL 10 m*M* Tris-Cl, pH 7.6. Spin as in previous step.
12. Save the supernatant and combine with the original 15 mL lysate from **step 17**, for a total amount of 25 mL lysate. Mix well, and save a 10–100-µL aliquot for the in vitro assay for rAAV production.
13. Add 2000 U DNaseI and 10 mg/mL RNase A (Ambion), to a final concentration of 0.2 mg/mL to the lysate and incubate in a 37°C water bath for 30 min.
14. Add 25 mL (equal volume) of saturated 4 *M* ammonium sulfate to the lysate, incubate on ice for 20 min and spin at maximum speed for 15 min.
15. Discard supernatant and resuspend pellet in 10 mL HEPES buffered saline (HBS), pH 7.8. Add and dissolve 6 g CsCl to the resuspended pellet in HBS and place it into an SW-41Ti centrifuge tube or similar apparatus.
16. Spin gradient at 41,000 rpm, 6°C for 24–48 h. Harvest gradient in 1-mL fractions, and perform an in vitro functional assay and dot blot on each fraction. rAAV is difficult to produce in titers that will constitute a visible band after ultracentrifugation; thus, the need for a diagnostic test for the presence of rAAV should be established.
17. The in vitro assay consists of 1–2 µL of gradient fractions on 293 cells seeded in a 6-, 12-, or 24-well plate, in DMEM with 2% fetal bovine serum, in the presence of adenovirus, at a multiplicity of infection of 2. The cells should be examined under a fluorescence microscope for GFP expression 24–48 h later.
18. Proceed to combine the positive rAAV fractions, based on the in vitro/dot blot results for a second SW-41Ti ultracentrifugation. Combine the fractions into one 12-mL ultracentrifuge tube, and fill the tube with a stock of 1.37 g/mL CsCl.
19. Ultracentrifuge the second gradient for 24–48 h at 41,000 rpm, 6°C. The virus can be more concentrated by allowing the spin to continue longer than 24 h.
20. Harvest the gradient fractions in 1-mL vol, and assay again for the presence of rAAV.
21. Dialyze the positive fractions against HBS, pH 7.8. Add 10% glycerol, then store at –80°C. Proceed to determine genomic titers by dot blot, using a kit such as the HybQuest system (Mirus), according to the manufacturer's recommendations. Comparison of the intensity of signals resulting from a known volume of viral stock to those of a known mass of p*sub*201.EGFP.stop or WPRE plasmid used as a hybridization control allows the calculation of viral genomes per unit volume of stock.

3.3. Transduction of Cells with rAAV and Quantitation of GFP Expression

1. Seed 10-cm plates with 293 cells in DMEM supplemented with 10% fetal calf serum so that they will be 30% confluent at the time of transfection. Prior to transduction, replace the media with DMEM supplemented with 2% fetal calf serum. The reduced-serum media should be added at a volume just large enough to cover the cells; 4–6 mL should be sufficient in a 10-cm plate.
2. Add equivalent MOI of the viral stocks with and without the WPRE to their respective plates, agitating gently. The authors found that the addition of 5000 viral genomes per target cell of stock was sufficient to render more than 90% of target 293 cells infected. After addition of viral stock, incubate the cells at 37°C (5% CO_2) for 4 h with occasional agitation, after which the media should be supplemented with additional fetal calf serum, to bring the serum content up to 10%.
3. Depending on the goals of the experiment, the cells may be harvested and assayed for transient GFP expression at 48 h following transduction, or they may be selected in media containing 200 µg/mL G418 (Gibco) for 3–4 wk to establish stably transduced populations.
4. To prepare for quantification of GFP expression by flow cytometry, wash the cells once with PBS, and harvest in PBS 5 m*M* EDTA at a density of approx 2×10^6 cells/mL. To permit the exclusion of dead cells from the analysis, propidium iodide may be added to a final concentration of 5 m*M*.

4. Notes

1. Expression plasmids should be selected using standard criteria for the generation of fusion proteins with GFP; the plasmid should contain a multiple cloning site 5' or 3' of GFP as appropriate to allow in-frame insertion of the gene to be studied within the message. An additional consideration when using the WPRE is that there must be a restriction enzyme site or sites suitable to allow the insertion of the 600 bp WPRE sequence 3' of the gene of interest and the GFP open reading frames and their stop codon, but 5' of the polyadenylation sequence for the message.
2. These primers anneal to the first and last 24 bases of the WPRE sequence, encompassing bases 1093–1684. Desired restriction enzyme sequences should be appended to the 5' end of each primer, to allow subsequent subcloning into expression vectors according to the criteria discussed in Methods.
3. The following general considerations should be taken into account when designing a cloning strategy that incorporates the WPRE into a plasmid expressing GFP fusion proteins:
   a. The WPRE acts at the level of RNA, so it must be included in *cis* within the message it is to enhance. The WPRE must be in the sense orientation to function.
   b. The WPRE is a roughly 600 base-long sequence; it should thus be inserted 3' of the GFP fusion protein gene to avoid reducing the efficiency of translation. Long 5' untranslated regions and structured RNA can reduce expression by perturbing ribosome function.

Fig. 3. Schematic illustrating correct placement of the WPRE in a construct expressing a GFP fusion protein. The WPRE should be included in the sense orientation within the message between the stop codon and polyadenylation signal of the gene of interest.

c. In its viral context the WPRE is located 139 bases from the noncanonical polyadenylation sequence. Constructs that place the WPRE 100–500 bases from the polyadenytion signal work well.
d. The WPRE should not be translated in order to avoid interference with the function of the fusion protein of interest. This requires that it be positioned 3' of the stop codon that terminates the translation of the fusion protein, but 5' of the polyadenylation signal and related sequences that facilitate the 3' end formation of the message (**Fig. 3**).

4. The WPRE has been shown by this and other labs to enhance gene expression in the context of multiple genes, promoters, and polyadenylation sequences. The degree of enhancement will vary depending on the fusion protein expressed, the cells used, and other parameters of the expression system.
5. The full-length, functionally defined WPRE encompasses nearly 600 bp. If the size of the message is an issue, the WPRE sequence may be shortened by deleting bases 1504–1684, to yield a minimally diminished enhancement that varies by system.
6. If one intends to generate a fusion protein in which GFP is positioned N-terminal to the protein of interest, it will be useful to consider commercially available plasmids designed for this purpose. Such plasmids generally contain multiple cloning sites, following the GFP reading frame, permitting a choice of restriction enzyme sites sufficient to allow the insertion of the gene of interest and the WPRE, as well as an artificial stop codon between the gene to be studied and the WPRE. Such a stop codon can be synthesized with the appropriate overhanging ends, to take advantage of restriction sites available at this position.

References

1. Antoniou, M., Geraghty, F., Hurst, J., et al. (1998) Efficient 3'-end formation of human beta-globin mRNA in vivo requires sequences within the last intron but occurs independently of the splicing reaction. *Nucl. Acids Res.* **26,** 721–729.
2. Nesic, D., Cheng, J., and Maquat, L. E. (1993) Sequences within the last intron function in RNA 3'-end formation in cultured cells. *Mol. Cell Biol.* **13,** 3359–3369.
3. Ryu, W. S. and Mertz, J. E. (1989) Simian virus 40 late transcripts lacking excisable intervening sequences are defective in both stability in the nucleus and transport to the cytoplasm. *J. Virol.* **63,** 4386–4394.
4. Donello, J. E., Beache, A. A., Smith, G. J., 3rd, et al. (1996) The hepatitis B virus posttranscriptional regulatory element is composed of two subelements. *J. Virol.* **70,** 4345–4351.

5. Donello, J. E., Loeb, J. E., and Hope, T. J. (1998) Woodchuck hepatitis virus contains a tripartite posttranscriptional regulatory element. *J. Virol.* **72,** 5085–5092.
6. Fischer, U., Meyer, S., Teufel, M., et al. (1994) Evidence that HIV-1 Rev directly promotes the nuclear export of unspliced RNA. *EMBO J.* **13,** 4105–4112.
7. Huang, Z. M. and Yen, T. S. (1994) Hepatitis B virus RNA element that facilitates accumulation of surface gene transcripts in the cytoplasm. *J. Virol.* **68,** 3193–3199.
8. Malim, M. H., Bohnlein, S., Hauber, J., et al. (1989) The HIV-1 rev trans-activator acts through a structured target sequence to activate nuclear export of unspliced viral mRNA. *Nature* **338,** 254–257.
9. Pasquinelli, A. E., Ernst, R. K., Lurd, E., et al. (1997) The constitutive transport element (CTE) of Mason-Pfizer monkey virus (MPMV) accesses a cellular mRNA export pathway. *EMBO J.* **16,** 7500–7510.
10. Saavedra, C., Felber, B., and Izaurralde, E. (1997) The simian retrovirus-1 constitutive transport element, unlike the HIV-1 RRE, uses factors required for cellular mRNA export. *Curr. Biol.* **7,** 619–628.
11. Huang, Z. M. and Yen, T. S. (1995) Role of the hepatitis B virus posttranscriptional regulatory element in export of intronless transcripts. *Mol. Cell Biol.* **15,** 3864–3869.
12. Smith, G. J., III, Donello, J. E., Luck, R., et al. (1998) The hepatitis B virus post-transcriptional regulatory element contains two conserved RNA stem-loops which are required for function. *Nucl. Acids Res.* **26,** 4818–4827.
13. Harris, M. E., et al. (2001) A cis-acting RNA element that increases expression by promoting longer polyadenylated tails, in preparation.
14. Huang, Y., Wimler, K. M., and Carmichael, G. G. (1999) Intronless mRNA transport elements may affect multiple steps of pre-mRNA processing. *EMBO J.* **18,** 1642–1652.
15. Loeb, J. E., Cordier, W. S., Harris, M. E., et al. (1999) Enhanced expression of transgenes from adeno-associated virus vectors with the woodchuck hepatitis virus posttranscriptional regulatory element: implications for gene therapy. *Hum. Gene Ther.* **10,** 2295–2305.
16. Xiao, X., Li, J., and Samulski, R. J. (1998) Production of high-titer recombinant adeno-associated virus vectors in the absence of helper adenovirus. *J. Virol.* **72,** 2224–2232.
17. Zufferey, R., Donello, J. E., Trono, D., et al. (1999) Woodchuck hepatitis virus post-transcriptional regulatory element (WPRE) enhances expression of transgenes delivered by retroviral vectors. *J. Virol.* **73,** 2886–2892.
18. Samulski R. J., Chang L. S., and Shenk, T. (1987) A recombinant plasmid from which an infectious adeno-associated virus genome can be excised in vitro and its use to study viral replication. *J. Virol.* **61,** 3096–3101.

28

Construction of Infectious Simian Varicella Virus Expressing Green Fluorescent Protein

Ravi Mahalingam and Donald H. Gilden

1. Introduction

Determining the precise location of pathogenic events inside living cells is critical to the understanding of infectious and other biological processes. Molecular cloning and expression of the green fluorescent protein (GFP) from the jellyfish *Aequorea victoria,* which emits bright green fluorescence at 509 nm *(1)*, has enabled visualization of events within eukaryotic cells. GFP is stable with minimal photobleaching *(1)*, and the gene-encoding GFP has been modified to enhance the fluorescence several-fold *(2)*. The authors have used GFP to label simian varicella virus (SVV) by homologous recombination *(3)*. SVV infection in primates resembles varicella zoster virus (VZV) infection in humans, clinically, pathologically, immunologically, and virologically, including features of latency in ganglia. VZV causes chickenpox (varicella) in children, becomes latent in dorsal root ganglia, and reactivates decades later to produce shingles (zoster). SVV causes a similar disease in monkeys, enters ganglia by hematogenous spread *(4)*, and remains latent in their ganglia for the lifetime of the animal. SVV expressing GFP (SVV-GFP) allows ready identification of cells infected with virus, both in vitro and in vivo, and is potentially useful for further analysis of varicella pathogenesis and latency in experimentally infected animals: such studies are not possible in humans.

2. Materials

1. SVV seronegative African green monkeys, African green monkey kidney cells (BSC-1), and SVV-infected BSC-1 cells in culture.
2. Restriction enzymes (*Sac*I, *Sma*I, *Bam*HI, *Hin*dIII, *Not*I and *Eag*I) (Life Technologies, Bethesda, MD).

From: *Methods in Molecular Biology, vol. 183: Green Fluorescent Protein: Applications and Protocols*
Edited by: B. W. Hicks © Humana Press Inc., Totowa, NJ

3. Cloned SVV DNA fragments containing the desired sequences.
4. Phenol, phenol–chloroform (1:1), chloroform, ethanol.
5. DNASIS software (Hitachi Software, South San Francisco, CA).
6. TE buffer: 10 m*M* Tris-HCl, 1 m*M* EDTA, pH 8.0.
7. Oligonucleotide primers for polymerase chain reaction (PCR) amplification (Sigma-Genosys, The Woodlands, TX) are as follows:

Primer	Sequence
1	5'-TGTCTGCTTAGGAGATTTTGGC-3'
2	5'-TAAAAAACGTCCTCGGATAGATGCATC-3'
3	5'-CCCGGGAGTGATAAGCGTT-3'
4	5'-CCCGGGGAATATACCGTAAC-3'
5	5'-GATATACCGGACCCATATCCCAACCC-3'
6	5'-GACGGCAGAACAAAACAAAATCCA-3'
7	(^{32}P labeled) 5'-CAACCGGGCTTCTGTTTTATCTTCAA-3'

8. Opti-MEM and Dulbecco's modified Eagle's medium (DMEM) (Life Technologies) lacking serum and antibiotics.
9. DNA plasmids: pGem11Zf (+) vector (Promega, Madison, WI), pRC/RSV plasmid (from Invitrogen, Carlsbad, CA), pEGFP-1 vector (enhanced, red-shifted GFP, from Clontech, Palo Alto, CA), pGem3Z containing an 11.9-kb SVV sequence.
10. *Escherichia coli* strain HB101.
11. GenecleanII kit (Bio101, Vista, CA).
12. T4 DNA ligase (Life Technologies).
13. T7 DNA polymerase (Life Technologies).
14. Mung bean nuclease (Life Technologies).
15. 90- and 60-mm tissue culture dishes.
16. Lipofectamine transfection reagent (Life Technologies).
17. Orthoplan Universal large-field Leitz fluorescence microscope with a KP490 filter, 2X interference blue excitation filter, and a K530 suppression filter or equivalent.
18. 2% Agarose gel, zetaprobe membrane.
19. 4% Paraformaldehyde.
20. Harris hematoxylin.
21. PBS: 25 m*M* sodium phosphate, pH 7.5, 150 m*M* NaCl.
22. TBS: 200 m*M* Tris-HCl, pH 7.5, 150 m*M* NaCl.
23. TEN: 10 m*M* Tris-HCl, pH 7.5, 1 m*M* EDTA, 100 m*M* NaCl.
24. Cell sonifier (Heat Systems Ultrasonics, Plainview, NY).
25. Freund's complete and incomplete adjuvants.
26. Normal monkey liver powder.
27. Normal sheep serum (Sigma).
28. Rabbit anti-SVV antiserum.
29. Normal sheep serum (Sigma).
30. Biotinylated goat antirabbit immunoglobulin (Dako, Carpenteria, CA).
31. Alkaline phosphatase-conjugated streptavidin (Dako).
32. Fuchsin substrate system (Dako).

3. Methods

3.1. Propagation of SVV in BSC-1 Cells

1. Prepare 60-mm Petri dishes containing BSC-1 cells (50% confluent) in Opti-MEM, without serum.
2. Infect cells with SVV by co-cultivation at a 3:1 ratio of uninfected to virus-infected cells.
3. SVV is propagated serially in BSC-1 cells by co-cultivation of infected cells with uninfected cells at a ratio of 1:4.

3.2. Construction of Recombinant EGFP with Flanking SVV Sequences

In the protocols below, all restriction enzyme digests of viral and recombinant DNA (5–10 μg) are carried out at final enzyme concentrations of 50–100 ng/μL, in a reaction volume of 20–100 μL. After a minimum 1 h incubation at the temperature suggested by the enzyme manufacturer, the DNA is extracted once with phenol, once with phenol–chloroform (1:1), and once with chloroform. The product is then ethanol-precipitated and dissolved either in water or TE buffer.

We selected the region between SVV genes *US2* and *US3* (both located in the U_S segment of the SVV genome) to insert the GFP gene (**Fig. 1**). The actual site was identified using DNASIS software to rule out hairpins and 3' complementarity in sequences.

3.2.1. Introduction of a SmaI Site into SVV Sequences by PCR

1. Introduce a *Sma*I site into a DNA fragment containing a portion of the U_S segment of the SVV genome, which includes a part of the US2 open reading frame, the poly(A) addition signal for *US2*, the putative promoter (TATA box) for *US3*, and part of the *US3* open reading frame (**Fig. 1**).
2. To generate such a DNA fragment, use oligonucleotide primers 1–4 corresponding to known SVV U_S sequences *(5)*. A *Bam*HI site is present downstream from primer 1 and a *Hin*dIII site upstream from primer 2 (**Fig. 1**).
3. Primers 3 and 4 are designed with SVV-specific sequences attached at their 3' and 5' -ends, respectively, and to a *Sma*I restriction site sequence (GGGCCC).
4. Using DNA from SVV-infected cells with primers 1 and 3, amplify a 578-bp DNA fragment by PCR. All PCRs are carried out by denaturing for 1 min at 94°C, annealing for 2 min at 55°C, and elongating for 3 min at 72°C, for a total of 34 cycles, as described *(6)*. For the final cycle we used denaturation, annealing, and elongation times of 1, 2, and 7 min, respectively.
5. Using DNA from SVV-infected cells with primers 2 and 4, amplify a 354-bp fragment by PCR.

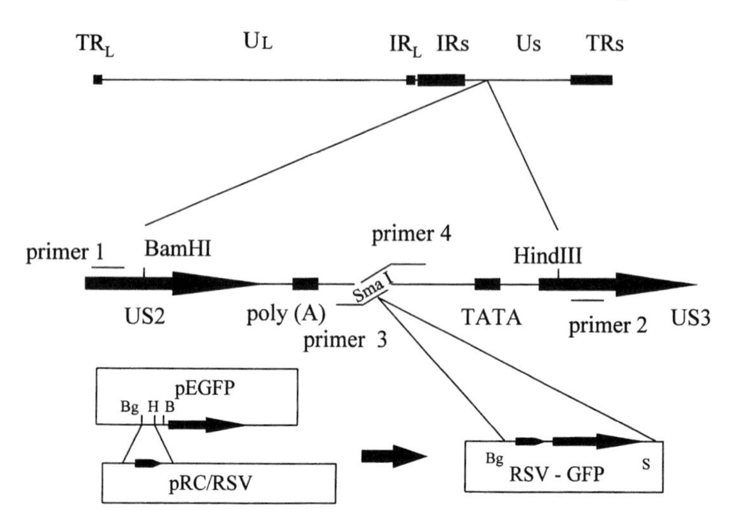

Fig. 1. Construction of a recombinant clone containing RSV-GFP in the U_S segment of the SVV genome. The location of the unique long (U_L) and unique short (U_S) segments bounded by internal and terminal repeat sequences (IR_S, TR_S, and IR_L), the *US2* gene, the poly(A)-addition site for *US2*, the putative TATA box for the *US3* gene, *US3*, and primers 1–4, are indicated. The *Sma*I sequences at the 3'- and 5'- ends of primers 3 and 4 are shown. pEGFP and pRC/RSV vectors, with the location of *Bgl*II (Bg), *Hin*dIII (H), *Bam*HI (B) and *Ssp*I (S) restriction sites, are also identified. The DNA fragment containing RSV-GFP was cloned into the *Sma*I site. (For optimal, color representation please see accompanying CD-ROM.)

6. In a third PCR amplification, mix the two PCR products (578- and 354-bp fragments) and primers 1 and 2, to amplify a 936-bp product. The annealing temperature for the third PCR was 40°C for 2 min for 7 cycles without any primers followed by 55°C for 2 min for 30 cycles using outside primers 1 and 2.
7. The final 936-bp product had a *Sma*I site at the desired location. Digest the fragment with *Bam*HI and *Hin*dIII to obtain a 735-bp DNA fragment.
8. Likewise, digest pGem11Zf (+) vector with *Bam*HI and *Hin*dIII.
9. Ligate the 735-bp PCR fragment to the digested pGem11Zf (+) vector fragment with T4 DNA ligase using the manufacturer's instructions.
10. Use the ligation mixture to transform *E. coli* HB101 cells.
11. Identify the recombinants containing the 735-bp insert, by digestion with *Bam*HI and *Hin*dIII.

3.2.2. Cloning the RSV Promoter into pEGFP-1 Vector

1. Digest the pRC/RSV plasmid with *Bgl*II and *Hin*dIII, and gel-purify the 398 bp DNA fragment containing the RSV promoter, using GenecleanII kit.
2. Digest the pEGFP-1 vector with *Bgl*II and *Hin*dIII.
3. Ligate the 398-bp fragment to the cut EGFP-1 vector with T4 DNA ligase, according to the manufacturer's instructions.

4. Use the ligation mixture to transform HB101 cells, and identify recombinants containing the 398-bp insert, by restriction digest with *Bgl*II and *Hind*III.

3.2.3. Insertion of RSV-GFP-DNA into Generated SmaI Site

1. Digest the recombinant clone described above, with *Bgl*II and *Ssp*I, and gel-purify the 1638-bp fragment containing the RSV-EGFP insert.
2. Fill in the ends using T7 DNA polymerase per the manufacturer's instructions.
3. Digest and dephosphorylate the pGEM11Zf (+) vector containing SVV insert (described above) with *Sma*I.
4. Ligate the blunt-ended DNA fragment to the *Sma*I digested pGEM11Zf (+) vector.
5. Use the ligation mixture to transform HB101 cells.
6. Analyze the resulting recombinants, using *Bam*HI and *Nco*I digestion. Identify clones containing the insert in the orientation *Bam*HI-GFP-RSV-*Hind*III, and 470 and 270 bp of SVV sequences from the U_S segment of the SVV genome flanking the 5' and 3' ends, respectively, of the GFP-RSV insert. In this SVV-GFP construct, the RSV promoter is used to drive the expression of GFP. Other promoters could also be used (*see* **Note 1**).

3.2.4. Extension of SVV Sequences Flanking the RSV-GFP

1. Abolish the *Not*I site and a *Eag*I site (part of the *Not*I site) located immediately outside the sequences encoding GFP in the recombinant clone, by linearizing it with *Not*I and blunting the ends with mung bean nuclease, per the manufacturer's instructions.
2. Religate the DNA and digest with *Not*I to reduce the background.
3. Use the *Not*I-digested ligation mixture to transform HB101 cells and select recombinants that do not contain a *Not*I site.
4. Digest the recombinant clone with *Eag*I and *Hind*III and gel-purify a 2377-bp DNA fragment containing the RSV-GFP sequences.
5. Digest a recombinant clone containing the 11.9-kb SVV *Bgl*II-D fragment (cloned into the unique BamHI site in pGem3Z) with *Eag*I and *Hind*III and gel-purify the 9523-bp fragment.
6. Ligate the 9523 bp SVV fragment with the 2377-bp *Eag*I-*Hind*III fragment containing the RSV-GFP sequences, effectively increasing the size of SVV sequences flanking RSV-GFP to 5240 and 6590 bp in the circular recombinant clone, as shown in **Fig. 2**. (**Note:** *Eag*I and *Hind*III ends cannot religate, since they have incompatible sticky ends.)
7. Use the ligation mixture to transform HB101, and select recombinants that contain a 13,467-bp insert by restriction digest with *Bam*HI .

3.3. Transfection of SVV-Infected Cells and Selection of Recombinant SVV Expressing GFP

1. Prepare 60-mm Petri dishes containing BSC-1 cells (50% confluent), and infect them with SVV by co-cultivation at a 3:1 ratio of uninfected to virus-infected cells.
2. Transfect the cells 2 d later. Add lipofectamine (9 µL) to Opti-MEM (150 µL), and incubate at room temperature for 45 min. Dilute 1–2 µg (per 60-mm Petri

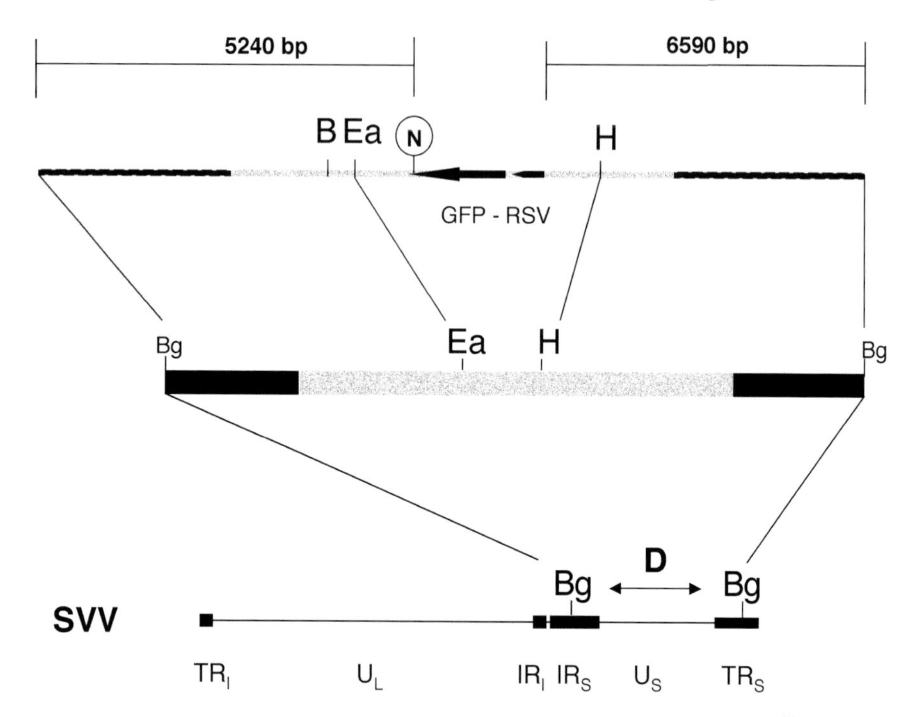

Fig. 2. Extension of SVV sequences flanking RSV-GFP. The location of various segments of the SVV genome, the SVV *Bgl*II-D fragment, and the *Bgl*II (Bg), *Eag*I (Ea), *Bam*HI (B), and *Hin*dIII (H) restriction sites, including the *Not*I (N) site within the RSV-GFP insert are indicated. The *Not*I site was mutated, and replaced the *Eag*I-*Hin*dIII fragment in the SVV *Bgl*II-D recombinant clone with the *Eag*I-*Hin*dIII fragment containing the RSV-GFP insert. The final 13.5-kb recombinant clone contained RSV-GFP sequences flanked by 5240 and 6590 bp of SVV sequences. Unique *Bam*HI site is located within the flanking sequences. (For optimal, color representation please see accompanying CD-ROM.)

dish), of the 16.6-kb recombinant clone containing RSV-GFP, with 150 μL Opti-MEM, and add this to the lipofectine mixture. Incubate at room temperature for 15 min, then add 1.2 mL DMEM (without serum or antibiotics) to the lipofectine–DNA solution.

3. Rinse the cells twice with 5 mL DMEM (without serum or antibiotics), then replace the DMEM with the Lipofectine-DNA-Opti-MEM mixture (1.5 mL/60-mm dish). Incubate the cells in the transfection mixture for 16 h at 37°C.
4. Identify cells expressing GFP using a fluorescence microscope.
5. Carefully transfer the foci displaying green fluorescence to 60- or 100-mm Petri dishes containing monolayers of uninfected BSC-1 cells.
6. Repeat this procedure several times until all cells exhibiting a cytopathic effect are green under fluorescence illumination light (**Figs. 3** and **4**; *see* **Notes 2** and **3**).

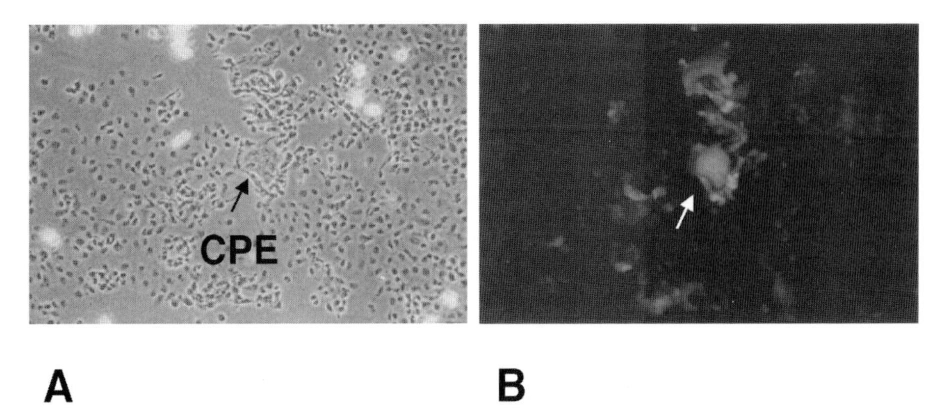

Fig. 3. BSC-1 cells infected with SVV-expressing GFP. SVV-infected cells in tissue culture were transfected with the 13.5-kb recombinant clone containing RSV-GFP flanked by SVV sequences (**Fig. 2**). The black arrow shows the cytopathic effect (CPE) in cells infected with SVV-GFP under normal light **(A)**. The white arrow indicates green fluorescence in the same area of CPE shown in A **(B)**. (For optimal, color representation please see accompanying CD-ROM.)

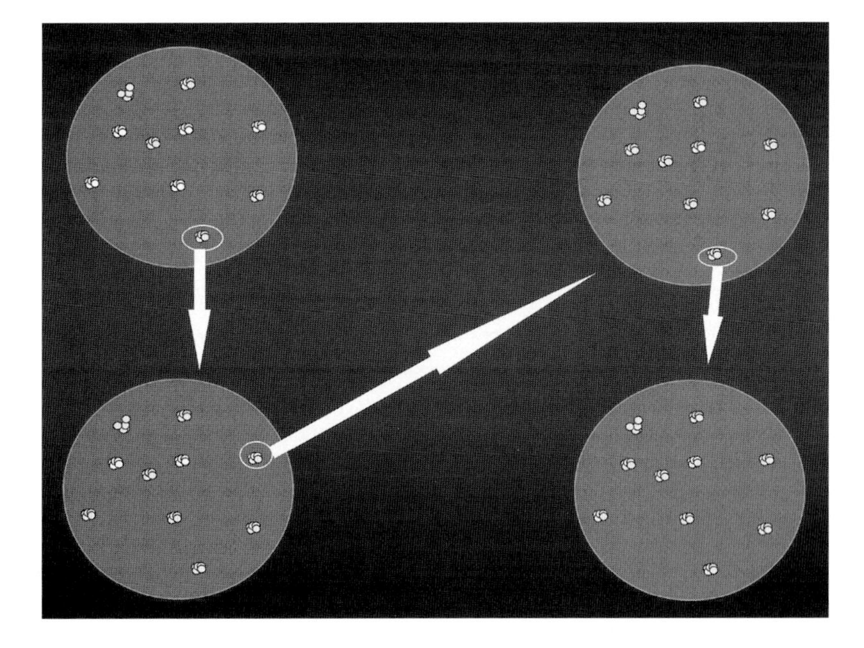

Fig. 4. Colony purification of SVV-GFP. After the initial transfection of SVV-infected-cells with the recombinant clone, 50% or less of the infectious centers emit green fluorescence. Sequential transfer of green fluorescing foci to uninfected cells produces an increase in infectious centers emitting green fluorescence. (For optimal, color representation please see accompanying CD-ROM.)

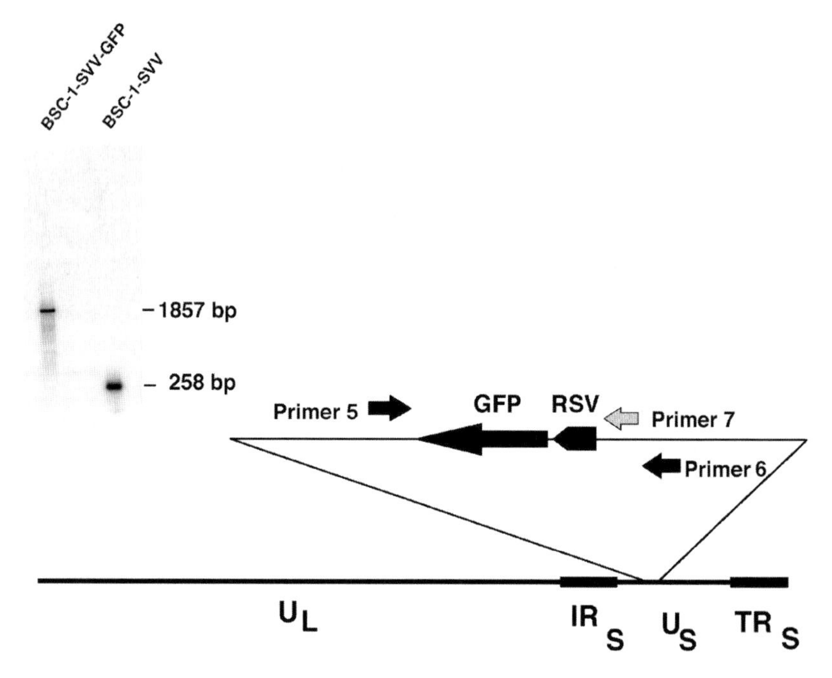

Fig. 5. PCR analysis of DNA from SVV-GFP-infected cells. Total DNA extracted from SVV-GFP- and SVV-infected BSC-1 cells was used along with SVV-specific primers 5 and 6 in PCR amplification. The products were analyzed as described in Methods, using a ^{32}P-labeled SVV-specific internal oligonucleotide (primer 7). The location and direction of GFP, RSV, and primers 5, 6, and 7 sequences on the SVV genome and the sizes of the PCR products, are indicated.

7. For Southern blot analysis, extract total DNA from SVV-GFP-infected cells, by lysis in 0.5% SDS and 100 μg/mL proteinase K at room temperature for 16 h.
8. Extract the DNA with phenol, precipitate with ethanol, and dissolve in TE buffer at concentration of 1 μg/μL.
9. Use 1 ng SVV-GFP in a PCR reaction using primers 5 and 6 (**Fig. 5**).
10. Separate the products by electrophoresis on a 2% agarose gel, transfer the DNA to a zetaprobe membrane and hybridize it to an internal ^{32}P-labeled oligonucleotide (primer 7).

3.4. Infection of Monkey with SVV-GFP and Processing of Tissue

1. Infect a 3-mo-old SVV-seronegative African green monkey intratracheally with 10^4 pfu of SVV-GFP.
2. Sacrifice the monkey 7 d later.
3. Fix the lungs in 4% paraformaldehyde, and stain 5-μm sections with Harris hematoxylin and eosin and observe by light microscopy (*see* **Note 4**).

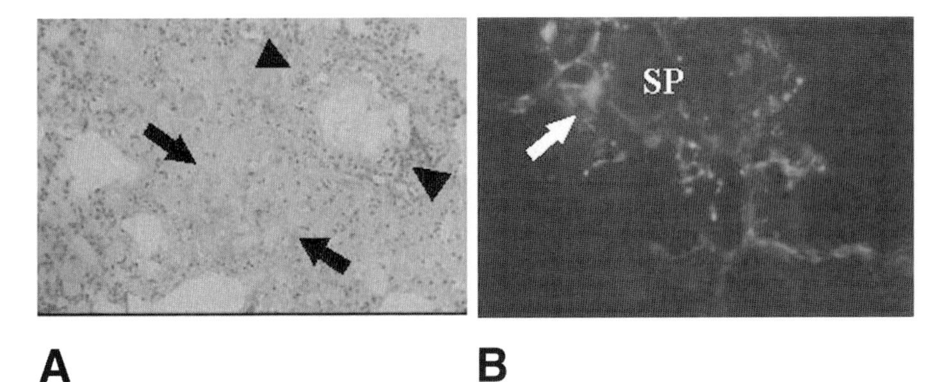

Fig. 6. SVV-infected monkey lung. Lung from a 3-mo-old monkey 10 d after infection with SVV-GFP. Photomicrograph showing necrotizing pneumonitis (*arrowheads*) and destruction of alveolar walls with proteinaceous and hemorrhagic exudate (*arrow*). H&E ×175. **(A)** Green fluorescence under 490 nm ultraviolet light (*arrow*), in an area of lung with intact alveoli indicates productive SVV-GFP infection. Alveolar space (SP) **(B)**. (For optimal, color representation please see accompanying CD-ROM.)

4. Analyze adjacent, unstained lung with an Orthoplan Universal large-field Leitz fluorescence microscope using a KP490 filter, 2× interference blue excitation filter, and a K530 suppression filter (*see* **Fig. 6**).

3.5. Preparation of Rabbit Anti-SVV Antiserum and Immunohistochemistry

1. Prepare 10× 150-mm tissue culture flasks containing BSC-1 cells (50% confluent), and infect them with SVV, by co-cultivation at a 1:1 ratio of uninfected to virus-infected cells. Rinse the virus-infected cells with 10 mL TBS and discard the wash.
2. Scrape the cells into 10 mL TBS, wash once with TBS, and resuspend the pellet in 3 mL TBS. Sonicate on ice 3× for 15 s.
3. Place the sonicated cells over a 10–50% sucrose gradient in TBS and centrifuge for 1 h at 150,000*g* at 4°C. Remove the top and bottom bands, and dilute 10-fold with TEN.
4. Pellet virions by centrifugation at 275,000*g* for 1 h at 4°C, then resuspend the pellet in 1 mL TE buffer.
5. Mix the pelleted virion suspension with an equal volume of Freund's complete adjuvant, for subcutaneous inoculation into rabbits. The authors boosted the rabbits, once every 2 wk for 10 wk, with a mixture of SVV nucleocapsids and Freund's incomplete adjuvant.
6. At the end of 10 wk, obtain serum and determine titer of anti-SVV antibodies.
7. Absorb a 1:10 dilution of the rabbit antiserum with normal BSC-1 cells at 37°C for 1 h and at 4°C for 16 h. Then adsorb the serum with normal monkey liver

LUNG

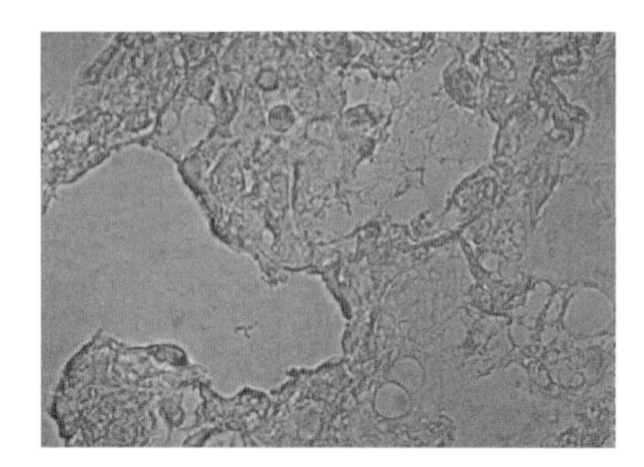
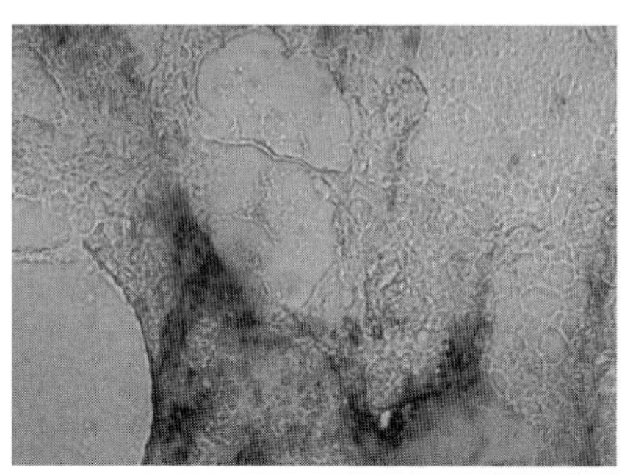

NRS ANTI-SVV

Fig. 7. Immunohistochemistry. SVV-specific antigen is seen in monkey lung with rabbit anti-SVV antiserum (ANTI-SVV) but not with normal rabbit serum (NRS). ×216. Results in **Figs. 6** and **7** were all from adjacent 50-µm sections. (For optimal, color representation please see accompanying CD-ROM.)

powder for 30 min at 37°C, and again for 20 h at 4°C. We found that the antiserum thus obtained reacted specifically with SVV-infected cells, but not with uninfected BSC-1 cells.

8. After processing sections to remove paraffin, incubate the sections with a 10% solution of normal sheep serum for 1 h followed by 1:10,000 dilution of either normal rabbit serum or rabbit anti-SVV antiserum in PBS for 30 min (both rabbit sera had been preabsorbed with normal monkey liver powder for 30 min, and again for 20 h at 4°C).
9. Rinse the sections with PBS.
10. Incubate for 20 min with a 1:300 dilution of biotinylated goat antirabbit IgG in PBS containing 5% normal sheep serum.
11. Wash 3× with PBS.
12. Incubate for 20 min with a 1:100 dilution of alkaline phosphatase-conjugated streptavidin.
13. Wash 3× with PBS.
14. Develop the color reaction for 5–30 min with fresh fuchsin substrate system, and mount sections, using an aqueous medium.

4. Notes

1. Selection of the most effective promoter and location on the target genome are the two most important criteria for long-term expression. The mouse Moloney

leukemia virus long terminal repeat (MMLV-LTR) promoter has been used to drive β-gal at four different locations on the herpes simplex virus type 1 genome. Carpenter and Stevens *(7)* observed β-gal expression for 18 mo, when they used the MMLV-LTR directly upstream from the latency-associated transcript promoter region, and expressed mRNA from the DNA strand opposite to that expressing latency-associated transcript.

2. For efficient transfection and homologous recombination, Petri dishes containing different ratios of uninfected to SVV-infected cells should be used.
3. Even after several rounds of colony purification, the recombinant virus may contain small amounts of the wild-type virus. To purify recombinant virus further, a stock of cell-free virus should be used to infect uninfected cells and to purify foci displaying green fluorescence as described above.
4. We have noticed that prolonged storage (months) in fixatives such as buffered formalin or 4% paraformaldehyde reduces the GFP signal in tissues, perhaps because of the high solubility of GFP in an aqueous environment. Quick-freezing tissues in OCT, and using frozen sections for GFP analysis, can overcome this problem. Recently, GFP was detected in frozen sections of rabbit ganglia that had been fixed in 4% paraformaldehyde at 4°C overnight, and mounted in OCT compound *(8)*.
5. GFP expression can also be used to study promoter activity. The authors have recently used GFP expression to identify the promoter sequences for *SVV* gene 21 *(9)*.
6. A recombinant virus that expresses GFP (as a fusion to the virus capsid protein) can be made and used to study virus infection.

References

1. Chalfie, M., Tu, Y., Euskirchen, G., Ward, W. W., and Prasher, D. C. (1994) Green fluorescent protein as a marker for gene expression. *Science* **263,** 802–805.
2. Zolotukhin, S., Potter, M., Hauswirth, W. W., Guy, J., and Muzyczka, N. (1996) A "humanized" green fluorescent protein cDNA adapted for high-level expression in mammalian cells. *J. Virol.* **70,** 4646–4654.
3. Mahalingam, R., Wellish, M., White, T., Soike, K., Cohrs, R., Kleinschmidt-DeMasters, B. K., and Gilden, D. H. (1998) Infectious simian varicella virus expressing the green fluorescent protein. *J. NeuroVirol.* **4,** 438–444.
4. Mahalingam, R., Wellish, M., Soike, K., White, T., Kleinschmidt-DeMasters, B. K., and Gilden, D. H. (2000) Simian varicella virus infects ganglia before rash in experimentally infected monkeys. *Virology*, in press.
5. Gray, W. L., Gusick, N. J., Ek-Kommonen, C., Kempson, S. E., and Fletcher T. M., III (1995) The inverted repeat regions of the simian varicella virus and varicella-zoster virus genomes have a similar genetic organization. *Virus Res.* **39,** 181–193.
6. Mahalingam, R., Smith, D., Wellish, M., Wolf, W., Dueland, A. N., Cohrs, R., et al. (1991) Simian varicella virus DNA in dorsal root ganglia. *Proc. Natl. Acad. Sci. USA*, **88,** 2750–2752.

7. Carpenter, D. E. and Stevens, J. G. (1996) Long-term expression of a foreign gene from a unique position in the latent herpes simplex virus genome. *Hum. Gene Ther.* **7,** 1447–1454.
8. Perng, G., Slanina, S. M., Yukht, A., Ghiasi, H., Nesburn, A. B., and Wechsler S. L. (2000) The latency-associated transcript gene enhances establishment of herpes simplex virus type1 latency in rabbits. *J. Virol.* **74,** 1885–1891.
9. Mahalingam, R., Wellish, M., White, T., and Gilden D. H. (2000) Identification of simian varicella virus gene 21 promoter region using green fluorescent protein. *J. Virol. Meth.* **86,** 95–99.

29

Green Fluorescent Protein in Retroviral Vector Constructs as Marker and Reporter of Gene Expression for Cell and Gene Therapy Applications

Nicoletta Eliopoulos and Jacques Galipeau

1. Introduction

The green fluorescent protein (GFP) was first observed in 1962, by Shimomura et al., as co-existing with the photoprotein, aequorin, purified from the bioluminescent jellyfish, *Aequorea victoria* ***(1a)***. Aequorin, activated through complex formation with Ca^{2+} and the cofactor coelenterazine, emits blue light that is absorbed by GFP, a 238-amino acid polypeptide. In response to this energy, GFP emits green light via the excitation of its fluorophore.

The cloning of the GFP gene was reported in 1992 *(2)*. Two groups of investigators in 1994 *(3,4)* determined that this cloned GFP gene can be successfully expressed in a diversity of living beings, other than jellyfish, as assessed by the production of fluorescence. GFP has a major excitation peak at 395 nm, with a secondary peak at 475 nm, and an emission spectrum that reaches 508 nm *(2,5)*. An uncommon and important characteristic of GFP is that it does not necessitate cofactors or substrates for its light-emitting feature *(6)*. Consequently, the fluorescence of GFP is valuable in a variety of biotechnological purposes, such as in noninvasive detection in intact living cells and tissues.

GFP mutants have been engineered with enhanced features, compared to the wild-type GFP, by modifications to its primary structure, facilitating fluorescence identification in prokaryotic and eukaryotic cells *(7)*. Improved fluorescence and photostability are achieved with the red-shifted variant generated by a mutation of amino acid serine at position 65 to threonine (S65T) *(8)*. The enhanced GFP (EGFP) (Clontech, Palo Alto, CA) with mutations that include F64L (mutation of Phe64 to Leu), to enhance protein folding, over and

From: *Methods in Molecular Biology, vol. 183: Green Fluorescent Protein: Applications and Protocols*
Edited by: B. W. Hicks © Humana Press Inc., Totowa, NJ

above mutation S65T, produces yet superior fluorescence *(9)*. EGFP, which shows an excitation peak at 488 nm and an emission peak at 507–509 nm, is the GFP variant employed in the procedures described in this chapter.

Genetic reprogramming of cultured cell lines with recombinant DNA is routinely carried out as a means to decipher the molecular mechanisms of disease. Gene transfer and expression is a powerful tool that may also be exploited for therapeutic purposes. Strategies can be devised in which the introduction of synthetic genetic information will alter the biological behavior and phenotype of cultured cell lines. If cultured, immortalized cell lines can be engineered, the same could be done in normal or diseased tissue of patients *(10)*. The development of fluorescent reporter transgenes, such as GFP, has markedly enhanced our ability to track the fate of genetically engineered cells in the experimental setting. The GFP cDNA may be introduced on its own in target cells for the expressed purpose of gene marking. With the assumption that GFP-expressing cells maintain the biological properties of parental cells, marked cells and their progeny can thereafter be readily identified and their function and fate observed in vitro and in vivo.

A closely related use of GFP cDNA would involve its expression in tandem with a second, linked transgene. In this setting, GFP would serve as a genetic marker, reporter, and selectable agent for engineered cells concurrently expressing the second, linked transgene. A double gene or bicistronic vector can be designed by the insertion of the encephalomyocarditis virus internal ribosomal entry site (IRES) between a therapeutic gene and the GFP reporter gene *(11,12)*. The expression cassette contains the two cDNAs and a single promoter, which, in combination with IRES, allows the translation of the two open reading frames from one mRNA. GFP would permit the noninvasive assessment of therapeutic gene transfer efficiency since GFP expression would be easily determined by fluorescence microscopy *(3)*. Alternatively, GFP/transgene fusion proteins can be devised, although functional integrity of the fusion protein needs to be validated, a step unnecessary in bicistronic constructs, in which the second linked transgene is expressed in its native form.

Genetic engineering consists of three components: target cells, synthetic genetic information, and delivery vehicle. In cell biology, and especially in gene therapy applications, transgene delivery often remains the most important variable for high-efficiency gene transfer. Various kinds of vectors are used for gene transfer into target cells. Although there is considerable research on expression vectors, an ideal vector does not yet exist.

Each method of gene delivery has advantages and disadvantages. Alternative gene delivery methods include transfection of naked plasmid DNA and the use of replication-defective viral vectors. Transfection of traditional eukaryotic expression vectors into cultured cells is usually inefficient, requiring prolonged drug selection to enrich for engineered cells. Viral vectors, such as adenovectors and retrovectors, will readily transduce cultured cells and are the means of choice for high-efficiency engineering of normal primary cells.

A variety of commercially available, GFP-expressing adenovectors are currently marketed. They are characterized by broad tissue and species tropism, high titers, and, usually, high levels of expression ***(13)***. Adenovectors do not integrate in the genome of the target cells. If the engineered cells proliferate subsequent to viral gene transfer, the number of replication-defective adenovectors per cell will decrease as cells multiply and reporter expression will be lost. If long-term expression in proliferating cultured or primary cells is sought, integrating vectors, such as murine oncoretrovectors or lentiviral vectors, are more desirable ***(14,15)***.

For experimental work, the authors have chosen to use a retroviral vector for gene transfer. The advantages of employing retroviral vectors include the simplicity of their genetic architecture, the size of the gene of interest (up to ~9 kb) that can be inserted, and the efficiency of gene transfer into target cells. Moreover, the foreign gene stably integrates into the chromosomal DNA of the host cell, and subsequently into progeny cells, and these vectors may evoke, if any, only a minor immune response in the host ***(16–19)***.

The disadvantages and risks associated with the use of retroviral vectors encompass their capacity to infect only cells actively dividing, their susceptibility to inactivation by serum complement, the risk of replication-competent virus arising in large-scale preparations of retroviral vectors, their inability to infect certain cell types which may result from the absence of specific receptors on these target cells, and considering that retroviral particles are generated in cell culture, that some cellular contaminants may thus coexist ***(10,16–21)***. In addition, the integration of the retrovirally transferred gene into the host cell genome is random and consequently there is theoretical risk of insertional mutagenesis.

This chapter outlines the methodology for cloning therapeutic genes into retroviral vectors comprising the GFP, the transient and stable transfection of resulting constructs into retroviral packaging cell lines, the titering of infectious retroviral particles produced, and the stable gene transfer by transduction of target cell lines and primary cells. Also discussed are the methods of detection and measurement of GFP expression, and the isolation by fluorescence activated cell sorting (FACS) of high transgene-expressing genetically engineered cells.

2. Materials

1. Restriction enzymes and buffers (Pharmacia, Baie d'Urfe, PQ).
2. Retroviral vector plasmid DNA, such as AP2 ***(22)***.
3. T4 DNA ligase and buffer (Gibco-BRL, Gaithersburg, MD).
4. Large fragment of DNA polymerase I, buffer, and deoxynucleoside triphosphate (Gibco-BRL).
5. Agarose, Tris-borate EDTA (TBE) 1X buffer; ethidium bromide: 10 mg/mL stock solution.
6. QiaQuick Gel Extraction Kit (Qiagen, Mississauga, ON).
7. Luria-Bertoni (LB) broth, LB agar, and ampicillin: 50 mg/mL in distilled water as a 1000X stock solution.
8. QIAquick Spin Miniprep Kit (Qiagen).
9. Phosphate buffered saline (PBS) (Gibco-BRL).
10. Trypsin solution : 0.05% trypsin and 0.53 m*M* EDTA (Wisent, St. Bruno, PQ).
11. Cell culture media (DMEM) (Wisent). The authors recommend using media, PBS, and trypsin following warming at 37°C or at room temperature.
12. Fetal bovine serum (FBS) (Wisent), heat-inactivated: Aliquot, and store at –20°C.
13. Penicillin/streptomycin (Pen/Strep) (Wisent): 50 U/mL penicillin and 50 μg/mL streptomycin.
14. Retrovirus packaging cell lines, such as GP+E86, and GP+envAM12 ***(23,24)***.
15. Sodium acetate: 2 *M* solution maintained at room temperature.
16. Glycogen (Gibco-BRL): 20 mg/mL stock solution at –20°C.
17. Drug resistance plasmid for co-transfection and the corresponding agent for selection. For instance, pJ6Ωbleo ***(25)*** generously provided by Richard C. Mulligan (Children's Hospital, Boston, MA), and Zeocin (Invitrogen, San Diego, CA): 100 mg/mL stock solution.
18. Lipofectamine (Gibco-BRL): 2 mg/mL stock solution, and polybrene (Sigma, St. Louis, MO): 6 mg/mL stock 1000X solution prepared in distilled water, and stored at 4°C for up to ~1 mo, or at –20°C, good for several months.
19. Target cell lines (NIH3T3 from American Type Culture Collection [ATCC]).
20. Polypropylene round-bottomed tubes (5 mL), 12 × 75 mm (Falcon, VWR Canlab, Mississauga, ON).
21. Low-protein binding filters (0.45 μm) (Gelman, Ann Arbor, MI), 10 and 20 mL syringes, and 18, 21, and 25 gage needles (Beckton Dickinson, Mississauga, ON).
22. Fluorescence microscope (inverted preferred) and flow cytometry apparatus.
23. Trypan blue (Gibco-BRL).
24. Mice (C57Bl/6) and/or rats (Lewis).
25. Dissection instruments.
26. 3% paraformaldehyde: Store at –20°C.
27. 22-mm square microscope cover glasses (Corning, Cambridge, MA).
28. Precleaned frosted-end microscope slides (Fisher, Nepean, ON).

3. Methods

3.1. Cloning of Therapeutic Genes into GFP-Encoding Retroviral Vector Plasmids

Bicistronic plasmid retroviral expression vectors, incorporating GFP and yellow fluorescent protein, have been reported, and are commercially available (Clontech). This laboratory has designed a bicistronic MSCV/MFG hybrid retrovector plasmid, primarily MSCV-derived, termed "AP2" (**Fig. 1**; *22*). This construct can express the EGFP reporter gene unaccompanied, or concomitant with, an inserted cDNA upstream of an IRES, and by means of a cytomegalovirus promoter in transfected cells and long terminal repeat (LTR) promoter element in transduced cells.

For the introduction of a transgene cDNA into the bicistronic AP2 backbone, a cloning strategy must be devised based on the restriction enzyme maps of the AP2 plasmid and of the construct encompassing the desired cDNA. Whenever feasible, generate compatible ends by selecting enzymes that cleave once or twice in the multiple cloning site of AP2 and near the perimeters of the coding region of the sought gene excluding its polyadenylation signal. Final construct will be structured as follows: promoter-cDNA-IRES-GFP-LTR. Penultimately, transfected and retrovirally transduced cells will express a bicistronic mRNA structured as follows: 5' cap-cDNA-IRES-GFP-poly(A). The level of GFP translation is dependent on the IRES, which in turn can be influenced by the secondary structure conferred to the whole RNA molecule by the cDNA sequence *(26)*. Hence, different AP2-derived constructs will lead to varying levels of GFP expression. Yet, in the authors' experience, virtually all AP2-derived constructs will express GFP at a level detectable by FACS, although some may not be detectable by fluorescence microscopy (which the authors find to be much less sensitive for GFP detection). Therefore, when possible, FACS analysis is utilized to measure GFP expression derived from these constructs, although, often, fluorescence microscopy alone will suffice.

1. Execute a restriction enzyme digest of >1 μg of each of vector, and of insert-containing plasmid DNA. Utilize 1–3 U enzyme(s)/μg DNA/h at 37°C (or at other temperature required for a particular enzyme).
2. Subsequently, terminate the enzymatic activity as recommended for the enzyme(s) used and as instructed on the specific manufacturer's information sheet.
3. Proceed, if required for the particular cloning strategy, with the ensuing DNA modifications, to generate compatible ends.
4. Perform agarose gel electrophoresis of the restriction-enzyme-digested DNA samples, including nondigested DNA as control for comparison purposes, to reveal the completion of the cleavage reactions.

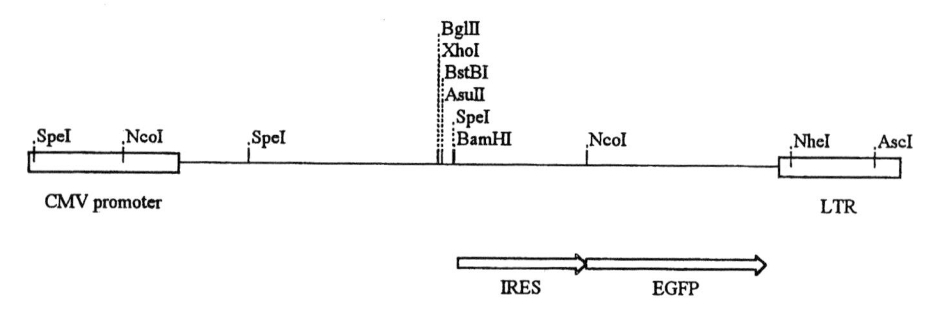

Fig. 1. Schematic representation of the AP2 retroviral construct. The bicistronic vector contains a multiple cloning site, and the EGFP reporter separated by an IRES. The CMV promoter drives transgene expression in cells transfected with the retroviral plasmid; the LTR controls expression in cells transduced (or infected) with the retroviral particles derived from producer cell lines. (Also on CD-ROM.)

5. Using a scalpel, harvest from the gel the desired bands, corresponding to the vector and insert fragments. Maneuver expeditiously, with adequate skin and eye protection and over low-intensity ultraviolet radiation, in order to minimize consequent DNA damage.
6. Purify the DNA from the gel slices by employing e.g., the QiaQuick Gel Extraction Kit (Qiagen), and quantitate the resulting DNA.
7. Ligate the vector to the insert in accordance with the T4 DNA ligase manufacturer's protocol. Prepare several ligation reactions comprising, initially, 40 ng vector, with a 3, 5, and 10 molar excess of insert, and also taking into account the dissimilar fragment sizes (*see* **Note 1**).
8. Transform DH5α-competent bacteria (subcloning efficiency or maximum efficiency) (Gibco-BRL), as indicated on information leaflet. For cloning realized with the AP2 vector, spread bacteria on LB agar plates containing ampicillin at a final concentration of 50 µg/mL. Position plates inverted in a 37°C incubator.
9. The following day, touch emerged colonies, such as with sterile toothpicks, and release them into tubes (14-mL polypropylene tubes [Falcon]), containing 4 mL LB broth and 50 µg/mL ampicillin (*see* **Note 2**). Place tubes at 37°C for ~12 h with ~200 rpm shaking.
10. Isolate plasmid DNA from 2 mL of each bacterial culture, utilizing, e.g., the QiaQuick Spin Miniprep Kit (Qiagen). Aliquot the resulting DNA and store at –20°C.
11. Confirm correct sequence of the recombinant constructs by a series of restriction enzyme digests and later, by DNA sequencing.
12. Once the DNA is validated, prepare 15% glycerol stocks of the remaining 2 mL bacterial culture, freeze rapidly as in a bath of dry ice and ethanol, and store at –80°C.

3.2. Stable Transfection of Retroviral Packaging Cells with Retroviral Expression Vectors Containing GFP

Infectious retroviral particles can be generated with the use of retroviral packaging cell lines *(27)*. These cell lines have been engineered to express all the protein components of a retroparticle (Gag, Pol, and Env), except for the RNA viral genome. Transfection of these packaging cells with a retroviral expression vector construct, bearing the necessary retroviral *cis*-acting sequences (such as the packaging signal, polypurine tract and a LTR), leads to a retroviral producer cell line. The possibility of recombination between the vector and the helper viral genome present in packaging cells has been lowered by the utilization of the packaging cells in which the genome of the helper virus is divided on two separate plasmids *(23,24,28)*. From these retroviral producer cells, retroviral particles budding from the cell membrane will bear two (+) single stranded RNA viral genomes derived from the retroviral expression vector construct. The released retroparticles can be collected and subsequently utilized for gene transfer experiments. Packaging cell lines are defined by the retroviral envelope protein they express, which in turn dictates the species tropism of sythesized retroparticles. The ATCC provides at least four packaging cell lines, including: GP+E86 (ecotropic, useful for gene transfer in murine cells only), GP+envAM12 and PA317 (amphotropic, useful for gene transfer in most mammalian species, including mice), and PG13 (useful for gene transfer in some mammalian species, excluding mice). Other retroviral packaging cell lines can be obtained directly from investigators or from commercial sources. GFP expressing retrovectors have significant added benefit when the isolation of retroviral producer cells is feasible by cell sorting (based on green fluorescence), and where measurement of retroviral titer can be easily done with a straightforward cytometry-based assay as described in **Subheading 3.5.**

Although generally employing lipid-mediated transfection, we have had success as well with the calcium phosphate protocol of transfection. This procedure is described with ecotropic GP+E86 packaging cells. GP+E86 cells stably transfected with a retroviral expression vector such as AP2 will continuously produce infectious retroviral particles. Described below is a detailed transfection protocol to generate retroviral producer cells:

1. Prepare the retroviral plasmid DNA for stable transfection into GP+E86 cells by linearizing 10 µg with a restriction enzyme that cleaves only in the plasmid backbone and not in the region confined by the 5' and 3' promoters (*see* **Note 3**).
2. Precipitate the linearized DNA, by adding to it in an Eppendorf tube, one-tenth vol 2 *M* sodium acetate, 2 vol 100% ethanol, and 1 µL glycogen (stock: 20 mg/mL). Mix and place at –20°C for a sufficient minimum of 1 h.

3. Pellet the DNA in a microcentrifuge at maximum speed for 15 min at a recommended but not essential temperature of 4°C.
4. Decant the supernatant, air-dry the pellet ~5 min and resuspend the DNA in 300 μL DMEM without serum. Withhold 10 μL for agarose gel electrophoresis to ascertain successful linearization of the DNA.
5. 1 d prior to transfection, plate GP+E86 cells at a cell density of 500,000 cells in 3 mL complete media (DMEM supplemented with 10% FBS and 50 U/mL Pen/Strep), in each of two 60-mm tissue culture plates, one to serve for the nontransfected control and one for the transfection.
6. For the reason that selection of stably transfected cells is desired, co-transfection with a drug-resistance plasmid is imperative. In the tube holding the resuspended DNA, add 1 μg (i.e., 10-fold less) of a drug-resistance plasmid such as pJ6ΩBleo ***(25)***, or any other drug selection plasmid that would be suitable.
7. In a separate Eppendorf tube, mix without vortexing, 25 μL lipofectamine Reagent (Gibco-BRL) with 275 μL DMEM.
8. Combine the DNA and the Lipofectamine solutions from the two tubes into one 15-mL-capacity sterile tube, swirl gently to mix and incubate at room temperature for 30 min.
9. Blend 2.4 mL DMEM into the 600 μL DNA–Lipofectamine mixture, and dispose over one 60-mm plate of packaging cells previously rinsed with serum-free media.
10. Place cells in a humidified 37°C incubator with 5% CO_2 for 6–8 h.
11. In the aftermath, add 2 mL complete media to cells, and return them to the incubator.
12. The following day, dissociate using trypsin solution, both the transfected and the control cells from the 60-mm plates, resuspend in ~25 mL complete media and replate each in one 150-mm plate.
13. 2 d later, when peak level of GFP expression is expected, commence drug selection of stable transfectants. Aspirate the media from the control and the transfected cells and replace with complete media containing selection drug (*see* **Note 4**).
14. Replace the media every 3–4 d with fresh drug-containing complete media up to the end of selection, i.e., until the nontransfected control cells have all succumbed from drug toxicity (~3–4 wk). Throughout this period, cells may necessitate passaging, if overconfluency is reached.
15. Colonies of stable transfectants may have arisen, and those containing the retroviral vector will be readily recognizable by the expression of GFP. If single colonies cannot be discerned, and are sought, plate cells at a low density. For the generation of monoclonal populations of virus-producing cells, pick colonies using, e.g., sterile cloning cylinders or, utilize, based on GFP expression, FACS (*see* **Subheading 3.8.**).
16. For a polyclonal population of stable transfectants, simply trypsinize and replate, consequently pooling all cells and colonies.
17. Subject monoclonal or polyclonal cell populations to flow cytometry analysis for GFP expression (*see* **Note 5**). Producers may be sorted based on GFP fluores-

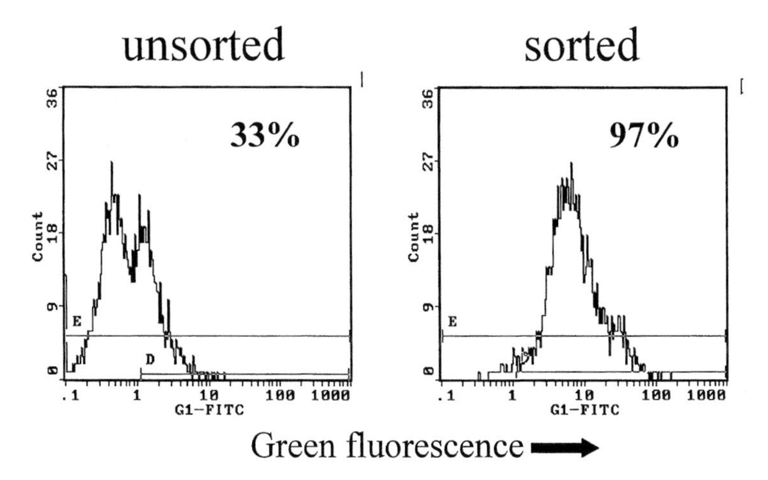

Fig. 2. Flow cytometry analysis of GFP expression in AP2 stably transfected GP+E86 retrovirus-producing cells (AP2-GP+E86 cells), following dominant selection by cell sorting. *Left*, unsorted AP2-GP+E86 retroviral producers. *Right*, sorted producers. The percent GFP-positive cells is indicated in upper right of panel. (Also on CD-ROM.)

cence if the proportion of GFP positive cells or the mean intensity of expression is lower than optimal or desired (*see* **Subheading 3.9.** and **Fig. 2**). We recommend a population of at least 50% GFP-positive cells for virus harvesting and utilization.

18. Freeze, as soon as possible, the virus-producing cell lines generated, in many vials and at various instances. Early-passage cells should be mostly utilized for frozen stock preparations.

3.3. Transient Transfection of Retroviral Packaging Cell Line with GFP-Encoding Plasmids

Small quantities of retrovector can be generated quickly by transient transfection of retroviral packaging cells. Supernatant from the transiently transfected producer cells will contain retroviral particles in the 48–72 h following transfection. Retroparticle production is usually self-limited to no more than a few days, and is usually 1–2 orders of magnitude less than seen in stably transduced producers. This protocol comprises **steps 4–12** of **Subheading 3.2.** It is essentially identical except for the variations listed below.

1. The plasmid construct need not be linearized.
2. A drug-resistance plasmid is not utilized, since it is only required for the selection of stable transfectants.
3. To transduce cells, the viral supernatant is employed at 48–72 h posttransfection.

3.4. Harvesting of Retroviral Particles Generated by Virus Producing Cells

Budding retroviral particles are released continuously in the culture media overlaying the retroviral producer cells. These distribute randomly in the media. Media enriched with particles can be used "as is," to gene-modify target cells. However, free-floating cells and subcellular debris must be removed from retroviral supernant prior to use on target cells.

1. Plate retroviral producer cells in a 10-cm tissue culture plate. When over 80% confluent, replace the media with a lower volume of complete media (~6 mL), thus enhancing the concentration of the released retrovirus.
2. The next day, collect and pass the virus-containing media through a syringe-mounted 0.45-μm, low-protein-binding filter so as to remove any cells or cellular debris present. Retroparticles have a diameter of ~100 nm, and will not be affected by filter process.
3. Place complete media over the producers and repeat previous step 10–24 h later, for subsequent transduction round.
4. For the genetic engineering of target cells, utilize viral supernatant immediately (*see* **Subheading 3.6.**), or store at –20°C for later use. Repeat freeze–thaw cycles will inactivate a substantial amount of retroviral particles and are not recommended. Transduction efficiency is greatest when the retroparticles from GP+E86 or GP+envAM12 cells are employed at time of harvest.

3.5. Assay for Titer Determination of Viral Particles Based on GFP Expression in Target Cells

A polyclonal population of stably transfected GP+E86 producer cells will continuously generate retroparticles. The amount of retroparticles made (titer) will be dependent on many variables, including retrovector RNA stability, volume:surface area ratio of media/producer cells, and confluency of producer cells in tissue culture plate. In most instances, subconfluent cells (~80%) with ~1 mL media/cm^2 surface area of adherent cells will yield a titer range of $0.5–15 \times 10^5$ infectious retroparticles/mL. The GFP reporter incorporated in retroviral constructs greatly facilitates the measurement of titer and a method utilizing flow cytometric analysis is described below. For titering the ecotropic GP+E86 producer, NIH3T3 murine fibroblast cells may be utilized.

1. Plate target cells (NIH3T3 or A549) in each well of two 6-well plates, at a density of $1.5–4 \times 10^4$ cells in DMEM + 10% FBS + 50 U/mL Pen/Strep.
2. The next day, just preceding transduction, enumerate cells in two wells, and calculate the average, to provide the approximate number of cells/well at time of virus addition.

3. For transduction, aspirate the media from eight other wells, and add the following amounts of viral supernatant diluted in complete media in a final volume of 1 mL/well, and in the presence of polybrene or lipofectamine at a final concentration of 6 µg/mL: 1000, 100, 10, 5, 1, 0.5, 0.1, and 0 µL for the control well.
4. The next day, add 2 mL complete media to each well.
5. 2 d later, corresponding to 72 h since virus addition and when GFP expression has reached maximum levels, visualize cells using a fluorescence microscope to estimate transduction efficiency.
6. For a precise quantitation of the percentage of GFP-positive cells, trypsinize cells in each well, resuspend in media, position half in tubes (e.g., 5 mL polypropylene tubes [Falcon]), and analyze by flow cytometry. **Figure 3** illustrates the flow cytometry read-out for titer determination of AP2-GP+E86 stable transfectants.
7. Calculate the titer with the equation below by considering the dilution of virus that leads to 10–50% of cells expressing GFP (*see* **Note 6**).

[(Percentage GFP-positive cells) × (Number of cells at time of transduction)]/(Volume of virus added in mL)

3.6. Transduction of Cell Lines and/or Primary Cells with GFP-Expressing Retroviral Particles In Vitro

Replication-defective retroparticles will adhere and fuse to target cells expressing the appropriate retroviral receptor. In the hour following fusion, reverse transcription of the retroviral RNA to DNA will occur and the integration complex consisting of retroviral DNA and the Pol protein will remain in the cytoplasm. The integration complex is incapable of penetrating the intact nuclear membrane and will remain transcriptionally silent. If, in the 8–12 h following fusion, the target cell undergoes mitosis, the integration complex will access the genomic DNA and integrate. Transcription of the retroviral vector will then follow in a continuous manner as dictated by the cluster of enhancer elements contained within the LTR *(29,30)*. Retroviral vectors are very good at transducing cycling cells with efficiencies approaching 100% if the ratio of viral particles to target cells is >1. This high-efficiency gene transfer is particularly useful if the goal is to genetically engineer short-lived cycling cells, such as normal primary cells, in which long-term drug selection is not feasible or desirable. The added benefit of incorporating the GFP reporter as part of the retroviral construct is the unambiguous confirmation and identification of gene-modified cells. Further, since retrovectors integrate in the genome of target cells, all daughter cells arising from subsequent mitosis will also bear the synthetic retroviral expression vector, (a desirable feature, if one seeks to establish a transgenic cell line). The following details the use of retroviral particles for gene transfer into generic immortalized murine cell lines commonly used in cell biology applications, and an example of gene transfer into primary rodent bone marrow stromal tissue is also given.

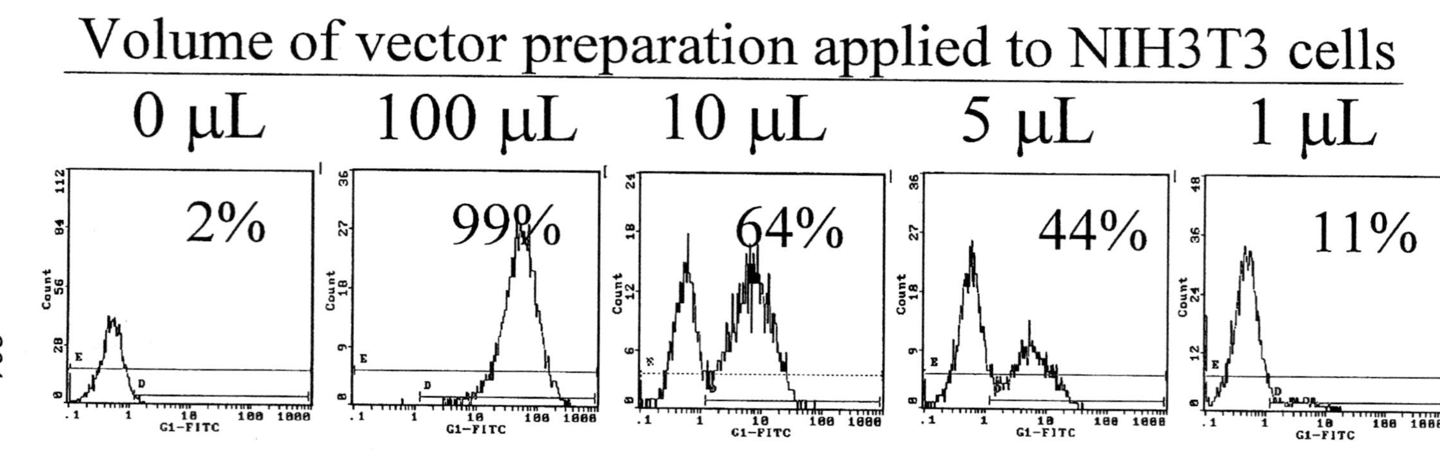

Fig. 3. Titer determination of AP2-GP+E86 producers by flow cytometry evaluation of GFP fluorescence. Left, Untransduced NIH3T3 mouse fibroblast cells and gene-modified 3T3 cells generated by exposure to 100, 10, 5, and 1 µL, respectively, of media containing AP2 retroviral particles. GFP expression was evaluated 72 h posttransduction. The percent GFP-positive cells is shown, and titer was assessed, based on the dilution giving GFP-expression ranging between 10 and 50% (*see* **Note 6**). (Also on CD-ROM.)

3.6.1. Adherent Cell Lines

This procedure is similar in principle to that utilized to determine the titer of virus-producing cells.

1. Plate target cells in a 10-cm tissue culture plate. Once cells are ~60% confluent, aspirate the media, and replace with viral supernatant freshly harvested from GP+E86 or GP+envAM12 producers (depending on target cell type). Cell division is required for retrovirus integration. Hence, if target cells are too confluent or if contact inhibition occurs, gene transfer is lessened.
2. Add polybrene or lipofectamine for a final concentration of 6 µg/mL to enhance transduction efficiency. However, these agents may exert some toxicity on certain cell types although the authors have not yet encountered any serious occurrence of this phenomenon.
3. Repeat the procedure once (or twice) a day for 3 consecutive d (*see* **Note 7**).
4. 1 d after the final round of transduction, discard the added viral supernatant, rinse cells twice with media and expose to fresh complete media.
5. 2 d later, visualize transduced cells by fluorescence microscopy to detect GFP and evaluate gene-transfer efficiency through flow cytometry analysis for GFP expression.
6. Continue the transduction (**steps 2** and **3** above) if high gene-transfer efficiency is not yet achieved.

3.6.2. Rodent Primary Bone Marrow Stromal Cells

Marrow stromal cells are a primary cell type that can be harvested and maintained in culture for many weeks. They proliferate well, and are highly receptive to retroviral gene transfer ***(31)***. Fibroblasts, hematopoietic precursors, and lymphocytes are also primary cells reported to be susceptible to retroviral gene transfer. Described below is a methodology we have developed for stromal cell retroviral gene transfer.

1. Sacrifice mouse or rat, and collect bone marrow cells by flushing the hind leg femurs and tibias with DMEM supplemented with 10% FBS and 50 U/mL Pen/Strep utilizing a needle (25- or 21-gauge, for mouse or rat, respectively) mounted syringe.
2. Prepare a single-cell suspension by multiple (5–10) passages through a needle (21- or 18-gage)-mounted syringe.
3. Plate mouse marrow in four 10-cm tissue culture dishes, and rat marrow in four 175-cm^2 flasks, in complete media, and incubate at 37°C with 5% CO_2.
4. 5 d later, discard the nonadherent hematopoietic cells, trypsinize the stromal cells and replate in an equivalent number of plates or flasks (*see* **Note 8**). Rinse the cells with PBS prior to applying trypsin solution (0.05% trypsin, 0.53 m*M* EDTA) for ~4 min.

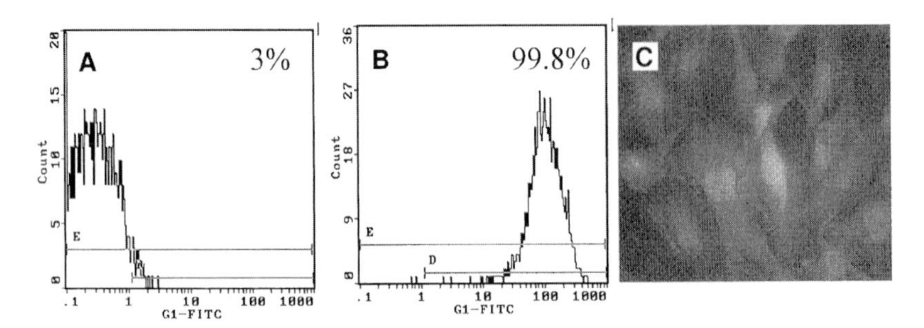

Fig. 4. GFP expression in murine bone marrow stromal cells transduced with retroviral particles from AP2-GP+E86 producers. **(A)** Nontransduced marrow stroma. **(B)** AP2-transduced stromal cells. The percent GFP-positive cells is indicated in top right of panels. Identical settings were employed for both test samples. **(C)** Fluorescence microscopy analysis of GFP expression in mouse marrow stroma. Photomicrograph exposure time utilized was 50 ms, and magnification is ×400. (For optimal, color representation please see accompanying CD-ROM.)

5. The following day, twice a day (~8–10 h intervals) for 3 consecutive d, replace the media over the stromal cells with viral supernatant.
6. For each round of transduction, add lipofectamine for a final concentration of 6 µg/mL. We have obtained significantly enhanced gene-transfer efficiencies in marrow stroma with the inclusion of lipofectamine.
7. 3 d after transduction completion, evaluate cells for GFP expression, and, if superior results are desired, repeat **steps 5** and **6. Figure 4** illustrates the GFP fluorescence in mouse bone marrow stromal cells, generated by transduction with retroparticles twice per day for 3 consecutive d and for each of 3 wk.

3.7. Fluorescence Microscopy for Noninvasive Detection of GFP Expression in Intact Cells

Often, the simplest way to determine GFP expression is by visualization of live cells directly from the incubator using an inverted fluorescence microscope equipped with a mercury lamp for GFP excitation, thereby not demanding plating of cells on glass cover slips and their fixation *(32)*. For visualization of GFP expression, ample excitation light is critical, in order to yield an easily perceivable fluorescence signal. Although novel filter sets exist, which are conceived especially for optimal GFP detection, GFP can be readily seen by using a standard fluorescein isothiacyanate (FITC) filter set if the extent of gene expression is sufficient as has been the case in this lab *(9,33)*. A beneficial aspect of GFP is its utilization in living cells as a real-time reporter. Of significance for superior detectability of GFP, is to enhance the intensity of the GFP signal and diminish the greenish yellow background signal of cellular autofluorescence. The methods by which this may be achieved encompass utilizing a wavelength of excitation that reduces autofluorescence, employing a narrower bandpass emission filter using the GFP variant best-suited for a sample *(33)*.

Cell culture media may cause autofluorescence and thus some investigators have reported discarding and replacing the media with PBS prior to assessing fluorescence with an inverted microscope *(32)*. The phenol red that is included in numerous growth media has been found to interfere with GFP fluorescence *(34)*. Nevertheless, our experience has been to view the cells primarily with their media.

GFP brightness is influenced by pH. The S65T and F64L GFP variant has a stronger, more intense fluorescence signal at pH 7.0 vs pH 6.0 *(33,35)*. Another major obstacle in fluorescence microscopy is photobleaching, which although not of principal importance here, because of the low rate of GFP photobleaching, it can nonetheless restrict the intensity of the GFP signal achieved *(33)*.

GFP not only fluoresces in live cells and tissues, but also after fixation with glutaraldehyde or formaldehyde *(1,3)*. Sometimes, however, fixation may extinguish GFP fluorescence by possible protein denaturation caused by utilization of strong organic solvents, high acidity of the fixative solution, or lengthy exposures *(1,35a)*. The authors have had problems visualizing the GFP signal in glutaraldehyde-fixed cells, especially in fixed tissue sections, and therefore recommend the exploration of numerous fixation conditions for one deemed satisfactory.

We have observed that GFP-expressing cells well withstand the following method of fixation.

1. Place a sterile 22-mm square microscope cover glass (Corning) into a 35-mm tissue culture plate.
2. Plate cells over cover glass and once subconfluency is attained, rinse cells 3× with PBS and expose to 2 mL 3% paraformaldehyde for 15 min at room temperature.
3. Remove the paraformaldehyde and wash cells 3× with PBS.
4. Using forceps, carefully lift the cover glass, and drop ~50 µL 37°C warmed gelvatol over cells.
5. Mount cover glass onto a microscope slide such as precleaned frosted-end slides (Fisher).
6. Visualize or photograph cells under fluorescence microscopy.

3.8. Flow Cytometry Analysis and FACS of Gene-Modified Cells, Based on GFP Expression

Flow cytometry analysis for assessment of the proportion of GFP-positive cells, and of their level of fluorescence is useful in gene transfer strategies. The brighter variants of GFP, such as EGFP, generate a sufficient degree of fluorescent signal to serve as selectable markers with flow cytometry and cell sorting *(7)*. As shown in **Fig. 4**, stromal cells transduced with EGFP-containing retroparticles (AP2) are discernable by the level of GFP expression from the unmodified controls.

By FACS, cells genetically engineered with recombinant retroviruses containing on a bicistronic construct the genes for a therapeutic protein and GFP may be selected for those expressing GFP. Moreover, these genetically engineered cells may be selected noninvasively for those displaying the desired intensity of GFP fluorescence, because it is expected to be proportional to the degree of therapeutic gene product expression in these cells *(7)*. **Figure 2** displays the flow cytometry analysis of GFP expression in stably transfected virus-producers, prior to and subsequent to FACS.

An important application of FACS is as a means to obtain single clones of GFP-expressing stable, virus-producing cells other than picking colonies using cloning cylinders *(32,36)*. We have succeeded in generating stable GFP-expressing producers with this method, which places one highly fluorescent cell from a heterogeneous population into each well of a 96-well plate in which 150 μL complete media has been previously placed. 1–3 wk later, adding fresh media once per week, populations of cells were noted in over 25% of the wells.

4. Notes

1. The quantity of insert relative to vector to be utilized in a ligation reaction may be calculated based on the following equation:

   Amount of insert = (Size of insert/Size of vector in base pairs) × Molar excess of insert × Amount of vector

   For example, for 40 ng 6700-bp vector and a 3-molar excess of an insert of 700 bp, 12 ng insert is required for the ligation.

2. If no colonies or incorrect constructs are obtained following transformation of the recombinant plasmid DNA into bacteria, it may be wise to repeat the ligation procedure using higher amounts of vector and insert, in addition to different ratios of the fragments.
3. Linearization of the plasmid DNA prior to transfection is not absolutely necessary since it will occur automatically, although randomly, for its integration into the host genome. This step accords certain control over the way the proviral DNA will integrate into the chromosomal DNA of the host cell.
4. The media used for drug selection of transfected cells varies according to a marker plasmid co-transfected with the GFP-containing construct or to a selectable marker existing on the plasmid vector. Moreover, the concentration of drug utilized for drug selection is a factor of the sensitivity of the particular unmodified cells to the cytotoxicity of the agent *(32)*. Consequently, we deem it important to expose in parallel with genetically engineered cells untransfected cells to the same selective media to confirm an earlier assessment of drug concentration and expo-

sure duration for cell eradication and to indicate the completion of the drug selection process. If many cells are still alive in the control plate after 4 wk of drug exposure, selection may not be complete, because of high cell density, which would have necessitated higher drug exposure.

5. GFP, particularly in mammalian cells, appears to necessitate a powerful promoter to permit a level of gene expression adequate for detection. Vectors utilizing the CMV promoter are, in the majority of cells, well-expressed. Numerous variables influence the extent of expression and detectability of GFP *(1)*. For instance, there will certainly be a higher amount of protein produced in a cell that contains many copies of the GFP gene and more powerful transcriptional promoters and enhancers controlling transcription.
6. For instance, for the titer determination of AP2-GP+E86 stable transfectants, a circumstance in which 5 μL viral supernatant transduced 44% of cells, numbered at 17,500 cells/well at time of virus addition, as determined by GFP expression (**Fig. 3**), the titer of these producers is ~1.5×10^6 infectious particles/mL volume

$$0.44 \times 17{,}500 \text{ cells/ } 0.005 \text{ mL virus}$$

7. Primary cells, much more so than cell lines, often require a multiplicity of infection of >50 infectious particles/target cell. An alternative is to expose cells to virus for several days, and even twice per day (8–12 h interval), if possible. Some cell types tolerate virus exposure more than others. It is therefore recommended to monitor cells, in order to estimate the number of times they may be subjected to a transduction round.
8. Working with stroma we have observed that, following trypsinization and replating of marrow stromal cells in a same-size plate, the transduction efficiency is greater, and probably results from enhanced cell distribution of colonies arisen in the 5 d after marrow harvest by dissociation.

Acknowledgments

Nicoletta Eliopoulos is a fellow of the Leukemia Research Fund of Canada, and Jaques Galipeau is a recipient of the Medical Research Council of Canada Clinician-Scientist award.

References

1. Tsien, R. Y. (1998) The green fluorescent protein. *Annu. Rev. Biochem.* **67,** 509–544.

1a. Shimoura, O., Johnson, F. H., and Saiga, Y. (1962) Extraction, purification and proporties of Aequorin, a bioluminescent protein from the luminous hydromeduson, Aequorea. *J. Cell Comp. Physiol.* **59,** 223–239.

2. Prasher, D. C., Eckenrode, V. K., Ward, W. W., Prendergast, F. G., and Cormier, M. J. (1992) Primary structure of the *Aequorea victoria* green-fluorescent protein. *Gene* **111,** 229–233.

3. Chalfie, M., Tu, Y., Euskirchen, G., Ward, W. W., and Prasher, D. C. (1994) Green fluorescent protein as a marker for gene expression. *Science* **263,** 802–805.

4. Inouye, S. and Tsuji, F. I. (1994) Aequorea green fluorescent protein. Expression of the gene and fluorescence characteristics of the recombinant protein. *FEBS Lett.* **341,** 277–280.
5. Cubitt, A. B., Heim, R., Adams, S. R., Boyd, A. E., Gross, L. A., and Tsien, R. Y. (1995) Understanding, improving and using green fluorescent proteins. *Trends Biochem. Sci.* **20,** 448–455.
6. Naylor, L. H. (1999) Reporter gene technology: the future looks bright. *Biochem. Pharmacol.* **58,** 749–757.
7. Galbraith, D. W., Anderson, M. T., and Herzenberg, L. A. (1999) Flow cytometric analysis and FACS sorting of cells based on GFP accumulation. *Meth. Cell Biol.* **58,** 315–341.
8. Heim, R., Cubitt, A. B., and Tsien, R. Y. (1995) Improved green fluorescence. *Nature* **373,** 663–664.
9. Baumann, C. T. and Reyes, J. C. (1999) Tracking components of the transcription apparatus in living cells. *Methods* **19,** 353–361.
10. Mulligan, R. C. (1993) The basic science of gene therapy. *Science* **260,** 926–932.
11. Ghattas, I. R., Sanes, J. R., and Majors, J. E. (1991) The encephalomyocarditis virus internal ribosome entry site allows efficient coexpression of two genes from a recombinant provirus in cultured cells and in embryos. *Mol. Cell. Biol.* **11,** 5848–5859.
12. Morgan, R. A., Couture, L., Elroy-Stein, O., Ragheb, J., Moss, B., and Anderson, W. F. (1992) Retroviral vectors containing putative internal ribosome entry sites: development of a polycistronic gene transfer system and applications to human gene therapy. *Nucl. Acids Res.* **20,** 1293–1299.
13. Seth, P. (2000) Adenoviral vectors. *Adv. Exp. Med. Bio.* **465,** 13–22.
14. Miller, D. G., Adam, M. A., and Miller, A. D. (1990) Gene transfer by retrovirus vectors occurs only in cells that are actively replicating at time of infection [published erratum appears in *Mol. Cell. Biol.* **12,** 433]. *Mol. Cell. Biol.* **10,** 4239–4242.
15. Hurford, R. K. J., Dranoff, G., Mulligan, R. C., and Tepper, R. I. (1995) Gene therapy of metastatic cancer by in vivo retroviral gene targeting. *Nat. Genet.* **10,** 430–435.
16. Whartenby, K. A., Abraham, G. N., Calabresi, P. A., Abboud, C. N., Calabresi, P., Marrogi, A., and Freeman, S. M. (1995) Gene-modified cells for the treatment of cancer. *Pharmacol. Ther.* **66,** 175–190.
17. Weichselbaum, R. R. and Kufe, D. (1997) Gene therapy of cancer. *Lancet* **349(Suppl. 2),** SII10–SII12.
18. Richter, J. (1997) Gene transfer to hematopoietic cells: the clinical experience. *Eur. J. Haematol.* **59,** 67–75.
19. Sandhu, J. S., Keating, A., and Hozumi, N. (1997) Human gene therapy. *Crit. Rev. Biotechnol.* **17,** 307–326.
20. Kavanaugh, M. P., Miller, D. G., Zhang, W., Law, W., Kozak, S. L., Kabat, D., and Miller, A. D. (1994) Cell-surface receptors for gibbon ape leukemia virus and amphotropic murine retrovirus are inducible sodium-dependent phosphate symporters. *Proc. Natl. Acad. Sci. USA* **91,** 7071–7075.

21. Roth, J. A. and Cristiano, R. J. (1997) Gene therapy for cancer: what have we done and where are we going? *J. Natl. Cancer Inst.* **89,** 21–39.
22. Galipeau, J., Li, H., Paquin, A., Sicilia, F., Karpati, G., and Nalbantoglu, J. (1999) Vesicular stomatitis virus G pseudotyped retrovector mediates effective in vivo suicide gene delivery in experimental brain cancer. *Cancer Res.* **59,** 2384–2394.
23. Markowitz, D., Goff, S., and Bank, A. (1988) A safe packaging line for gene transfer: separating viral genes on two different plasmids. *J. Virol.* **62,** 1120–1124.
24. Markowitz, D., Goff, S., and Bank, A. (1988) Construction and use of a safe and efficient amphotropic packaging cell line. *Virology* **167,** 400–406.
25. Morgenstern, J. P. and Land, H. (1990) A series of mammalian expression vectors and characterisation of their expression of a reporter gene in stably and transiently transfected cells. *Nucl. Acids Res.* **18,** 1068.
26. Mizuguchi, H., Xu, Z., Ishii-Watabe, A., Uchida, E., and Hayakawa, T. (2000) IRES-Dependent second gene expression is significantly lower than cap-dependent first gene expression in a bicistronic vector. *Mol. Ther.* **1,** 376–382.
27. Miller, A. D. (1990) Retrovirus packaging cells. *Hum. Gene Ther.* **1,** 5–14.
28. Ory, D. S., Neugeboren, B. A., and Mulligan, R. C. (1996) A stable human-derived packaging cell line for production of high titer retrovirus/vesicular stomatitis virus G pseudotypes. *Proc. Natl. Acad. Sci. USA* **93,** 11,400–11,406.
29. Miller, A. D. (1992) Retroviral vectors. *Curr. Top. Microbiol. Immunol.* **158,** 1–24.
30. Miller, A. D., Miller, D. G., Garcia, J. V., and Lynch, C. M. (1993) Use of retroviral vectors for gene transfer and expression. *Meth. Enzymol.* **217,** 581–599.
31. Jaalouk, D. E., Eliopoulos, N., Couture, C., Mader, S., and Galipeau, J. (2000) Glucocorticoid-inducible retrovector for regulated transgene expression in genetically engineered bone marrow stromal cells. *Hum. Gene Ther.* **11,** 1837–1849.
32. Sullivan, K. F. (1999) Enlightening mitosis: construction and expression of green fluorescent fusion proteins. *Meth. Cell Biol.* **61,** 113–135.
33. Piston, D. W., Patterson, G. H., and Knobel, S. M. (1999) Quantitative imaging of the green fluorescent protein (GFP). *Meth. Cell Biol.* **58,** 31–48.
34. Goodwin, P. C. (1999) GFP biofluorescence: imaging gene expression and protein dynamics in living cells. Design considerations for a fluorescence imaging laboratory. *Meth. Cell Biol.* **58,** 343–367.
35. Patterson, G. H., Knobel, S. M., Sharif, W. D., Kain, S. R., and Piston, D. W. (1997) Use of the green fluorescent protein and its mutants in quantitative fluorescence microscopy. *Biophys. J.* **73,** 2782–2790.
35a. Ward, W. W., Cody, C. W., Hart, R. C., and Coumier, M. J. (1980) Spectrophotometric identity of the energy-transfer chromophores in Kenill and Aequorea green fluorescent proteins. *Photochem. Photobiol.* **31,** 611–615.
36. Lybarger, L., Dempsey, D., Franek, K. J., and Chervenak, R. (1896) Rapid generation and flow cytometric analysis of stable GFP-expressing cells. *Cytometry* **25,** 211–220.

Index

C

G

R

S

CD-ROM Contents

This CD-ROM contains GIF, JPEG, PDF, MPEG, and AVI files for *Green Fluorescent Protein: Applications and Protocols*, edited by Barry W. Hicks. These files are mainly color versions of the artwork found in the book. Animation files are included in some chapters. The Editor gratefully acknowledges the generous support of the Universal Imaging Corporation for this CD-ROM.

System Requirements

This self-launching CD-ROM is compatible with both PC and Macintosh operating platforms on which Internet Explorer or Netscape Navigator version 4.0 or later is installed. Users should note that their browser settings should be configured to playback files in MPEG format (Quicktime 4.0 or later is recommended). Users can re-launch the CD-ROM by double-clicking the icon which reads "GFPhome.htm" in the "Hicks GFP" folder.

Limited Warranty and Disclaimer

Humana Press Inc. warrants the CD-ROM contained herein to be free of defects in materials and workmanship for a period of thirty days from the date of the book's purchase. If within this thirty day period Humana Press receives written notification of defects in materials or workmanship, and such notification is determined by Humana Press to be valid, the defective disk will be replaced.

In no event shall Humana Press or the contributors to this CD-ROM be liable for any damages whatsoever arising from the use or inability to use the software or files contained therein.

The authors of this book have used their best efforts in preparing this material. Neither the authors nor the publisher make warranties of any kind, express or implied, with regard to these programs or the documentation contained within this book, including without limitation, warranties of merchantability or fitness for a particular purpose. No liability is accepted in any event, for any damages including incidental or consequential damages, lost profits, costs of lost data or program material, or otherwise in connection with or arising out of the furnishing, performance, or use of the programs on this CD-ROM.